Methods in Molecular Biology

Series Editor
John M. Walker
School of Life Sciences
University of Hertfordshire
Hatfield, Hertfordshire, AL10 9AB, UK

For further volumes:
http://www.springer.com/series/7651

Metalloproteins

Methods and Protocols

Edited by

Juan C. Fontecilla-Camps

Metalloproteins Unit, Institut de Biologie Structurale (IBS), Grenoble, France

Yvain Nicolet

Metalloproteins Unit, Institut de Biologie Structurale (IBS), Grenoble, France

Humana Press

Editors
Juan C. Fontecilla-Camps
Metalloproteins Unit
Institut de Biologie Structurale (IBS)
Grenoble, France

Yvain Nicolet
Metalloproteins Unit
Institut de Biologie Structurale (IBS)
Grenoble, France

ISSN 1064-3745 ISSN 1940-6029 (electronic)
ISBN 978-1-4939-6328-7 ISBN 978-1-62703-794-5 (eBook)
DOI 10.1007/978-1-62703-794-5
Springer New York Heidelberg Dordrecht London

Printed on acid-free paper

Humana Press is a brand of Springer
Springer is part of Springer Science+Business Media (www.springer.com)

Preface

The fast development of genome sequencing has accumulated a huge amount of information concerning the nature and binding properties of the proteins coded by the genetic material of many different organisms. Thus, a high-throughput X-ray absorption spectroscopy (HT-XAS) metalloproteomic study performed on 3,879 proteins produced by the New York SGX Research Center for Structural Genomics indicated that about 9 % of them contained transition metals such as Zn, Cu, Ni, Co, Fe, or Mn. These metalloproteins, which are the subject of this book, are involved in many key biological processes such as gas transport and metabolism, photosynthesis, cell respiration, the Krebs cycle, and many other vital redox reactions.

One very important characteristic of many of the transition metal-containing enzymes is the sensitivity of their catalytic and redox centers to oxygen attack. This feature complicates their study because it often requires strict anaerobic conditions, as any exposure to O_2 may result deleterious. This book addresses the multiple aspects of metalloenzyme research: after an introductory chapter by Professor Perry A. Frey the following four chapters deal with the production and purification of metalloproteins using both standard recombinant techniques and a cell-free system. The possibility of not using standard equipment, such as a glove box, is addressed.

Once homogenous intact metalloproteins have been obtained they can be subjected to their functional characterization. One of the best techniques to explore the redox changes and catalytic properties of a great number of metalloenzymes is electrochemistry. This technique is discussed in Chapter 6. The electrochemical analysis of metalloproteins can be effectively coupled to infrared (IR) absorption spectroscopy as explained in Chapter 7. IR techniques can also be applied to monitor changes in the protein structure by using a technique called surface-enhanced infrared absorption or SEIRA. The use of resonance Raman (RR) spectroscopy to investigate the nature of short-lived intermediates formed during reactions of metalloproteins is addressed in Chapter 8. This technique is generally well suited to study enzymes carrying chromophoric prosthetic groups. Similar to IR and RR spectroscopies is a technique called nuclear resonance vibrational spectroscopy or NRVS. As discussed in Chapter 9 NRVS requires access to a synchrotron X-ray source. Data collection strategies are similar to those used in extended X-ray absorption fine structure (EXAFS) spectroscopy.

Many metalloproteins display paramagnetic states of their catalytic or redox centers when poised at relevant potentials. Such paramagnets are well suited to be studied by electron paramagnetic resonance (EPR) spectroscopy, which is described in Chapter 10. A particular advantage of this technique is that molecules with fully filled electron shells, including the solvent and the protein backbone, are EPR silent. Chapter 11 is dedicated to Mössbauer spectroscopy. This technique, specific to iron, has the advantage of being able to detect all forms of this metal ion, and their coordination sphere, found in a protein. Like in the case of NRVS, the sample should be enriched in ^{57}Fe.

Although paramagnetism is indeed a very convenient feature of transition metals some of them, such as the common Zn^{2+}, are diamagnetic and consequently spectroscopically silent in their biologically relevant redox state. In this case, the best-adapted technique to study the local electronic and physical structure around the metal site in a protein is X-ray absorption spectroscopy. Of course, XAS, which comprises X-ray absorption near-edge structure (XANES) and EXAFS, can also be used to study paramagnetic centers and provides very accurate stereochemical parameters in general. Like NRVS this technique requires access to a synchrotron X-ray source, and beam time should be requested accordingly. Because the overall signal for a given atom will include those from other metals present in the sample XAS is not well suited for the study of bulk cell preparations or cell lysates. Chapter 12 shows how to overcome this problem. It describes the use of XAS for the characterization of protein bands after electrophoretic separation and Western blotting.

The wealth of data provided by spectroscopic analyses can be ideally coupled to the structural data that can be obtained using X-ray crystallography (although nuclear magnetic resonance (NMR) is also a powerful technique for protein structure determination it is not well suited for the study of paramagnetic species). Chapter 13 discusses the use of X-rays not only to determine the three-dimensional structure of a crystallized metalloprotein but also to establish the nature of the metal(s) that it contains. Phasing using the multi-wavelength anomalous dispersion (MAD) method is described in some detail. Metalloprotein crystals can also be analyzed by X-ray fluorescence (to identify metals) or UV/visible spectroscopy (to monitor redox changes). Like in the other cases where tunable X-rays are required, beam time should be requested at a suitable synchrotron facility.

The last two chapters of this book deal with the theoretical interpretation of metalloenzyme catalytic and redox processes. Chapter 14 reviews the main procedures and computational methods that use quantum mechanics (QM) to study the reactivity and electronic properties of transition metal-containing enzymes. The continuous increase in computing power has recently allowed scientists to apply QM methods to the study of relatively large systems such as the active site of enzymes, including complex metal components. Finally, Chapter 13 describes in detail a particular domain of density functional theory (DFT) called broken symmetry (BS). DFT-BS is a very powerful albeit complex technique used to compute the energetic and spectroscopic properties of bioinorganic clusters (especially those containing transition metal ions). It is especially well suited to calculate coupling constants and g tensors as well as isomer shifts and quadrupolar tensors to be respectively compared to those obtained from EPR and Mössbauer spectroscopic experiments.

In summary, this book, which collects the contributions from the laboratories of renowned scientists, covers most of the relevant aspects of the metalloprotein field. It is the hope of the editors that the practical experience acquired in those laboratories and now contained in this compilation will be useful for both scientists who have already worked in the field for many years as well as those that are just joining in.

Grenoble, France *Juan C. Fontecilla-Camps*
Grenoble, France *Yvain Nicolet*

Contents

Contributors

ESEN ERCAN ALP • *Argonne National Laboratory, Advanced Photon Source, Argonne, IL, USA*

FRASER A. ARMSTRONG • *Inorganic Chemistry Laboratory, University of Oxford, Oxford, UK*

SIMON ARRAGAIN • *Centro de Biotecnología y Genómica de Plantas, Universidad Politécnica de Madrid, Madrid, Spain*

LUCA BERTINI • *Department of Biotechnology and Biosciences, University of Milano-Bicocca, Milan, Italy*

GENEVIÈVE BLONDIN • *Laboratoire de Chimie et Biologie des Métaux, Institut de Recherche en Technologies et Sciences du Vivant, Grenoble, France*

NICK E. LE BRUN • *Centre for Molecular and Structural Biochemistry, School of Chemistry, University of East Anglia, Norwich, UK*

MAURIZIO BRUSCHI • *Department of Environmental Science, University of Milano-Bicocca, Milan, Italy*

MARTIN CLÉMANCEY • *Laboratoire de Chimie et Biologie des Métaux, Institut de Recherche en Technologies et Sciences du Vivant, Grenoble, France*

UGO COSENTINO • *Department of Environmental Science, University of Milano-Bicocca, Milan, Italy*

JASON C. CRACK • *Centre for Molecular and Structural Biochemistry, School of Chemistry, University of East Anglia, Norwich, UK*

STEPHEN P. CRAMER • *Physical Biosciences Division, Lawrence Berkeley National Laboratory, Berkeley, CA, USA*

CARLOS ECHAVARRI-ERASUN • *Centro de Biotecnología y Genómica de Plantas, Universidad Politécnica de Madrid, Madrid, Spain*

RHIANNON M. EVANS • *Inorganic Chemistry Laboratory, University of Oxford, Oxford, UK*

LYDIA A. FINNEY • *Argonne National Laboratory, X-Ray Science Division, Advanced Photon Source, Argonne, IL, USA*

PERRY A. FREY • *Department of Biochemistry, Institute for Enzyme Research, University of Wisconsin-Madison, Madison, WI, USA*

SERGE GAMBARELLI • *Laboratoire de Chimie Inorganique et Biologique, Institut de Nanosciences et Cryogénie, Grenoble, France*

RICARDO GARCIA-SERRES • *Laboratoire de Chimie et Biologie des Métaux, Institut de Recherche en Technologies et Sciences du Vivant, Grenoble, France*

LUCA DE GIOIA • *Department of Biotechnology and Biosciences, University of Milano-Bicocca, Milan, Italy*

CLAUDIO GRECO • *Department of Environmental Science, University of Milano-Bicocca, Milan, Italy*

JEFFREY GREEN • *Department of Molecular Biology and Biotechnology, University of Sheffield, Sheffield, UK*

OSCAR GUTIÉRREZ-SANZ • *Laboratorio de Bioelectrocatalisis, Departamento de Biocatalisis, Instituto de Catalisis y Petroleoquimica, Madrid, Spain*

JON M. KUCHENREUTHER • *Department of Chemical Engineering, Stanford University, Stanford, CA, USA*

Antonio L. De Lacey • *Laboratorio de Bioelectrocatalisis, Departamento de Biocatalisis, Instituto de Catalisis y Petroleoquimica, Madrid, Spain*
Jean-Marc Latour • *Laboratoire de Chimie et Biologie des Métaux, Institut de Recherche en Technologies et Sciences du Vivant, Grenoble, France*
Hirotoshi Matsumura • *Oregon Health & Science University, Institute of Environmental Health, Portland, OR, USA*
Vincent Maurel • *Laboratoire de Chimie Inorganique et Biologique, Institut de Nanosciences et Cryogénie, Grenoble, France*
Evan Maxey • *Argonne National Laboratory, X-Ray Science Division, Advanced Photon Source, Argonne, IL, USA*
Pierre Moënne-Loccoz • *Oregon Health & Science University, Institute of Environmental Health, Portland, OR, USA*
Giorgio Moro • *Department of Biotechnology and Biosciences, University of Milano-Bicocca, Milan, Italy*
Jean-Marie Mouesca • *Laboratoire de Chimie Inorganique et Biologique, Institut de Nanosciences et Cryogénie, Grenoble, France*
Emily Ollmann • *Argonne National Laboratory, X-Ray Science Division, Advanced Photon Source, Argonne, IL, USA*
Luis M. Rubio • *Centro de Biotecnología y Genómica de Plantas, Universidad Politécnica de Madrid, Madrid, Spain*
Olaf Rüdiger • *Laboratorio de Bioelectrocatalisis, Departamento de Biocatalisis, Instituto de Catalisis y Petroleoquimica, Madrid, Spain*
Alessandro A. Scandurra • *Centro de Biotecnología y Genómica de Plantas, Universidad Politécnica de Madrid, Madrid, Spain*
Stacey A. Shiigi • *Department of Bioengineering, Stanford University, Stanford, CA, USA*
James R. Swartz • *Departments of Chemical Engineering and Bioengineering, Stanford University, Stanford, CA, USA*
Andrew J. Thomson • *Centre for Molecular and Structural Biochemistry, School of Chemistry, University of East Anglia, Norwich, UK*
Anne Volbeda • *Metalloproteins Unit, Institut de Biologie Structurale, UMR 5075 CEA-CNRS-UJF, Grenoble, France*
Hongxin Wang • *Department of Chemistry, University of California, Davis, CA, USA*
Jesse Ward • *Argonne National Laboratory, X-Ray Science Division, Advanced Photon Source, Argonne, IL, USA*
Yoshitaka Yoda • *JASRI, SPring-8, Hyōgo Prefecture, Japan*
Giuseppe Zampella • *Department of Biotechnology and Biosciences, University of Milano-Bicocca, Milan, Italy*

Chapter 1

Advances in Research on Metalloproteins

Perry A. Frey

Abstract

Metal ions play essential roles in biological processes. Ions such as K^+ and Na^+ are important in ion transport, and Mg^{2+}, Ca^{2+}, and Zn^{2+} are important chelators in many processes, including phosphotransfer and harvesting of light for energy metabolism. The transition metals readily undergo one-electron chemistry, and in this capacity they function uniquely in biological processes such as long-range and inner sphere electron transfer. They also facilitate many one-electron chemical reactions involving free radical intermediates. Iron, being the dominant element in the earth and a transition metal, most frequently participates in biological one-electron chemistry.

Key words Transition metals, Metalloenzymes, Metallocofactors, One-electron bioinorganic chemistry

The chemistry of metal ions equips them to carry out biological functions essential to life. Metal ions such as K^+, Mg^{2+}, and Zn^{2+} are biologically essential because of their electrostatic and chelation properties. For example, Mg^{2+} in chlorophyll serves an important function as a chelator that does not undergo reduction upon photoexcitation. A transition metal ion would be an equally good chelator but would undergo photoreduction, thereby derailing the energy transduction pathway in photosynthesis. However, the chemical properties unique to transition metal ions enable them to contribute in many other ways to life processes by engaging directly in biochemical transformations. A central functional property of transition metals in to all life forms is the propensity to undergo one-electron processes under physiological conditions. This property arises from the mid-range, one-electron reduction potentials of metals such as iron, cobalt, nickel, manganese, molybdenum, tungsten, and copper. Transition metals also bind to proteins or multidentate chelators in stable complexes and engage chemically in many biological molecules, especially enzymes.

All biological energy metabolism depends on electron transfer. Long-range electron transfers through distances of 10–14 Å link

Juan C. Fontecilla-Camps and Yvain Nicolet (eds.), *Metalloproteins: Methods and Protocols*, Methods in Molecular Biology, vol. 1122, DOI 10.1007/978-1-62703-794-5_1,

sites in multienzyme complexes of electron transfer pathways and within protein complexes that catalyze essential processes such as nitrogen fixation. Transition metals uniquely relay single electrons through long distances [1] linking electron transfer cofactors. Such electron transfer pathways energize the ATP synthases that generate biological energy in the form of ATP [2]. Anaerobic and aerobic life forms all depend on these processes.

Life in aerobes depends on oxygen metabolism at several levels. Transition metal ions display special reactivity with molecular oxygen, a paramagnetic molecule, in part because of the ability to undergo one-electron chemistry. Transition metal ions may exist in either diamagnetic or paramagnetic complexes, and so can interact with molecular oxygen in either reversible binding or in chemical transformations involving one-electron chemistry. Iron in hemoglobin reversibly binds molecular oxygen and serves to transport it within and among cells. Oxygenases are metalloproteins that facilitate one-electron chemistry between transition metals and molecular oxygen, leading to oxygenation of substrates. Prominent examples include the cytochrome P450s that catalyze oxygenation of alkanyl-sites in organic molecules through the chemistry of iron protoporphyrin IX. Many other oxygenases are mononuclear iron or diiron or copper proteins in which the transition metal engages directly in the oxygenation chemistry.

Few organic molecules readily undergo one-electron chemistry under physiological conditions. Heterocyclic organic compounds with conjugated π-electron systems often engage in one-electron chemistry, even with molecular oxygen. Examples include quinones; flavins [3], including the coenzymes FMN and FAD, biopterins, and under special conditions NAD [4]. In contrast, many transition metals function in multiple oxidation states.

Alkanes and chemically isolated alkanyl groups within organic molecules do not undergo one-electron chemistry under physiological conditions, except with the involvement of transition metal ions. According to the following rule put forward in 1990: "...any enzymatic reaction in which C–H cleavage occurs at an unactivated and unfunctionalized carbon will be found to require the action of a metal-containing cofactor that provides a paramagnetic hydrogen-abstracting species; and the mechanism will involve the formation of a substrate-derived organic radical..." [5]. This is a durable rule, and the participating metals are uniformly transition metal ions capable of one-electron chemistry. (In an apparent exception, reports held that phenylalanine monooxygenase from *Chromobacterium violaceum* was not a metalloenzyme [6] and that the metal-free enzyme catalyzed the nonstereospecific oxygenation of cyclohexyl-alanine [7]. Subsequent research showed that phenylalanine monooxygenase from *C. violaceum*, like all other species of phenylalanine monooxygenase, requires Fe^{2+} in the oxygenation of substrates [8]).

Prior to the enunciation of this rule, the metallocofactors chemically engaged in one-electron bioinorganic reactions included iron protoporphyrin IX, diiron complexes, mononuclear iron complexes, and the Vitamin B12 coenzyme adenosylcobalamin. More recently, the complex of *S*-adenosyl-L-methionyl-[4Fe–4S] in Radical SAM enzymes has been found to initiate highly diverse free radical mechanisms in more than 40 families of enzymes [9].

Most transition metalloenzymes are iron metalloenzymes because iron is the most plentiful element by mass in the earth and the most plentiful transition metal in the earth's crust [10]. Other transition metals are chemically as capable as iron but far less abundant. Nevertheless, despite the lesser abundance of cobalt, nickel, manganese, copper, molybdenum, and tungsten, all of them function in essential enzymatic systems.

The chemical properties of transition metal ions fuel the current focus on transition metal ions in bioinorganic chemistry. Progress in research on metalloenzymes depends critically on the availability of methods for investigating the structures and chemical properties of the enzymes, especially of the integral metal ion-complexes. The collection of articles in this volume from internationally prominent researchers review the most modern and innovative methodologies that can be employed to dissect the actions of metalloproteins. Included are articles on purification, thin film electrochemistry. Mossbauer spectroscopy, FTIR spectroscopy, EPR spectroscopy, nuclear magnetic vibrational spectroscopy, X-ray absorption spectroscopy, X-ray crystallography, transient phase Raman spectroscopy, density functional theory, and quantum mechanical methods. The complete analysis of metalloenzyme function requires all of these methods, making this volume an essential resource for researchers in bioinorganic chemistry.

References

1. Gray HB, Winkler JR (2005) Long range electron transfer. Proc Natl Acad Sci U S A 102:3534–3539
2. Abrahams JP, Leslie AG, Lutter R, Walker JE (1994) Structure at 2.8 Å resolution of F1-ATPase from bovine heart mitochondria. Nature 370:621–628
3. Williams RF, Shinkai S, Bruice TC (1975) Radical mechanisms for 1,5-dihydroflavin reduction of carbonyl compounds. Proc Natl Acad Sci U S A 72:1763–1767
4. Wong S, Frey PA (1976) UDP-galactose 4-epimearase-catalyzed oxygen-dependent reduction of a free radical substrate analogue by two-electron reducing agents. J Am Chem Soc 98:7886–7887
5. Frey PA (1990) The importance of organic radicals in enzymatic cleavage of unactivated C–H bonds. Chem Rev 90:1343–1357
6. Carr RT, Benkovic SJ (1993) An examination of the copper requirement of phenylalanine hydroxylase from *Chromobacterium violaceum*. Biochemistry 32:14132–14138
7. Carr RT, Balasubramanian S, Hawkins PC, Benkovic SJ (1995) Mechanism of metal-independent hydroxylation by Chromobacterium violaceum phenylalanine hydroxylase. Biochemistry 34:7525–7532
8. Chen D, Frey PA (1998) Phenylalanine hydroxylase from Chromobacterium violaceum: uncoupled oxidation of tetrahydropterin and the role of iron in hydroxylation. J Biol Chem 273:25594–25601
9. Frey PA, Hegeman AD (2008) The Radical SAM superfamily. Crit Rev Biochem Mol Biol 43:63–88
10. Frey PA, Reed GH (2012) The ubiquity of iron. ACS Chem Biol 7:1477–1481

Chapter 2

Purification of O_2-Sensitive Metalloproteins

Carlos Echavarri-Erasun, Simon Arragain, and Luis M. Rubio

Abstract

The most dependable factor to perform successful biochemical experiments in an O_2-free environment is the experience required to set up an efficient laboratory, to properly manipulate samples, to anticipate potential O_2-related problems, and to maintain the complex laboratory setup operative. There is a long list of O_2-related issues that may ruin your experiments. We provide here a guide to minimize these risks.

Key words Schlenk line, Glove box, Anaerobic protein purification, Metalloprotein, Metal clusters, O_2-free device, Sodium dithionite (DTH), Liquid nitrogen (LN_2), Liquid oxygen (LO_2), Radical-SAM enzyme

1 Introduction

It is estimated that 30 % of all proteins in a given organism contain associated metal(s) [1]. Metalloproteins are central to prokaryotic metabolic diversity, being essential to the global biogeochemical cycles of N, C, O, S, and H [2]. Arguably, the most important metalloenzymes in this context are the following: the [Ni–Fe] and [Fe–Fe] hydrogenases, involved in hydrogen production and consumption [3]; the molybdenum (Mo), vanadium (V), and iron-only (Fe) nitrogenases that fix atmospheric N_2 into NH_3 [4]; the [Ni–Fe] carbon monoxide dehydrogenase (CODH) carrying out a biological water-shift reaction [5, 6]; and more generally, [Fe–S] cluster-containing proteins involved in other fundamental biochemical processes such as respiratory and photosynthetic electron transport chains [7, 8].

Metal centers of proteins may be involved in catalysis, electron transfer, metal trafficking, sensing environmental conditions, or may simply have structural roles [9]. The catalytic activities of many metal-containing proteins, including all examples listed above, are sensitive to O_2 [10]. In some cases, O_2 reversibly inactivates the enzyme, which might be able to recover activity after its

Juan C. Fontecilla-Camps and Yvain Nicolet (eds.), *Metalloproteins: Methods and Protocols*,
Methods in Molecular Biology, vol. 1122, DOI 10.1007/978-1-62703-794-5_2,

removal (e.g., [Ni–Fe] hydrogenase [11]). In some other cases, O_2 irreversibly destroys the enzyme metal center(s), as it occurs in the [Fe–Fe] hydrogenase [12] and nitrogenase [13]. Other well-studied O_2 sensitive metal clusters are those of the iron–sulfur proteins aconitase and Fumarate Nitrate Reductase regulatory protein (FNR, *see* Chapter 4). Active aconitase carries a [4Fe–4S] cluster that is partially degraded into an inactive [3Fe–4S] form when exposed to O_2. Interconversion of both types of clusters occurs and is dependent on redox potential and Fe availability [14]. FNR [4Fe–4S] to [2Fe–2S] cluster conversion, caused by oxidation, inactivates FNR as transcriptional regulator and is used to regulate gene expression in response to cellular O_2 levels [15] (*see* Chapter 4). Finally, another broad example of metalloprotein sensitivity to O_2 can be found in the Radical-SAM protein superfamily of which more than 48,000 member proteins have been identified in databases [16–18]. Radical-SAM enzymes are involved in many different thermodynamically unfavorable reactions and require redox-active [4Fe–4S]-SAM conserved cluster for activity.

Specific laboratory setup and special techniques are required to work with O_2 sensitive proteins. We detail here material and methods used in anaerobic protein purification experiments, taking into account that not all laboratories have access to O_2-free glove boxes. A description of Schlenk line assembly and operation, O_2-free buffer preparation, sample manipulation, anaerobic purification tricks, protein concentration and storage methods will guide the reader through the virtual setup of an anaerobic biochemistry laboratory.

2 Materials

2.1 Carrier Gases

Ultrahigh purity gases with O_2 content lower than 3 ppm are required. High-pressure cylinders (usually 50 l at 200 atm) are probably the most popular solution for gas delivery. The cylinders have to be attached to a proper pressure reductor, and they are mobile and cheap, but laborious to move, may occupy essential laboratory space, and present safety concerns (*see* **Note 1**). Therefore, it is recommended that gas cylinders should be stored in a dedicated space on the outside of the research building (Fig. 1g). In this external space, the cylinders should be secured against a wall and connected to pressure reductors from where stainless steel or copper tubing should conduct gases to the required labs, but at a much lower pressure (12 atm or lower). In these laboratories, fine-pressure reductors (0–10 atm) should be located in the pipeline-end, in which attachment of an O_2-trapping catalyst is recommended (*see* **Note 2**).

The most popular inert gases for anaerobic biochemistry are nitrogen (N_2) and argon (Ar). N_2 is normally cheaper and a more

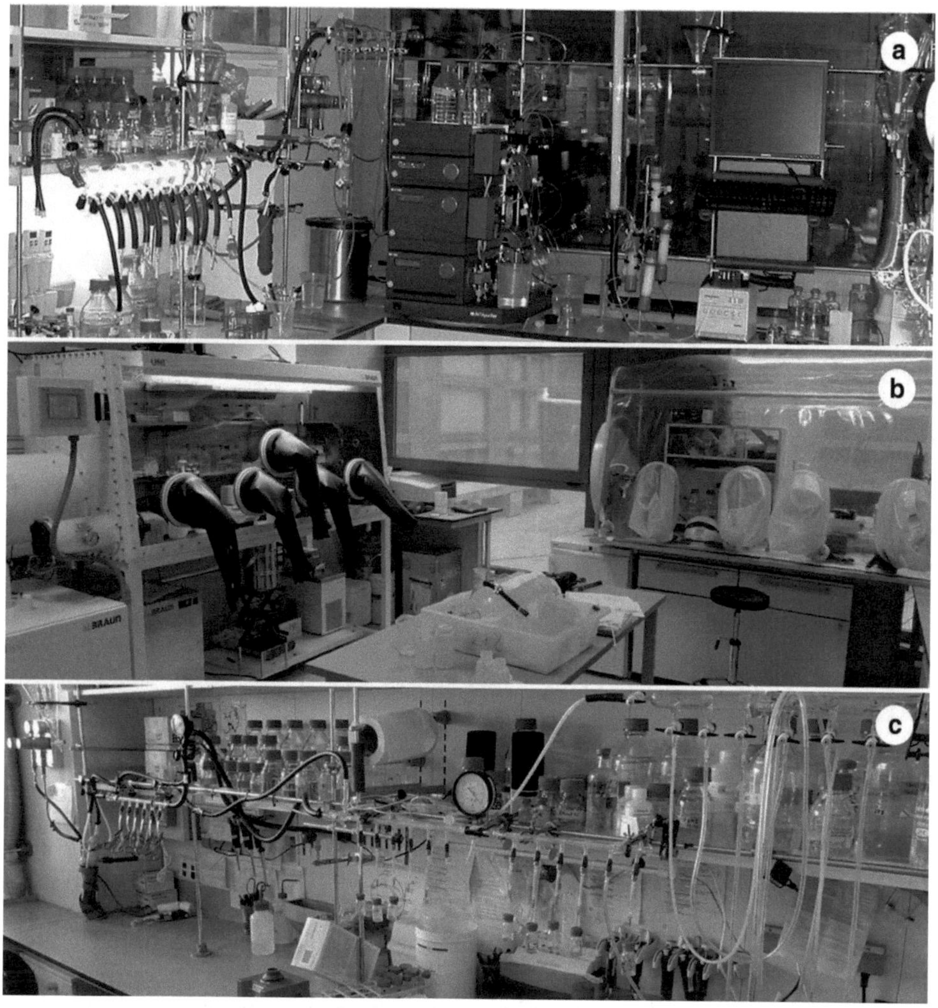

Fig. 1 (**a**) FPLC system connected to a Schlenk line and to a gas distributing manifold. System is connected to building N_2 and Ar lines through homemade O_2 traps (in *red* at the *center* of the picture). The vacuum line contains a water trap between the Schlenk line and the vacuum pump (hidden in the picture). Individual valves allow flowing gas into many flasks and vials independently. A number of chromatography columns are displayed in the *center* of the picture. (**b**) MBraun glove box to the *left* and Coy Labs vinyl glove box to the *right*. (**c**) Schlenk line specifically devoted to acetylene reduction assays. It is composed of two attached glass manifolds controlled by a multi-position valve from which it is possible to apply gas (Ar or N_2) or vacuum. Connected "dummy" vials are shown. (**d**) Co^{2+} affinity purification of His-NifDK performed outside a glove box. (**e**) NifB-co purification carried out outside a glove box in a jacketed column with double-sealed rubber caps. Note N_2 flow via needle lines on top of the column. (**f**) N_2 and H_2 gas end lines inside a laboratory. For safety reasons, the H_2 line is properly labeled and built in dual-tubing setup open at the building outside. Two gas mixers are installed: N_2/H_2 (*top*) and air/H_2 (*bottom*). (**g**) Research facilities should have their highly pressurized gas tanks located outside the building in a dedicated space, with restricted access. Picture includes an N_2 generation system with dual deposit capacity of 3,000 l total (<3 ppm O_2) and ten bar maximum pressure. (**h**) Schlenk line fed with highly purified N_2 (<2 ppm O_2) from a gas line controlled with valves and a pressure reductor. Homemade trap with a heater to scavenge O_2 traces in N_2. Vacuum line connected to an Edwards RV3 rotatory pump via a dry-ice water trap. Along the manifold, delivery valves are present, each connected to air-free hermetic vials of known volume through needles puncturing rubber-sealed caps. Additionally, three lines on the right side are dedicated to preparing anaerobic buffers in Büchner flasks. (**i**) MBraun glove box with FPLC apparatus and anaerobic columns inside

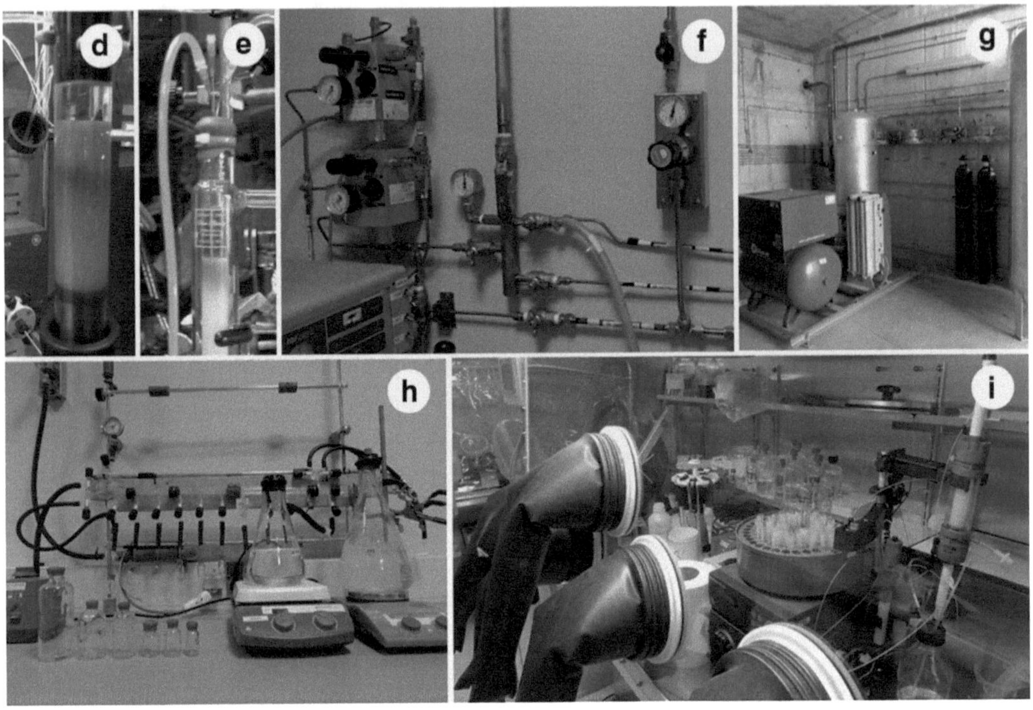

Fig. 1 (continued)

popular alternative. However, the use of Ar gas is required in specific nitrogenase-related experiments when the presence of N_2 might interfere with nitrogenase activity determinations.

In-house N_2 generators are an increasingly popular alternative to gas cylinders because ease of use, N_2 quality output (<5 ppm), and lower running costs. These systems are based on adsorption technology that relies in an adsorbent solid matrix with pores in which H_2O, O_2, and CO_2 molecules are immobilized whereas N_2 molecules pass through. Because adsorption matrixes eventually saturate, N_2 generators present duplicated fixing vessels and a sensor mechanism that switches regularly between the two, allowing them to run continuously. Thus, while a vessel is operative producing high-quality N_2, the other vessel is under regeneration by using a controlled-heating process until, eventually, the system switches back again. Importantly, N_2 generators storage gases at lower pressure than cylinders do (normally below 10 atm); therefore, bigger gas reservoirs are required (Fig. 1g). Downie presented a useful guide about industrial gas production and facts [19].

2.2 Gas Mixers

Some experiments require specific gas mixtures of two or more gases. The easiest approach is to order premade mixtures to gas suppliers that check the composition formula accurately. However, this strategy is usually expensive and not recommended when high volumes are needed. In this case, research facilities can rely in the use of gas mixers and their own gas sensor systems, or other devices, such as gas chromatographers, to calibrate their own gas mixes with high accuracy at a lower cost.

A variety of systems are available in the market: (1) multi-gas flow meters are handy, versatile and cheap to operate; they work at low pressures and flows, but output mix requires external calibration and testing that could be time consuming. However, once the correct parameters are found, these devices provide mixes reliable for routine laboratory work. (2) Mechanical gas mixers allow for higher pressures and flows; they are accurate (±1 % error in the mix) but are gas specific and more expensive (Fig. 1f). (3) A later generation of gas mixers combine mechanics, sensors, and electronics to deliver accurate gas mixes and flows that are analyzed to control gas quality. However, this sort of high-end devices is costly.

2.3 O_2 Traps and Sensor Alarms

The need for the purest N_2 for experimentation may require control sensor systems and additional O_2-free measures beyond the quality of supplied gas [20]. It is recommended that each gas workstation should have an O_2 trapping system attached after the pressure regulator. Although there are commercial solutions from a variety of companies, including Alltech Inc, PerkinElmer, Agilent, Sigma-Aldrich, it is also feasible to build an O_2 trap yourself (Fig. 1a). Most O_2 traps rely on molecular copper-oxide based adsorbents that act as O_2 scavengers. These systems can reduce O_2 presence to ppb values. BASF produces a variety of adsorbents dedicated to eliminate O_2 traces in pure gases. Importantly, it is possible to regenerate these catalysts periodically by applying a stream of H_2 and/or controlled heat.

Monitoring O_2 traces in gas lines is strongly recommended as a control measure. Currently, there are three main O_2 sensor technologies present in the market: (1) zirconia based sensors, (2) electrochemical sensors, and (3) paramagnetic sensors. The zirconium and electrochemical sensors are the most prevalent and both can detect O_2 in ppb values. Zirconium sensors are more robust and expensive, whereas the electrochemical sensors are relatively inexpensive but have a short-life, usually lasting no longer than a year.

Once the basic setup to deliver pure inert gas into a laboratory has been completed, attention must focus on how gas should be delivered to perform experiments in O_2-free atmosphere.

2.4 Schlenk Line and Rubber Septa Containers

A Schlenk line (or vacuum gas manifold) is an air-free system that connects a line of inert gas and a vacuum line by a controlling valve. In addition, Schlenk lines have multiple valve-controlled ports to connect hermetic containers thus allowing the manipulation, under O_2-free conditions, of several samples simultaneously (Fig. 1c, h).

In a nitrogenase research laboratory, independent N_2 or Ar lines exiting O_2-traps are connected to Schlenk manifolds.

A rotary vane RV3 Edwards pump connected to another manifold-end provides the necessary vacuum. If the Schlenk line is used for anaerobic buffer production, a water-trap is placed in between the manifold and the pump to prevent water vapor from reaching the pump oil reservoir and corroding the internal mechanism. Simple water-traps are based on an external jacket surrounding an internal deposit filled with ethanol and dry ice. Water vapor present in the flowing gas freezes in the deposit outer-jacket.

Because vacuum gas manifolds are usually made out of glass and because their design is inherently delicate, careful attention must be placed on the pressure and vacuum conditions applied. Installing accurate pressure and vacuum gauge(s) as well as an overpressure valve system is recommended as control measures. Glass manifolds should not be manipulated at pressures higher than 0.5 bars (50 kPa) or at vacuum conditions below −0.9 bar. It is very important that glass connectors and lines be properly sealed using vacuum grease and Teflon tape. Keeping positive pressure in any container or manifold, while not under vacuum, helps avoiding O_2 contamination.

2.5 Anaerobic Jars

Also named as McIntosh and Filde's anaerobic systems, these are hermetic anaerobic containers in which a modified atmosphere can be applied using a Schlenk line. Anaerobic jars are mostly used for culturing microbes under air-free conditions in agar-based media, but may also be used to culture microbes in liquid media, or to set up biochemical reactions under a specific gas composition. However, anaerobic jars are very limiting because no air-free manipulation is possible once the experiment is set. To monitor gas transfer to and from the jar, these systems usually present a vacuum/pressure gauge. It is also possible to eliminate O_2 inside the jar by inserting an anaerobic generation bag, which minimizes the need for a Schlenk line (*see* **Note 3**). The use of anaerobic indicator strips to confirm O_2-free atmosphere within the jar is recommended (*see* **Note 4**).

2.6 Glove Box

Glove boxes allow manipulating objects and processes in a closed chamber with controlled gas composition, typically an O_2-free N_2 atmosphere. Basic anaerobic glove boxes are designed with a set of systems to ensure O_2-free operation, including (1) a hermetic chamber, (2) butyl gloves for manipulation, (3) chemical catalyzer device to trap O_2 traces, (4) O_2 sensor, and (5) a transfer box or airlock. Most of these systems work in a 0–5 ppm O_2 range.

A good example of traditional anaerobic system would be the Coy Labs vinyl polymer series (Grass Lake, Michigan, USA) (Fig. 1b). Coy Labs glove boxes are affordable, highly reliable, and easy to service. The use of vinyl allows for a clear visual access while an external aluminum structure holds the bag. A catalyzer device based on a palladium matrix circulates the internal atmosphere gas while trapping contaminating traces of O_2. This catalyzer device works optimally at 30 °C and in the presence of 3–6 % H_2 in the gas phase. If O_2 contamination occurs the catalyst rapidly fixes it generating H_2O. The O_2 sensor unit reads O_2 contamination in ppm values (0–2,000) and also determines H_2 presence in % values. Delivery of gas (N_2/H_2 mix) or objects inside the glove box is performed through the airlock using a manual or automatic vacuum–N_2–vacuum–N_2–vacuum–N_2/H_2 mix cycle process. Analog and electronic airlocks exist that have door alarm systems to minimize risk of contaminating the glove box with external O_2.

More demanding experimental needs have further refined the glove box concept. Manufacturers such as Jacomex (Dagneux, France), MBraun (Garching, Germany), and GloveBox Systems (Massachusetts, USA) build metal anaerobic glove boxes units with front (and back) panels made of crystal clear polymer (Fig. 1b). Importantly, these systems may work at positive and negative pressures and have dedicated sealed ports for connecting devices out and inside the glove box. Several glove box systems can be interconnected allowing larger working space. In these glove boxes the O_2-trapping systems are placed on the outside, saving valuable working space, and operate by circulating the atmosphere through the catalyst and back into the box in coordination with O_2 and H_2O sensors. Safety is also improved because H_2 is not used routinely during glove box operation. Electronics play a major role integrating all sensor data, i.e., temperature, O_2 (in ppb), and humidity (*see* **Note 5**). Most of these glove boxes present large airlocks that can be manipulated in automatic mode, and also present very popular manually managed small airlocks. These systems can incorporate external cooling systems to work at temperatures 10–15 °C below room temperature, a very important asset when conducting lengthy protein purifications (Fig. 1i). In summary, these glove boxes are versatile and can be customized to researcher needs, but at a higher price tag.

3 Methods

3.1 Operation of Schlenk Lines

Transfer of vacuum or O_2-free N_2 or Ar gas from the manifold to the experimental vials is carried out through airtight Schlenk lines (*see* **Note 6**).

1. Make the Schlenk line anaerobic by performing three cycles of vacuum/gas with dummy vials connected.
2. Seal the sample vials hermetically with rubber stoppers and connect them to Schlenk line ports via tubing and needles (Fig. 1h).
3. Start by applying −0.9 bar of vacuum for 30 s.

4. Switch Schlenk line main valve from vacuum to N_2 flow-in position to fill the vials with N_2
5. Repeat the vacuum/N_2 cycle at least three more times.
6. After the last cycle, the vials must be kept under slight positive pressure to minimize O_2 contamination through the stopper. Remove the vials from the manifold.
7. At the end of the day, make sure that the manifold has at least one port open to atmospheric pressure before switching off vacuum pump.

3.2 Preparation of Anaerobic Buffers

Total procedure takes 1–2 h per liter of buffer.

1. Connect a hermetically sealed Büchner flask containing aerobic buffer solution to a Schlenk line while flowing out N_2. Switch to vacuum mode (9.48 kPa or −0.9 bar) with gentle magnetic stirring. Vacuum mode forces dissolved gases out of the liquid phase.
2. Switch the Schlenk line main valve at least three times from the vacuum position to the N_2 flow-in position to refill the flask with gas to accelerate air removal.
3. Fill the Büchner flask with excess N_2 after the final cycle to keep positive gas pressure on the inside.
4. Büchner flasks can be stored in the laboratory bench or inside a glove box.

3.3 Addition of Sodium Dithionite as Reductant and O_2 Scavenger

Sodium dithionite ($Na_2S_2O_4$), abbreviated DTH, is an important component of buffers used for anaerobic protein purification. DTH stock solutions must be freshly prepared (*see* **Note 7**).

1. Store solid DTH stocks away from the light inside anaerobic jars with desiccant sieve at room temperature.
2. Weight DTH and degas it at the Schlenk line within serum vials hermetically closed with rubber stoppers.
3. Add anaerobic 50 mM Tris–HCl buffer, pH 8, to make a 100 mM DTH stock solution.
4. Use this stock solution to make 1–2 mM DTH-containing anaerobic buffers.
5. Check buffer reduction with methyl viologen-soaked paper strips (*see* **Note 8**).

3.4 Anaerobic Manipulation of Samples from Liquid N_2 into Vials

Liquid nitrogen (LN_2) is useful in sample manipulation outside a glove box because its low boiling temperature point (−196 °C) combined with its low O_2 content prevents samples from oxidation (*see* **Note 9**). O_2-sensitive protein samples are stored as droplets in cryogenic dewars under LN_2 to prevent oxidation. To thaw protein samples proceed as follows:

1. Place an empty glass vial into a beaker and fill both with LN_2.
2. Add frozen droplets to the vial using tweezers. Calculate the volume of sample needed and remove from the dewar the exact number of droplets (estimate 30–50 μl per droplet).
3. Approach the Schlenk line. Decant excess LN_2 from the vial while applying a flow of pure N_2 from the manifold and into the vial.
4. Seal the vial with a rubber stopper, connect it to the Schlenk line via a needle and switch immediately to vacuum for 1 min, or until LN_2 completely evaporates. Switch back to the N_2 position.
5. Repeat vacuum/N_2 gas cycle three more times.
6. Remove vial from manifold and place it on ice until protein is thawed.

3.5 How to Set Up an Anaerobic Protein Purification Method Without a Glove Box

It is possible to perform anaerobic protein purifications without a glove box if procedures described in this chapter are carefully applied. By ensuring that running buffer is anaerobic, and that tubing connections, peristaltic pump, chromatography column, UV/Vis detector, and sample collecting vials are airtight, there is a good chance that anaerobic protein purification will succeed.

3.5.1 Preparation of Chromatography Columns

1. Buffers must be O_2 free.
2. If possible, add 1–2 mM DTH to the buffers to scavenge O_2 traces and keep reducing conditions (*see* **Note 10**).
3. Peristaltic pump tubing and connectors to the sample flask and to the column must be airtight. Wash all tubing with anaerobic buffer before handling the sample. When disconnecting tubes, use clamps to stop buffer flow. Reassemble as fast as possible avoiding bubbles.
4. Apply 10–20 column volumes (CV) of anaerobic buffer to the column to turn stationary phase anaerobic. For instance, an FPLC system attached to a column with 50 ml of Q-sepharose resin may need around 0.6–0.8 l of DTH-containing anaerobic buffer for the methyl viologen strip test to indicate a reduced condition (blue) at the column exit (*see* **Note 11**).
5. Connect peristaltic pump tubing to flask containing protein sample using clamps to stop flow.
6. Load sample onto the column and run chromatography (Fig. 1d).

3.5.2 Collecting Fractions from a Chromatography Column

Sample collection has to be performed manually. Degassed vials with dual-sealed rubber caps are used (Subheading 3.1). Column outlet must be connected to tubing ending in a needle. To collect samples, simply insert needle into a degassed vial. Two additional needles must be in place in the vial rubber cap before inserting the collection needle: an entry N_2 line and a needle to release overpressure (Fig. 1e).

3.5.3 Concentrating Proteins in an Amicon Cell

It is important to carefully render the membrane and the Amicon cell anaerobic before applying the protein sample.

1. Assemble the Amicon cell and membrane on the bench and connect the former to the tubing that will deliver N_2 to pressurize the device.
2. In addition, apply a constant low flow of N_2, by using tubing and a bended needle, into the upper opening of the Amicon cell for 3–5 min. This gas should be supplied anytime the pressure valve is not in position.
3. Fill one half of the Amicon cell with DTH-reduced anaerobic buffer. Wash the cell walls by gentle twirling. Remove upper gassing needle and close pressure valve.
4. Place the Amicon cell on slow magnetic stirring and let all the buffer solution flow through the ultrafiltration membrane to reduce it. Check buffer reduction status with methyl viologen strips.
5. Once reduced, open the pressure valve and load the protein sample by using an anaerobic syringe and needle. Close the valve and place the cell back on magnetic stirrer until desired volume is left in the cell.
6. Remove sample with anaerobic syringe and needle (protected with soft rubber to avoid puncturing the ultrafiltration membrane), and transfer it to an anaerobic vial.

3.5.4 Storing Proteins Under Liquid N_2

O_2-sensitive proteins are routinely stored under LN_2 as droplets (*see* **Note 12**).

1. Place an empty polypropylene vial, with tiny holes in it, inside a beaker and fill both with LN_2. Use safety glasses.
2. Use anaerobic syringe and needle to slowly drop protein solution into LN_2. Do not place the needle too close to the LN_2 because it will freeze, thereby stopping the flow.

4 Notes

1. Safety and maintenance tips. Gas leaks are not just a menace to experimental work, but also expensive and potentially dangerous due to the nature of some gas mixes used for experimentation. Proper maintenance of gas systems and tubing must be periodically carried out. Maintenance of high-pressure systems attached to gas cylinders should be outsourced to professionals. Special attention should be paid to pressure regulators, valves, anti-return valves, tubing connections, and cylinder manifolds. Importantly, only certified professionals should perform modifications to these installations.

Laboratory personnel must be trained to manage the low-pressure gas lines connected to the different instruments. Importantly, proper tools should be available in any anaerobic biochemistry laboratory. A regular wrench, a grip adjustable wrench, a ratchet set, a drive socket set, screwdrivers, pliers and locking plier sets, a metric tape, and hummer tools should be available. Thread seal tape and pipe thread tube compound are essential to ensure that gas line components are properly sealed.

2. Screw thread, fittings and tubing. The two main screw thread options found in gas lines and laboratory equipment are the ISO Metric screw thread and the British Standard Pipe (BSP) thread, which is based on the pioneer British Standard Whitworth. Additionally, some USA-made pieces of equipment use National Pipe Thread Taper (NPT) standard. It is necessary to emphasize that, although these standard systems have—on occasions—quite similar thread patterns (for example, ISO Metric and NPT use an identical 60° angle), these should not be mixed (for example, the British Standard Whitworth uses a 55° angle).

   Problems emerge when a laboratory has pieces of equipment from different geographic origins, because manufactures usually rely in their home country thread standards. As a result, interconnecting equipment and gas lines may become a practical nightmare. Traditionally, gas lines used to be installed in stainless steel or copper tubing, which guarantee robustness at the cost of lacking flexibility, further complicating the compatibility issue. The proliferation of new fast-lock and semi-fast-lock systems, complemented with the use of better plastic tubing materials (polyamide,- polyurethane, and PTFE), has dramatically simplified this problem.

3. Fluka Analytical provides one fast and convenient solution to limit O_2 concentration to <1 % in anaerobic jars (up to a volume of 3.5 l). This technology is based on anaerobic atmosphere generation bags (Oxoid AnaeroGen developed by Mitsubishi Gas Company Inc.). Interestingly, this solution does not rely in the use of H_2 and a catalyst to produce water but in the use of ascorbic acid as an active component yielding an atmosphere with 9–13 % CO_2 content [21].

4. Fluka Analytical provides a simple solution to test anaerobiosis in enclosed compartments based on test strips saturated with resazurin solution. When the strips change color from pink to white anaerobic conditions have been achieved.

5. Most modern glove boxes are designed for processes that require a <0.1 % relative humidity (i.e., atomic applications, chemical synthesis, batteries, laser welding, etc.). Therefore, these systems are equipped with reliable zirconium sensors. However, biochemical anaerobic work normally requires some

humidity inside the glove box to avoid rapid desiccation of samples. Importantly, zirconium sensors are not recommended in those humid conditions and need to be replaced by electrochemical sensors.

6. An easy method to spot gas leaks is applying soapy water to the suspected leak area. If positive, bubbles immediately form. There are spray cans specifically manufactured for this task. However, these solutions only work well for significant leaks and are not effective to spot small leaks that can be as damaging to the experimental work. It is recommended to have one gas leak detector system in the laboratory to test gas lines and glove boxes when a leak is suspected. Ultrasonic detectors, infrared point sensors, and electrochemical gas detectors are commercially available. Ultrasonic detectors are probably the most prevalent because they are not gas specific and pinpoint the leak relying in the fact that most gas leaks are audible in the 25 kHz to 10 MHz range.
7. Although added to the buffers to prevent protein oxidative damage, DTH is, itself, susceptible to oxidation breaking down into sulfur compounds that lower buffer pH and might damage protein [Fe–S] clusters. Therefore, DTH must be added to the buffers only after they have been made anaerobic and just before their use. It is strongly recommended not to use DTH-containing buffers older than 1 day. However, there are some drawbacks when using DTH. For example, it over-reduces Ni and is not recommended in Ni^{2+} affinity chromatography protocols.
8. Methyl viologen detecting paper strips are used to test buffer anaerobiosis. Whatman filter paper is soaked in 250 mM Tris–HCl with 1 % (w/v) methyl viologen and then dried in the oven at 80 °C. The dried paper is cut into small strips of 8×5 mm^2 surface. When exposed to a DTH-reduced buffer solution, white paper strips turn blue.
9. LN_2 is generated after purifying N_2 from air in a size exclusion process followed by compression, and is stored in cryogenic pressurized reservoirs. The resulting LN_2 presents O_2 concentrations lower than 0.05 %. Commercial LN_2 with O_2 contents of 5 ppm of lower can also be obtained. A potential oxidation risk for metalloproteins immersed in LN_2 dewars occurs when O_2 from air condensates overtime to form liquid O_2 (LO_2, boiling point −183 °C). LN_2 contaminated with LO_2 turns pale blue. This situation should be avoided at all costs.
10. The anaerobic buffer container must be pressurized to avoid O_2 contamination. For this purpose, we use gas lines with needles to puncture Büchner flask caps. Slow N_2 or Ar flow is needed, but also a release needle valve of smaller diameter has to be in place. This system must be replicated in the collecting sealed tubes.

11. Because of the large volume required to reduced a resin, it may be convenient to turn the system anaerobic overnight, so that protein purification starts early next morning. Importantly, this procedure has to be repeated anytime a new purification procedure is planned.
12. If a glove box is available, it is possible to use it to store purified protein preparations in aliquots into hermetically sealed cryotubes. In glove boxes with installed cold traps, cryotubes can be frozen inside the glove box. Otherwise, freeze closed cryotubes in LN_2 just after removing them from the airlock.

Acknowledgments

This work was supported by ERC Starting Grant 205442 and by MICINN Grant BIO2009-12661.

References

1. Waldron KJ, Robinson NJ (2009) How do bacterial cells ensure that metalloproteins get the correct metal? Nat Rev Microbiol 7:25–35
2. Knoll AH, Canfield DE, Konhauser KO (2012) Fundamentals of geobiology. Wiley-Blackwell, New York
3. Fontecilla-Camps JC, Amara P, Cavazza C, Nicolet Y, Volbeda A (2009) Structure-function relationships of anaerobic gas-processing metalloenzymes. Nature 460:814–822
4. Hoffman BM, Lukoyanov D, Dean DR, Seefeldt LC (2013) Nitrogenase: a draft mechanism. Acc Chem Res 46:587–595
5. Drennan CL, Heo J, Sintchak MD, Schreiter E, Ludden PW (2001) Life on carbon monoxide: X-ray structure of *Rhodospirillum rubrum* Ni-Fe-S carbon monoxide dehydrogenase. Proc Natl Acad Sci U S A 98:11973–11978
6. Lindahl PA (2012) Metal-metal bonds in biology. J Inorg Biochem 106:172–178
7. Fontecave M (2006) Iron-sulfur clusters: ever-expanding roles. Nat Chem Biol 2:171–174
8. Johnson DC, Dean DR, Smith AD, Johnson MK (2005) Structure, function, and formation of biological iron-sulfur clusters. Annu Rev Biochem 74:247–281
9. Dos Santos PC, Dean DR (2008) A newly discovered role for iron-sulfur clusters. Proc Natl Acad Sci U S A 105:11589–11590
10. Imlay JA (2006) Iron-sulphur clusters and the problem with oxygen. Mol Microbiol 59:1073–1082
11. De Lacey AL, Fernandez VM, Rousset M, Cammack R (2007) Activation and inactivation of hydrogenase function and the catalytic cycle: spectroelectrochemical studies. Chem Rev 107:4304–4330
12. Stripp ST, Goldet G, Brandmayr C, Sanganas O, Vincent KA, Haumann M et al (2009) How oxygen attacks [FeFe] hydrogenases from photosynthetic organisms. Proc Natl Acad Sci U S A 106:17331–17336
13. Wong PP, Burris RH (1972) Nature of oxygen inhibition of nitrogenase from *Azotobacter vinelandii*. Proc Natl Acad Sci U S A 69:672–675
14. Kennedy MC, Emptage MH, Dreyer JL, Beinert H (1983) The role of iron in the activation-inactivation of aconitase. J Biol Chem 258:11098–11105
15. Sutton VR, Mettert EL, Beinert H, Kiley PJ (2004) Kinetic analysis of the oxidative conversion of the [4Fe-4S]2+ cluster of FNR to a [2Fe-2S]2+ cluster. J Bacteriol 186:8018–8025
16. Sofia HJ, Chen G, Hetzler BG, Reyes-Spindola JF, Miller NE (2001) Radical SAM, a novel protein superfamily linking unresolved steps in familiar biosynthetic pathways with radical mechanisms: functional characterization using new analysis and information visualization methods. Nucleic Acids Res 29:1097–1106

17. Marsh EN, Patterson DP, Li L (2010) Adenosyl radical: reagent and catalyst in enzyme reactions. Chembiochem 11:604–621
18. University of California-San Francisco Medical Center (2013) Structure-Function Linkage Database http://sfld.rbvi.ucsf.edu/django/superfamily/29/sequences/all/
19. Downie NA (2002) Industrial gases. Kluwer Academic, Boston, MA
20. McIlwrick CR, Phillips CS (1973) The removal of oxygen from gas streams: applications in catalysis and gas chromatography. J Phys E Sci Instrum 6:1208–1210
21. Beerens H (1998) Bifidobacteria as indicators of faecal contamination in meat and meat products: detection, determination of origin and comparison with *Escherichia coli*. Int J Food Microbiol 40:203–207

Chapter 3

Expression and Purification of NifB Proteins from Aerobic and Anaerobic Sources

Carlos Echavarri-Erasun, Simon Arragain, Alessandro A. Scandurra, and Luis M. Rubio

Abstract

NifB is the key protein in the biosynthesis of nitrogenase iron-molybdenum cofactor. Due to its extreme sensitivity to O_2 and inherent protein instability, NifB proteins must be purified under strict anaerobic conditions by using affinity chromatography methods. We describe here the methods for NifB purification from cells of the strict aerobic nitrogen-fixing bacterium *Azotobacter vinelandii*, the facultative anaerobic nitrogen-fixing bacterium *Klebsiella pneumoniae*, and the facultative anaerobic non-nitrogen fixing bacterium *Escherichia coli* recombinantly expressing a *nifB* gene of thermophilic origin.

Key words Anaerobic protein purification, *Azotobacter vinelandii*, *Klebsiella pneumoniae*, *Escherichia coli*, NifB, NifB-co, Nitrogenase, Immobilized metal affinity chromatography (IMAC), Glutathione-*S*-transferase (GST), Phenylmethylsulfonyl fluoride (PMSF), Sodium dithionite (DTH)

1 Introduction

NifB is a key protein in the biosynthesis of nitrogenase. Its importance relies on the fact that it not only catalyzes the unprecedented synthesis of the complex [6Fe-9S-C] central cage of the iron-molybdenum cofactor (FeMo-co) [1], but is also required for the biosynthesis of the FeV-co and FeFe-co of the V- and Fe-only-nitrogenases, respectively [2]. Thus, NifB is required for all biological nitrogen fixation activity in Nature.

Literature regarding NifB protein is scarce due to the difficulties of purifying this protein intact and in soluble form. Surprisingly, NifB catalytic product (NifB-co) was purified earlier than the NifB protein, and its role as intermediate in FeMo-co biosynthesis clearly established [3]. It was Curatti and collaborators who first succeeded in purifying NifB from *A. vinelandii* (AvNifB) demonstrating its essential role in the in vitro FeMo-co synthesis assay and its *S*-adenosylmethionine dependence, which confirmed

Juan C. Fontecilla-Camps and Yvain Nicolet (eds.), *Metalloproteins: Methods and Protocols*,
Methods in Molecular Biology, vol. 1122, DOI 10.1007/978-1-62703-794-5_3,

the involvement of radical chemistry in NifB-co synthesis [4]. GST-tagged NifB purified from *K. pneumoniae* (KpNifB) was different from AvNifB in that it carried bound NifB-co in the as isolated form. Thus, pure KpNifB was found to be readily active in the in vitro FeMo-co synthesis assays [5]. A third form of NifB has been purified from *A. vinelandii* by using a gene fusion with NifN [6]. This purified NifENB complex permitted tracking the origin of the central C atom of FeMo-co to the SAM molecule of NifB [7].

This chapter includes methods for the expression and purification of AvNifB and KpNifB proteins. In addition, a method to purify a shorter and more robust NifB protein from the thermophilic bacterium *Methanocaldococcus infernus* (MiNifB) in recombinant *E. coli* cells is provided (*see* **Note 1**).

2 Materials

Unless otherwise specified, all chemicals are from Sigma-Aldrich (St. Louis, Mo, USA). All buffers have been made O_2 free by degassing at Schlenk lines or by sparging N_2 for 60 min (*see* Chapter 1). Sodium dithionite (DTH) is added to degassed buffers when indicated to eliminate O_2 traces and to keep buffers under low redox potential conditions.

2.1 Cell Growth

A. vinelandii strain UW232 carries an engineered *nif*B_{his}-*nifHDK* operon under the control of the *nifH* promoter [4]. *K. pneumoniae* UC16 strain carries a glutathione-*S*-transferase (GST)-*nifB* chimeric gene under the control of the *tac* promoter in expression vector pGEX-4T-3 [5]. *E. coli* BL21 (DE3) pRHB511 strain carries the *M. infernus nifB* gene, codon optimized for *E. coli*, under the control of a T7 promoter in expression vector pET16b.

1. *A. vinelandii* Burk's nitrogen-free media consists of the BB and BS components that are autoclaved separately and mixed at room temperature in sterile conditions [8]. Burk's media is supplemented to 29 mM ammonium acetate final concentration, when growing cells in non-diazotrophic conditions.
2. BB component (10× stock): 0.14 M KH_2PO_4 and 0.45 M K_2HPO_4.
3. BS component (10× stock): 8 mM $MgSO_4$, 6.1 mM $CaCl_2$, 0.18 mM $FeCl_3$, 0.01 mM Na_2MoO_4, and 2 % (w/v) sucrose.
4. Ammonium acetate stock: 22 % (w/v).
5. *K. pneumoniae* minimal growth medium consists of solutions I and II [9]. Both solutions are autoclaved separately and mixed in sterile conditions. The medium is supplemented to 17.1 mM ammonium acetate, final concentration, when growing in non-diazotrophic conditions.

6. Solution I (10×): 170 mM Na_2HPO_4 and 22 mM KH_2PO_4.
7. Solution II (10×): 10 mM $MgSO_4$, 170 mM NaCl, 6.8 mM $CaCl_2$, 0.18 mM $FeCl_3$, 0.01 mM Na_2MoO_4, and 2 % (w/v) sucrose.
8. *E. coli* Luria Bertani (LB) medium: 10 g tryptone, 5 g yeast extract, and 10 g NaCl per liter supplemented with the required antibiotics (150 μg/ml ampicillin or 50 μg/ml chloramphenicol).

2.2 Buffers and Protease Inhibitors

1. PMSF: 100 mM phenylmethylsulfonyl fluoride (PMSF) stock in dimethyl sulfoxide (DMSO) is kept away from the light and at room temperature.
2. Leupeptin: 0.5 mg/ml leupeptin stock in 100 mM Tris-HCl, pH 8 solution is kept in 500 μl aliquots at −20 °C.
3. DTH stock solutions are always freshly prepared. DTH is weighted and degassed at the Schlenk line within serum vials hermetically closed with rubber stoppers. Anaerobic Tris–HCl, 50 mM pH 8, buffer is added to create a 100 mM stock DTH solution.

2.3 AvNifB Purification

1. 20 ml Co^{2+} His-tag affinity column (Talon resin, Clontech).
2. 20 ml Q-sepharose column (GE Healthcare).
3. Buffer A: 50 mM sodium phosphate, pH 7.4, 4 M glycerol, 1.7 mM DTH.
4. Buffer B: 50 mM sodium phosphate, pH 7.4, 20 mM imidazole, 0.1 % *n*-dodecyl-β-D-maltopyranoside, 5 mM β-mercaptoethanol, 0.5 mM DTH.
5. Buffer C: 50 mM sodium phosphate, pH 7.4, 400 mM NaCl, 10 % glycerol, 20 mM imidazole, 0.1 % *n*-dodecyl-β-D-maltopyranoside, 5 mM β-mercaptoethanol, 0.5 mM DTH.
6. Buffer D: 50 mM Tris–HCl, pH 8, 300 mM imidazole, 20 % glycerol, 0.2 % *n*-dodecyl-β-D-maltopyranoside, 5 mM β-mercaptoethanol, 1 mM DTH.
7. Buffer E: 50 mM Tris -HCl, pH 8, 75 mM imidazole, 5 % glycerol, 0.05 % *n*-dodecyl-β-D-maltopyranoside, 5 mM β-mercaptoethanol, 1 mM DTH.
8. Buffer F: 50 mM Tris -HCl, pH 8, 400 mM NaCl, 20 % glycerol, 0.2 % *n*-dodecyl-β-D-maltopyranoside, 5 mM β-mercaptoethanol, 1 mM DTH.

2.4 KpNifB Purification

1. 20 ml GST-sepharose FF column (GE Healthcare).
2. Buffer G: 100 mM Na_2HPO_4, 10 mM KH_2PO_4, pH 8.5, 280 mM NaCl, 5.4 mM KCl, 10 % glycerol, 5 mM β-mercaptoethanol, 1 mM DTH.

3. Buffer H: 50 mM Tris–HCl, pH 8, 280 mM NaCl, 10 % glycerol, 0.1 % *n*-dodecyl-β-D-maltopyranoside, 5 mM β-mercaptoethanol, 10 mM reduced glutathione (GSH), 1 mM DTH.

2.5 MiNifB Purification

1. 5 ml Ni^{2+}-HiTrap metal affinity column (GE Healthcare).
2. 53 ml HiPrep 26/10 desalting column (GE Healthcare).
3. Buffer I: Tris–HCl 50 mM pH 8, 200 mM KCl, 30 mM imidazole, 10 % glycerol, 5 mM β-mercaptoethanol, 0.05 % *n*-dodecyl-β-D-maltopyranoside, 0.1 % Triton X-100, and 0.1 % Tween 20.
4. Buffer J: 50 mM Tris–HCl, pH 8, 200 mM KCl, 30 mM imidazole, 10 % glycerol, and 5 mM β-mercaptoethanol.
5. Buffer K: 50 mM Tris–HCl, pH 8, 200 mM KCl, 100 mM imidazole, 10 % glycerol, and 5 mM β-mercaptoethanol.
6. Buffer L: 50 mM Tris–HCl, pH 8, 200 mM KCl, 250 mM imidazole, 10 % glycerol, and 5 mM β-mercaptoethanol.
7. Buffer M: 50 mM Tris–HCl, pH 8, 200 mM KCl, 10 % glycerol, and 5 mM β-mercaptoethanol

3 Methods

3.1 Expression and Purification of His-Tag NifB from *A. vinelandii* (AvNifB)

A standard *A. vinelandii* 32-1 fermentation process yields around 75 g frozen cell paste.

3.1.1 *A. vinelandii* UW232 Cell Growth Conditions

1. Four 100-ml baffled flasks containing 25 ml of Burk's medium supplemented with 29 mM ammonium acetate are inoculated with individual *A. vinelandii* colonies and grown overnight at 30 °C under vigorous shaking (250 rpm) until the culture reaches an OD of 2–3.
2. Autoclave four 10-l glass fermenters equipped with a propeller stirring system and a ceramic air-diffuser capable of delivering 5 l/min of very fine oil-free filtered air bubbles (*see* **Note 2**). Each fermenter contains 8 l of Burk's medium supplemented with limiting ammonium (5.7 mM ammonium acetate, 80 μg/ml final in N).
3. Inoculate fermenters with cells from the flasks cultures, in a laminar flow hood, to an initial OD of 0.02.
4. Place fermenters in a large water bath at 30 ° C, start air sparging and monitor OD_{600nm} changes and residual ammonium in the medium every hour (*see* **Note 3**). Ammonium depletion is programmed to occur around 8–9 h after fermenter inoculation when OD_{600nm} reaches 1.2.

5. Continue incubating cultures for 3 additional hours after ammonium depletion to derepress nitrogenase. Additional cell duplication may occur during this period.
6. In vivo nitrogenase (acetylene reduction) activity and OD_{600nm} changes are monitored at 1 h intervals during derepression by withdrawing 1-ml aliquots from each culture (*see* **Note 4**). Data obtained is very valuable to predict the levels of Nif protein production.
7. Collect cells 3 h after derepression (OD_{600nm} around 2) by using a tangential filtration hollow fiber device (*see* **Note 5**). Centrifuge to obtain cell paste, freeze 1–2 g chunks of paste in liquid N_2 and store at −80 °C. Frozen cells can be stored for years under these conditions.

3.1.2 Preparation of A. vinelandii UW232 Cell-Free Extracts

Cell lysis by anaerobic osmotic shock [8] followed by ultracentrifugation is used to prepare particle-free protein extracts. Samples must be kept cold whenever possible.

1. 75 g of frozen *A. vinelandii* cells are transferred into a glove box and are resuspended into 225 ml of anaerobic buffer A (*see* **Notes 6–8**), supplemented with 0.46 ml PMSF stock, 0.12 ml leupeptin stock, and 230 μg of freshly weighted DNAse I, with a magnetic stirring for 15 min.
2. Transfer cells to hermetically sealed centrifuge bottles, moved them out of the glove box and spin down cells for 10 min at 43,000 × *g* at 4 °C. Have special care in sealing centrifuge bottles properly and use caps that allow pressurizing bottles with N_2.
3. Transfer the bottles back into the glove box, discard supernatant, resuspend pellet in 225 ml of buffer B, and homogenize vigorously during 5 min using a homogenizer. Osmotic lysis occurs at this step.
4. Transfer cells to hermetically sealed centrifuge bottles, moved them out of the glove box and spin down cells for 90 min at 43,000 × *g* at 4 °C to remove cell debris.
5. Inside the glove box, transfer supernatant (cell-free extract) to a 250-ml serum vial, supplement with 400 mM NaCl and 10 % glycerol, and mix gently.

3.1.3 AvNifB Purification

This procedure is conducted under a N_2 atmosphere at 4–8 ° C and consists of a Co^{2+} affinity chromatography followed by a Q-sepharose anion exchange chromatography.

1. Pack and equilibrate 20-ml Co^{2+} resin into a 2.5-cm × 10 cm column.
2. Reduce resin with 10 column volumes (CV) of buffer C.

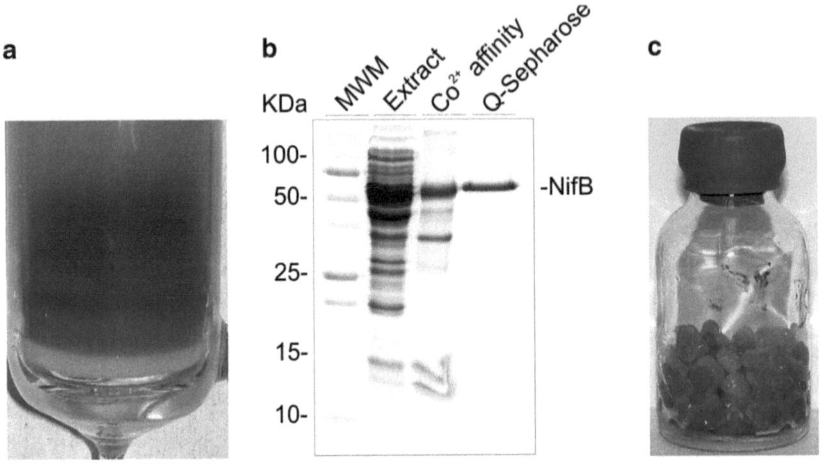

Fig. 1 (**a**) NifB elution from a Co^{2+} affinity chromatography. (**b**) SDS-PAGE analysis of AvNifB purification steps. (**c**) NifB droplets frozen on LN_2 for storage

3. Load freshly prepared cell-free extract onto the Co^{2+} column at a flow rate of 2 ml/min.
4. Wash with 5 CV buffer C.
5. Wash with 5 CV buffer C supplemented to 60 mM imidazole.
6. Elute NifB with 2 CV buffer D (Fig. 1a).
7. Dilute elution fractions fourfold with 50 mM Tris-HCl, pH 8, 5 mM β-mercaptoethanol, 1 mM DTH.
8. Pool eluted fractions and load them at a 2 ml/min flow rate onto a 20-ml Q-sepharose column previously equilibrated with 10 CV of buffer E.
9. Wash with 5 CV buffer E supplemented to 180 mM NaCl.
10. Elute NifB with 2 CV Buffer F.
11. Check NifB purity by SDS-PAGE (Fig. 1b).
12. Concentrate NifB by ultrafiltration through a 30-kDa membrane filter in an Amicon device.
13. Store concentrated NifB preparation in 30–50 μl drops under liquid nitrogen (Fig. 1c).

3.2 Expression and Purification of GST-NifB from *Klebsiella pneumoniae* (KpNifB)

A standard *K. pneumoniae* 180-l fermentation process yields around 250 g frozen cell paste.

3.2.1 *K. pneumoniae* UC16 Cell Growth Conditions

1. Four 500-ml baffled flasks containing 125 ml of *Klebsiella* growth medium supplemented to 17.1 mM ammonium acetate are inoculated with individual *K. pneumoniae* colonies and grown overnight at 30 °C under vigorous shaking (250 rpm).
2. Autoclave four 50-l polypropylene carboys equipped with a homemade air-diffusing system capable of delivering 25 l/min

of filtered gas. Each carboy contains 45 l of *Klebsiella* growth medium supplemented with 2.8 mM ammonium acetate.

3. Inoculate carboys with cells from the flasks cultures, in a laminar flow hood, to an initial OD of 0.02.
4. Place carboys inside a large water bath with temperature regulated to 30 °C, start air sparging, and monitor OD_{600nm} changes and residual ammonium in the medium every hour (*see* **Note 3**). Ammonium depletion is programmed to occur around 6–8 h after fermenter inoculation when OD_{600nm} reaches 1.4.
5. After ammonium depletion, switch sparging to 97 % N_2, 3 % CO_2 anaerobic mix. Add 0.95 mM L-serine and 5 μM IPTG to derepress nitrogenase and induce NifB expression, respectively, and incubate cultures for 4 additional hours.
6. In vivo nitrogenase (acetylene reduction) activity and OD_{600nm} changes are monitored at 30 min intervals during derepression by withdrawing 1-ml anaerobic aliquots from each culture (*see* **Note 9**).
7. Collect cells by using a tangential filtration hollow fiber device under anaerobic conditions (*see* **Notes 5** and **10**). Centrifuge at 7,000 × *g* and 4 °C for 10 min to obtain cell paste.
8. Freeze 1–2 g chunks of paste in liquid N_2 and store under liquid N_2. Frozen cells can be stored for years under these conditions.

3.2.2 Preparation of K. pneumoniae UC16 Cell-Free Extracts

Lysis of *K. pneumoniae* cells is achieved by French Press treatment (*see* **Note 11**).

1. Two hundred and fifty grams of frozen *K. pneumoniae* cells are transferred from a liquid N_2 dewar into a glove box and resuspended in 750 ml of buffer G supplemented with 1.5 ml PMSF stock, 0.38 ml leupeptin stock, and 750 μg of freshly weighted DNAse I.
2. Resuspension takes around 30 min and is performed at room temperature by using a magnetic stirrer and an IKA homogenizer.
3. Transfer cells to a hermetically sealed 1-l Pyrex flask, remove it from the glove box, and place it in ice. An additional, empty, sealed flask is needed for the lysis process.
4. Lysis is performed in a cold room by using French Press at 15,000 psi under anaerobic conditions (*see* **Note 12**). A N_2 gas line inside the cold room is used to constantly flow N_2 gas into the entry (full) and the exit (empty) flasks. Repeat sample lysis by switching entry and exit flasks.

5. Immediately transfer lysate into the glove box and check pH. If required, adjust pH to 7.4 by adding buffer G.
6. Supplement lysate with *n*-dodecyl-β-D-maltopyranoside to 0.1 % (v/w) and incubate for 10 min with gentle stirring.
7. Transfer to hermetically sealed centrifuge bottles and spin down for 60 min at 44,000 × *g* at 4 °C to remove cell debris.
8. Inside the glove box, transfer supernatant (cell-free extract) to a 1-l flask and supplement with Triton X-100 to 0.1 % (v/v).

3.2.3 KpNifB Purification

This procedure is conducted under N_2 atmosphere at 4–8 °C and consists of a GSH-affinity chromatography followed by concentration and desalting on an Amicon cell.

1. Pack and equilibrate 20-ml of GSH-Sepharose resin into a 2.5-cm × 10 cm column.
2. Reduce resin with 10 column volumes (CV) of buffer G.
3. Load freshly prepared cell-free extract onto the GSH-sepharose column at a flow rate of 5 ml/min.
4. Wash with 3 CV of buffer G supplemented with 1 % Triton X-100.
5. Wash with 10 CV of buffer G.
6. Elute GST-NifB with 3 CV of buffer H.
7. Inside a glove box, concentrate GST-NifB by ultrafiltration through a YM100 membrane, in an Amicon device, to 3–5 mg NifB/ml.
8. Buffer can be exchanged by repeating NifB concentration/dilution cycles on the Amicon.
9. Store GST-NifB preparation in 30–50 μl drops under liquid nitrogen.

3.3 Expression and Purification of M. infernus His-NifB from Recombinant E. coli Cells

3.3.1 E. coli Cell Growth Conditions

A standard *E. coli* 5-l fermentation process lasts 8–9 h and yields around 250 g frozen cell paste.

1. Inoculate a Sartorius Biostat fermenter containing 5 l of LB medium with a starter culture of *E. coli* BL21 (DE3) pRHB511cells to an initial OD of 0.10 (*see* **Note 13**).
2. When culture reaches a OD_{600nm} of 17, induce NifB expression by adding 20 ml of 20 % lactose solution and lowering the temperature to 26 °C (*see* **Note 14**).
3. Continue incubation for 2.5 additional hours after lactose addition to reach a OD_{600nm} value of 18.
4. Collect cells under anaerobic conditions in hermetically sealed centrifuge bottles and spin down at 7,000 × *g* and 4 °C for 10 min.

5. Freeze 1–2 g chunks of cell paste in liquid N_2 and store under liquid N_2. Frozen cells can be stored for years under these conditions.

3.3.2 Preparation of Anaerobic E. coli Cell-Free Extracts

Lysis of *E. coli* cells is carried out by sonication inside a glove box.

1. Transfer 100 g of frozen *E. coli* cells from the liquid N_2 dewar into the glove box and resuspended in 800 ml buffer I supplemented 1.5 ml PMSF stock, 0.38 ml leupeptin stock, and 750 μg of freshly weighted DNAse I. It takes around 15 min for the cells to resuspend properly by using a magnetic stirrer.
2. Lyse cells with a sonicator (UP 100H Ultraschallprozessor) equipped with 12 mm tip. Sonicate for 10 min at 90 % power output using 10 s pulse/15 s pause cycles.
3. Transfer lysed cells to hermetically sealed centrifuge bottles and centrifuge for 40 min at 47,000 × *g* at 4 °C.
4. Collect supernatant, transfer to clean hermetically sealed centrifuge bottles inside glove box and incubate bottles at 75 °C in a water bath outside the glove box for 30 min. Observe protein precipitation (*see* **Note 15**).
5. Centrifuge for 20 min at 30,000 × *g* at 20 °C to remove precipitated proteins.
6. Inside glove box, transfer supernatant to a clean flask to start chromatographic separations.

3.3.3 MiNifB Purification

Most steps in this protocol are conducted at room temperature inside a Coy Labs glove box and consist of Ni^{2+} affinity chromatography followed by desalting on a HiPrep 26/10 column.

1. Load a 5 ml HiTrap IMAC column with 2–3 CV of 200 mM of $NiSO_4$ solution. Remove metal excess by washing with 10 CV of degassed water.
2. Equilibrate HiTrap IMAC resin with 10 CV of buffer J.
3. Load freshly prepared cell-free extract, previously filtered through 45-μm pore filter, onto the Ni^{2+} affinity column at a flow rate of 1–2 ml/min.
4. Wash with 5 CV buffer J.
5. Wash with 2 CV buffer K.
6. Elute MiNifB with 5 CV Buffer L.
7. Elution fractions are quickly analyzed by SDS-PAGE. Choose fractions that carry higher amount of MiNifB and less contaminant proteins and pool them into a single vial.
8. Equilibrate HiPrep 26/10 desalting column with 5 CV of buffer M.

9. Load pooled MiNifB fractions onto a HiPrep column at a flow rate of 5 ml/min. Run desalting chromatography at a flow of 5 ml/min on buffer M and collect. This step is fundamental to remove imidazole from MiNifB preparations.
10. Freeze MiNifB preparation in 0.1 ml aliquots in cryotubes and store under liquid nitrogen.

4 Notes

1. Cell growth conditions described in this chapter have been carefully chosen, so that a variety of organisms expressing nitrogenase proteins is covered. *A. vinelandii* cells are cultured in 10-l glass reactors equipped with systems enhancing O_2 diffusion: variable-speed propellers, ceramic gas micro-diffusers, and 35 % O_2-enriched air. *K. pneumoniae* nitrogenase derepression requires lack of ammonium, presence of a poor nitrogen source (serine), strict anaerobic conditions, and, in the specific case shown here, addition of 5 μM IPTG to the medium. Homemade, inexpensive, 50-l carboys and slow gas diffusers are used. Heterologous, lactose-dependent, expression of *M. infernus* NifB in *E. coli* cells is achieved in a 5-l Sartorius bioreactor by carefully controlling medium O_2 concentration and pH.
2. *A. vinelandii* is unusual because it fixes nitrogen aerobically at a very high respiratory rate. Therefore, exceptional air transfer fermenter capability is needed. Fast propellers, ceramic micro bubble diffusers yielding 100–500 μm bubbles, and O_2-enriched air are used. All fermenter processes described here use oil-free air compressors.
3. Residual ammonium is determined by withdrawing 1-ml culture aliquots into a ceramic plate, adding one drop of Nessler's reagent to each aliquot and waiting 5–10 s for color development. Yellow-orange color indicates presence of ammonium; if ammonium is depleted from the culture the reaction is colorless.
4. The in vivo acetylene reduction assay is based on the nitrogenase ability to reduce acetylene to ethylene [10]. Nine-milliliter serum vials are loaded with 1-ml culture suspension and closed with rubber stoppers. Acetylene (0.5 ml pure gas) is injected into the vials yielding a 93 % air, 6 % acetylene gas phase, and vials are incubated at 30 °C in a water bath with vigorous shacking. Reactions are stopped after 15 min by injection of 100 μl of 4 N NaOH into the vial. Ethylene production is quantified by injecting 50 μl samples withdrawn from the vials gas phase into a Shimadzu GC-2014 gas chromatographer equipped with a

hydrogen flame ionization detector and N_2/air as carrier gases. Sample injection port and column temperature are 100 and 150 °C, respectively. An acetylene broad peak immediately follows the ethylene peak. Ethylene production is quantified based on peak height (μV) and transformed into nmoles ethylene min^{-1} OD^{-1} units. Well-derepressed *A. vinelandii* and *K. pneumoniae* cell cultures usually deliver activity values in the range 10–20 nmoles ethylene produced min^{-1} OD^{-1}.

5. The hollow fiber system is based on the GE Healthcare MAXCELL UFP-500-E-65 filtering cartridge (4.4 m^2 area) connected to a Berkeley water pump and a 50-l recirculating reservoir. The culture is transferred from the fermenter to the reservoir by using a second pump. Once in the deposit, it moves in a closed circuit, in which the Berkeley pump pushes forward the culture through the hollow fibers and back into the reservoir. The flow increases pressure inside the fibers, which are porous and permeable to water, but not to the cells in the culture. A culture of 180 l is concentrated down to 3–4 l in 30 min. The MAXCELL UFP-500-E-65 filtering cartridge is normally kept in sodium hydroxide solution to prevent bacterial growth. This unit should be washed intensively to remove sodium hydroxide and should not be used to collect cells until pH of washing water is neutral.
6. To make buffers anaerobic proceed as follows: hermetically-sealed Büchner flask containing aerobic buffer solution is connected to a Schlenk line using vacuum mode (9.48 kPa or −0.9 bar) with gentle magnetic stirring for 1 h (per liter). During this period, the Schlenk line is switched at least three times from vacuum position to N_2 flow-in position to accelerate air removal. After completing the process, N_2 is flowed inside the Büchner flask to keep positive gas pressure on the inside. Büchner flasks can be stored in the laboratory bench or inside a glove box.
7. Sodium dithionite ($Na_2S_2O_4$), abbreviated DTH, is an important component of buffers used for anaerobic protein purification. DTH solid stocks must be stored away from light inside anaerobic jars with desiccant sieve at room temperature. Freshly weighed DTH is normally added to the buffers in the 1–2 mM range to prevent oxidative damage to the protein. Importantly, DTH lowers buffer and protein redox potentials so it should be used with caution or not at all in certain experiments, such as electrochemical titrations. DTH itself is susceptible to oxidation and breaks down into S compounds that lower the pH and damage protein [Fe–S] clusters. Therefore, DTH must be added to the buffers only after they have been made anaerobic (*see* **Note 6**) and just before their use. It is strongly recommended not to use DTH-containing buffers older than 1 day.

8. Methyl viologen soaked (and dried) white paper strips can be used to detect buffer redox conditions. This method is useful to test reducing conditions of chromatography mobile phases. Apply a few buffer drops on top of the strip and wait a couple of seconds for color development. Blue color indicates that applied drops were able to reduce methyl viologen.
9. In vivo acetylene reduction assay under anaerobic conditions require an Schlenk line and 9-ml vials sealed under Ar atmosphere. Airtight syringe/needles are used to deliver 1 ml culture into the vial. Gas pressure inside the vial must be released by quickly puncturing the stopper before injecting 0.5 ml pure acetylene. Ethylene production reaction and determination proceed as described in **Note 4**.
10. Air-free cell collection requires using a constant N_2 flow via gas lines to the collecting reservoir and centrifuge bottles, as well as the use of liquid N_2 to store *K. pneumoniae* cell paste.
11. This chapter describes an osmotic shock method to lyse *A. vinelandii* cells, a French Press method to lyse *K. pneumoniae* cells, and a sonication method to lyse *E. coli* cells. Both, *A. vinelandii* and *E. coli* cell disruption protocols can be performed inside a glove box, whereas *K. pneumoniae* lysis is performed in a French Press located in a cold room with a N_2 gas line attached to ensure strict anaerobic conditions. Additional anaerobic lysis protocols using Avestin C3 and C5 systems have been successfully applied for *K. pneumoniae* and *E. coli*.
12. A French Press system equipped with a 1 in. 40 K cell, and upgraded with a Rapid Fill Kit, is suited for anaerobic lysis by adding proper air-tight tubing, connections and entry and exit sample flasks. The French Press is first made anaerobic by running 2–3 complete cycles with anaerobic buffer containing 2 mM DTH. Cell chamber volume is 35 ml; thus 100–150 ml of anaerobic reducing buffer must be used. Methyl viologen paper strips are used to check buffer reduction at the exit of French Press cell.
13. *E. coli* fermentation occurs at 37 °C with air sparging at 2 l/min and automatic adjustment of pH, temperature and dissolved O_2. OD_{600nm} is monitored at 30 min intervals. Dissolved O_2 concentration is key in successful recombinant NifB expression. Very high O_2 concentration levels can cause intracellular oxidation of [Fe–S] clusters in *E. coli*.
14. Lowering the temperature decreases the amount of inclusion bodies formed.
15. *Methanocaldococcus infernus* is a thermophile organism that grows at 55–91 °C with an optimal growth temperature of 85 °C [11]. Expressing *M. infernus* NifB in a mesophilic

bacterium, such as *E. coli*, can help the purification process because upon heating NifB remains soluble, while most *E. coli* proteins precipitate.

Acknowledgments

This work was supported by ERC Starting Grant 205442 and by MICINN Grant BIO2009-12661.

References

1. George SJ, Igarashi RY, Xiao Y, Hernandez JA, Demuez M, Zhao D et al (2008) Extended X-ray absorption fine structure and nuclear resonance vibrational spectroscopy reveal that NifB-co, a FeMo-co precursor, comprises a 6Fe core with an interstitial light atom. J Am Chem Soc 130:5673–5680
2. Joerger RD, Bishop PE (1988) Nucleotide sequence and genetic analysis of the nifB-nifQ region from *Azotobacter vinelandii*. J Bacteriol 170:1475–1487
3. Shah VK, Allen JR, Spangler NJ, Ludden PW (1994) In vitro synthesis of the iron-molybdenum cofactor of nitrogenase. Purification and characterization of NifB cofactor, the product of NIFB protein. J Biol Chem 269:1154–1158
4. Curatti L, Ludden PW, Rubio LM (2006) NifB-dependent in vitro synthesis of the iron-molybdenum cofactor of nitrogenase. Proc Natl Acad Sci U S A 103:5297–5301
5. Zhao D, Curatti L, Rubio LM (2007) Evidence for *nifU* and *nifS* participation in the biosynthesis of the iron-molybdenum cofactor of nitrogenase. J Biol Chem 282: 37016–37025
6. Wiig JA, Hu Y, Ribbe MW (2011) NifEN-B complex of *Azotobacter vinelandii* is fully functional in nitrogenase FeMo cofactor assembly. Proc Natl Acad Sci U S A 108:8623–8627
7. Wiig JA, Hu Y, Lee CC, Ribbe MW (2012) Radical SAM-dependent carbon insertion into the nitrogenase M-cluster. Science 337:1672–1675
8. Shah VK, Davis LC, Brill WJ (1972) Nitrogenase. I. Repression and derepression of the iron-molybdenum and iron proteins of nitrogenase in *Azotobacter vinelandii*. Biochim Biophys Acta 256:498–511
9. Yoch DC, Pengra RM (1966) Effect of amino acids on the nitrogenase system of *Klebsiella pneumoniae*. J Bacteriol 92:618–622
10. Stewart WD, Fitzgerald GP, Burris RH (1967) In situ studies on nitrogen fixation with the acetylene reduction technique. Science 158:536
11. Jeanthon C, L'Haridon S, Reysenbach AL, Vernet M, Messner P, Sleytr UB et al (1998) *Methanococcus infernus* sp. nov., a novel hyperthermophilic lithotrophic methanogen isolated from a deep-sea hydrothermal vent. Int J Syst Bacteriol 48:913–919

Chapter 4

Techniques for the Production, Isolation, and Analysis of Iron–Sulfur Proteins

Jason C. Crack, Jeffrey Green, Andrew J. Thomson, and Nick E. Le Brun

Abstract

Iron–sulfur clusters constitute a group of cofactors found in many proteins that play key roles in an exceptionally wide range of metabolic processes. The chemical reactivity of iron–sulfur clusters means that they can be particularly prone to damage when removed from the protective environment of the cell. In general, the key to obtaining an intact, biologically active iron–sulfur cluster-containing protein is to maintain a strictly anaerobic environment throughout the entire process of protein purification and analysis. For many proteins, particularly those with more labile clusters, it is essential.

Key words Anaerobic protein purification, [2Fe-2S], [4Fe-4S], Iron–sulfur clusters, Isotopic iron labelling

1 Introduction

Iron–sulfur cluster protein cofactors exhibit a wide range of functions, including electron transfer, catalysis and signal transduction, and play key roles in numerous biochemical pathways, including respiration and photosynthesis [1–5]. Underpinning this importance is the ability of iron–sulfur clusters to undergo redox reactions, the precise nature of which is significantly influenced by the protein environment surrounding the cluster. Generally, cluster stability tends to decrease as the overall oxidation level of the cluster increases. In addition to their capacity to undergo one electron redox reactions, leading in some cases to their participation in organic radical chemistry, they also carry out Lewis acid-based catalysis. Hence, iron–sulfur clusters are inherently reactive towards a variety of redox-active species, including small redox-active compounds, reactive oxygen and reactive nitrogen species. This property has been exploited during the evolution of bacterial iron–sulfur cluster-containing sensor regulators, which coordinate the cell's response to oxidative and nitrosative stress conditions, by

Juan C. Fontecilla-Camps and Yvain Nicolet (eds.), *Metalloproteins: Methods and Protocols*, Methods in Molecular Biology, vol. 1122, DOI 10.1007/978-1-62703-794-5_4,

specifically detecting, with high sensitivity, biologically important analytes (e.g., drugs, O_2, NO) [6–8]. However, this also means that iron–sulfur cluster cofactors are readily prone to damage. This can also be the case when clusters encounter strongly coordinating species, which may abstract iron/sulfide or displace thiolates ligands from the cluster [9–11]. Any of these interactions may result in a change in the redox state of the cluster, cluster conversion, or even complete cluster loss. While the cell usually strives to protect its complement of iron–sulfur proteins from the destructive effect of these species, they become particularly prone to damage when removed from the protective environment of the cell.

Here we describe a range of techniques and methods that enable the researcher to obtain intact and biologically active iron–sulfur proteins in pure and stable form. Methods for the generation of highly concentrated samples containing iron–sulfur clusters isotopically labelled with ^{57}Fe for Mössbauer and related spectroscopies are also described. Finally, with an emphasis on iron–sulfur regulatory proteins, we provide examples of experimental methods for studying reactivity with O_2 and NO.

2 Materials

All solutions (growth medium/buffers) should be prepared using 18 M Ω/cm water. All items (buffers, plastic/glass items, centrifuge tubes, etc.) should be placed into an anaerobic cabinet (<3 ppm O_2) a minimum of 24 h before they are required.

2.1 Routine Growth Medium

1. LB Broth: For each liter of medium required, add 10 g tryptone, 5 g yeast extract, 10 g sodium chloride, followed by the required volume of water. Mix well to dissolve.
2. Separate into portions of ~500 ml in 2 l Erlenmeyer flasks.
3. Sterilize by autoclaving at 121 °C for 20 min. Once cool, the medium is ready to use.

2.2 Minimal Growth Medium for ^{57}Fe Labelled Iron–Sulfur Proteins

1. M9 Salts: For each liter of medium required, add 12.8 g $Na_2HPO_4 \cdot 7H_2O$, 3 g KH_2PO_4, 1 g NH_4Cl, 0.5 g NaCl, followed by an appropriate volume of water. Mix well to dissolve.
2. Prepare 100 g of Chelex-100 resin (Bio-Rad), as per manufacturer's instructions, and place into a Whatman Klair-Flex 0.2 μm filtration device or pack into a column. Rinse with water.
3. Pass the M9 salts through the Whatman Klair Flex filtration device containing Chelex-100 resin. Separate into ~500 ml portions in 2 l Erlenmeyer flasks (*see* **Note 1**). This step removes polyvalent metal ions from the solution; i.e., $Fe^{2+/3+}$.

4. Autoclave: 121 °C for 20 min. Allow to cool.
5. For each liter of medium required, dissolve the following in ~8 ml of Chelex-100 treated water: 2 g glucose (to give 0.2 % w/v final concentration), 2 g casein hydrolysate (0.2 % w/v; *see* **Note 2**), 2 mg Thiamine HCl, 0.2 g $MgCl_2{\cdot}6H_2O$ (1 mM), 14.7 mg $CaCl_2{\cdot}2H_2O$ (100 μM). Filter-sterilize the solution (using a 0.2 μm filter), and combine with the M9 salt solution.
6. Add ^{57}Fe dissolved in concentrated HCl to give a final concentration of 20 μM. The medium is ready to use.

2.3 Preparation of $^{57}FeCl_3$ Solution

1. Weigh out approx. 100 mg of ^{57}Fe metal powder into a suitable container (*see* **Note 3**).
2. Add 8 ml of 37 % HCl and incubate at ~80 °C for ~16 h using an inert, sealed container.
3. Once cool, determine the iron content of the solution (*see* Subheading 3.7).

2.4 Preparation of Winkler Solutions for Dissolved O_2 Determinations

1. Solution 1: dissolve 3.55 g (2.1 M) of $MnSO_4{\cdot}H_2O$ in 100 ml of water and sparge with nitrogen gas (at least 99.998 % N_2) to remove dissolved O_2.
2. Solution 2: dissolve 20 g (20 % w/v) of NaI, 0.5 g NaN_3 (0.5 % w/v), and 40 g of NaOH (40 % w/v) in 100 ml of water and sparge with nitrogen gas to remove dissolved O_2.
3. Solution 3: dissolve 7.91 g of $Na_2S_2O_3$ (50 mM) in 1 l of water.
4. Indicator solution: dissolve 0.5 g of starch (1 % w/v) in 50 ml of water (*see* **Note 4**).

2.5 Preparation and Calibration of NONOate Solutions for Experiments with Nitric Oxide

1. To prepare a NONOate stock solution, weigh out an appropriate amount (typically 2 mg) of Proli-NONOate (*see* **Note 5**). Dissolve in 1 ml of anaerobic 50 mM NaOH.
2. Add 20 μl of this stock solution to 2 ml of 50 mM NaOH to give a total volume of 2.02 ml.
3. Record the absorbance at 252 nm. The concentration of the Proli-NONOate stock solution is calculated using an ε_{252nm} value of 8,400 M^{-1} cm^{-1}.
4. If necessary, the extent of NO release can be determined as follows: an aliquot of NONOate is allowed to decompose in the presence of ~182 μM reduced myoglobin (ε_{435nm} = 121,000 M^{-1} cm^{-1}). The concentration of NO released is determined using ε_{421nm} = 155,000 M^{-1} cm^{-1} for the myoglobin–NO complex.
5. Proli-NONOate was previously found to liberate 1.9 (±0.1) molecules of NO [12].

3 Methods

Although in vitro methods for iron–sulfur cluster insertion into apo-proteins exist (*see* Subheading 3.3), it is usually advantageous to combine protein expression with in vivo iron–sulfur cluster assembly. We describe below a generic set of conditions that have proved suitable for a variety of proteins (e.g., a number of different Wbl proteins [13]), as well as conditions suitable for the in vivo assembly of highly O_2-sensitive iron–sulfur proteins, as exemplified by FNR [14]. Both of these are likely to be applicable to other iron–sulfur proteins. For additional information see refs. [14–19].

3.1 Generic Growth Conditions for In Vivo Iron–Sulfur Cluster Assembly

1. Supplement the growth medium with an appropriate antibiotic and inoculate with a 400-fold dilution of the starting culture. We routinely use pET (Novagen) or pGEX (GE Healthcare) vectors in *E. coli* BL21 λDE3 (Invitrogen, New England Biolabs).
2. Incubate at 37 °C in an orbital shaker set at 200 rpm until A_{600nm} reaches 0.6–0.8.
3. Remove the flasks from the incubator and place on ice for 18 min. During this time allow the incubator to cool to 30 °C.
4. Add 96 mg/l (0.4 mM final concentration) isopropyl β-D-1-thiogalactopyranoside (IPTG) and return the flasks to the incubator. Incubate at 30 °C for 50 min with 110 rpm orbital shaking.
5. For routine growth medium: add 53 mg/l (200 μM final concentration) ammonium iron(III) citrate and 4.5 mg/l (30 μM final concentration) L-methionine; go to **step** 7.
6. For ^{57}Fe labelling: add 180–200 μM ^{57}Fe dissolved in concentrated HCl (*see* Subheading 2.3 for preparation instructions).
7. Incubate at 30 °C, with 110 rpm orbital shaking, for a further 3 h 30 min.
8. Harvest the cells by centrifugation and, where possible, freeze under anaerobic conditions (*see* **Note 6**).

3.2 In Vivo Assembly of the Highly O_2-Sensitive FNR [4Fe-4S] Cluster

1. For native [4Fe-4S] FNR expression, the minimal medium outlined in Subheading 2.2 is suitable (*see* **Note** 7). However, generic growth conditions outlined in Subheading 3.1 are not suitable for this highly O_2-sensitive cluster. Instead, anaerobic growth conditions are required [20].
2. Supplement the growth medium with 100 mg/l ampicillin and inoculate with *E. coli* BL21 λDE3 containing pGS1859 (*see* **Notes 8, 9**).
3. Incubate at 37 °C, with 200 rpm orbital shaking, until A_{600nm} reaches 0.5.

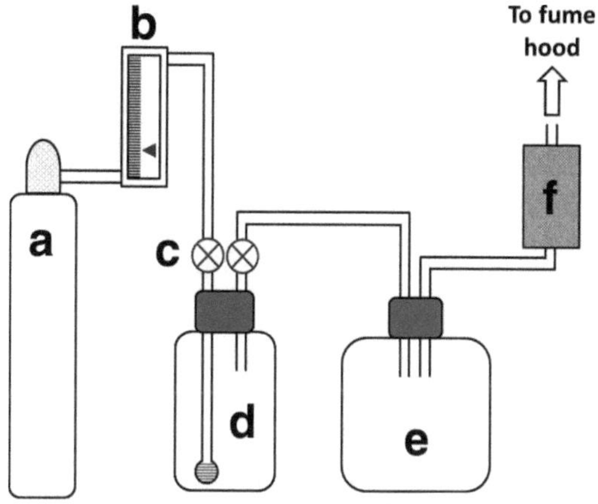

Fig. 1 Diagram of N_2-sparging apparatus for anaerobic in vivo assembly of [4Fe-4S] FNR. (**a**) Nitrogen cylinder. (**b**) Gas flow meter. (**c**) Taps allowing the flask to be sealed and removed from the gas line. (**d**) 1 l Duran bottle containing gas dispersion tube. (**e**) 2 l fluid trap. (**f**) 0.2 μm HEPA (High-Affinity Particulate Air) filter. All components except **a** and **f** are housed inside a fridge (modified to enable gas flow in and out) to permit incubation at 4 °C

4. Add 0.4 mM IPTG and incubate at 37 °C for a further 60 min.
5. Reduce the culture volume (typically 6–8 l) by 50 %. To do this, centrifuge half the total culture at 3,488 × *g* and discard the supernatant. Resuspend the resulting cell pellet in the remaining culture volume (3–4 l).
6. Separate the resulting concentrated culture into ~1 l portions using previously sterilized modified 1 l Duran bottles (see below) and supplement with 200 μM ammonium iron(III) citrate, 50 μM L-methionine, and 1 ml/l (0.1 % v/v) Antifoam 204 (Sigma-Aldrich).
7. Sparge the cultures in the modified Duran bottles with O_2-free nitrogen gas (1 l/min) at 4 °C overnight (*see* Fig. 1).
8. Transfer the sparged cultures to the anaerobic cabinet and decant the culture into O-ring sealed centrifuge tubes.
9. Harvest the cells, outside the anaerobic box, by centrifugation at 4 °C. Return immediately to the box. Decant the supernatant and discard. Where possible, freeze the cell pellets under anaerobic conditions (*see* **Note 6**).

3.3 In Vitro Assembly: NifS Catalyzed [4Fe-4S] Assembly

1. Previously prepared apo-protein is transferred to the anaerobic cabinet and allowed to equilibrate with the anaerobic atmosphere (*see* **Note 10**).
2. Freshly prepare the following under anaerobic conditions, using a suitable buffer to dissolve the solutes: Solution-1, 50 mM (12.8 mg/ml) L-cysteine, 125 mM (41.7 mg/ml)

dithiothreitol. Solution-2, 20 mM (78.5 mg/10 ml) $(NH_4)_2Fe(II)(SO_4)_2$.

3. Add an appropriate volume of solution-1 to the apo-protein solution to give a final concentration of 1 mM L-cysteine and 2.5 mM dithiothreitol.
4. Add an appropriate volume of solution-2 to the apo-protein, such that there is a sevenfold excess of Fe^{2+} ions over apo-protein.
5. To initiate iron–sulfur cluster reconstitution, add purified NifS (*see* **Note 11**) to a final concentration of 225 nM. The reaction mixture is sealed using a rubber stopper, stirred (using a magnetic follower), and incubated at 37 °C for up to 4.5 h (*see* **Note 12**).
6. A spectrophotometer and anaerobic cuvettes can be used to monitor the progress of the reconstitution. The reaction should be aborted if a black iron-sulfide precipitation becomes apparent.
7. Pass the reconstitution mixture through a 0.2 μm syringe filter to remove any particulate matter.
8. The reconstituted protein can be separated from low molecular weight reactants using one of two methods. The first, which is ideal for small scale preparations up to 2.5 ml, involves rapid gel filtration on a PD10 column (GE Healthcare). The second method, which is ideal for larger preparations up to ~25 ml, involves the use of a 1 ml ion exchange column to simultaneously clean and concentrate the reconstituted protein (*see* **Note 13** and Subheading 3.5).

The following sections are intended as a guide to purifying, concentrating and assaying the iron/sulfide content of iron–sulfur proteins. A detailed protocol for the isolation of native [4Fe-4S] FNR is provided, which could be suitably modified for other iron–sulfur proteins. Regardless of the culturing conditions (aerobic/anaerobic) used, iron–sulfur protein purification is exclusively carried out under anaerobic conditions. To achieve this, a dedicated Fast Protein Liquid Chromatography (FPLC) system, together with columns, located within an anaerobic cabinet operating at ~2.0 ppm oxygen should be used. Apo-proteins can usually be obtained by purifying under aerobic conditions on the bench.

3.4 Purification of Native [4Fe-4S] FNR

1. Resuspend bacterial cell pellets from 5 l of culture in 105 ml of buffer A (10 mM potassium phosphate, 100 mM KCl, 5 % (v/v) glycerol, pH 6.8).
2. Add 12 ml of CelLytic B lysis detergent (CelLytic B 10×; Sigma) (*see* **Note 14**), 0.3 mM phenylmethylsulfonyl fluoride

(PMSF) dissolved in ethanol, 2.5 mg lysozyme, and 20 μl of DNase I (5 mg/ml stock).

3. Incubate at ambient temperature for up to 20 min, with occasional agitation.
4. Transfer the cell lysate into O-ring sealed centrifuged tubes and centrifuge outside of the anaerobic cabinet at 40,000 × *g* for 45 min at 2 °C.
5. Return the cell lysate to the anaerobic cabinet and transfer the supernatant to a suitable container. Discard the pellet.
6. Load the supernatant onto a 10 ml HiTrap SP FF ion exchange column (GE Healthcare), previously equilibrated with buffer A.
7. Wash the column with 10 % (v/v) buffer B (10 mM potassium phosphate, 800 mM KCl, 5 % (v/v) glycerol, pH 6.8) until $A_{280nm} \leq 0.1$.
8. Bound proteins are eluted using a linear gradient (60 ml) of 10–100 % (v/v) buffer B. Fractions containing FNR are pooled (*see* **Note 15**).
9. Dilute the pooled fractions threefold with buffer C (10 mM potassium phosphate, 5 % glycerol, pH 6.8) and load onto a 2 ml HiTrap Heparin HP column (GE Healthcare) equilibrated with buffer A.
10. Wash the column with 5 column volumes of 10 % buffer B.
11. Elute bound proteins using a linear gradient (50 ml) from 10 to 100 % (v/v) buffer B. Fractions containing FNR are pooled, concentrated if necessary (*see* Subheading 3.6), and, where possible, should be frozen under anaerobic conditions (*see* **Note 6**).

3.5 Concentrating Iron–Sulfur Proteins

1. Equilibrate a suitable 1 ml ion exchange or affinity column via syringe using a buffer. The buffer should contain low salt (NaCl or KCl) concentrations (≤100 mM) to ensure tight binding (*see* **Notes 16, 17**).
2. Load the sample onto the column as depicted in Fig. 2 (*see* **Note 17**). In this orientation binding of the iron–sulfur protein to the column is clearly visible. If the column becomes saturated, repeat the process on a fresh 1 ml column.
3. Wash the column with 10–20 ml of the buffer used in **step 1**. This step only applies to reconstitution mixtures (*see* **Note 18**).
4. To elute, invert the column, as depicted in Fig. 2, and apply elution buffer containing ≥500 mM salt (NaCl or KCl). The eluent is collected in microcentrifuge tubes, five drops at a time to obtain the highest concentration (*see* **Note 19**).

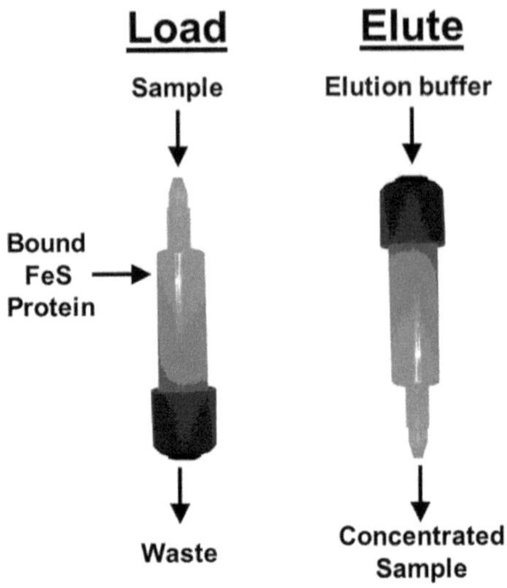

Fig. 2 Concentration of iron–sulfur proteins via ion exchange. Load; the sample is loaded, via syringe, onto the bottom of the column. Elute; the column is inverted and bound material is eluted using a high salt (≥500 mM) buffer. Manual fractionation into microcentrifuge tubes allows highly concentrated material to be obtained

5. Pool the most concentrated fractions; other, more dilute fractions may also be pooled together and retained.
6. In some cases, such as during the preparation of ^{57}Fe labelled proteins, it is advantageous to spin concentrate the dilute fractions using a suitable centrifugation concentrator device. This can then be combined with the most concentrated fractions. Using this approach, solutions containing ~2 mM iron–sulfur protein are achievable.
7. Where possible, freeze under anaerobic conditions (*see* **Note 6**).

3.6 Determination of Iron Content

1. Prepare a 200 μM solution (10 ml) of Fe(III) using an Fe(III) standard solution suitable for atomic absorption spectroscopy (*see* **Note 20**).
2. Using 2 ml screw cap tubes and the solution prepared in **step 1**, prepare a calibration curve from 0 to 200 μM according to Table 1 (*see* **Note 21**).
3. Using 2 ml screw cap tubes, prepare appropriate dilutions (total volume 0.1 ml) of the sample(s) for analysis (*see* **Note 21**).
4. Add 0.1 ml of ~22 % HNO_3 to each tube and incubate at 95 °C for 30 min.
5. Allow the samples to cool to ambient temperature and then briefly centrifuge.

Table 1
Standards for iron determination

Fe solution (μl)	Water (μl)	[Fe] μM
0	100	0
10	90	20
20	80	40
30	70	60
40	60	80
50	50	100
60	40	120
70	30	140
80	20	160
90	10	180
100	0	200

Solutions prepared using a stock solution of 200 μM iron

6. Add 0.6 ml of a 7.5 % (w/v) solution of ammonium acetate to each tube. Mix by inversion.
7. Add 0.1 ml of a freshly prepared 12.5 % (w/v) solution (0.25 g/ml) of ascorbic acid. Mix by inversion.
8. Add 0.1 ml of a freshly prepared 10 mM solution (4.94 mg/ml) of Ferene. Mix by inversion. Incubate at room temperature for 30 min.
9. The absorbance of each standard/sample is recorded at 593 nm.
10. A calibration curve is constructed by plotting the concentration of iron (x axis) versus A_{593nm} (y axis). The concentration of iron present in the protein sample is determined by reference to this calibration curve, taking into account any dilutions made to the sample.

3.7 Determination of Acid-Labile Sulfide Content (See Note 22)

1. Weigh out 0.25 g of sodium sulfide nonahydrate and dissolve in 500 ml of 10 mM (0.2 g/500 ml) NaOH. The resulting solution contains 2.08 mM sulfide.
2. Using 1.5 ml flip top microcentrifuge tubes and the solution prepared in **step 1**, prepare sulfide standards in the range 0–260 μM (according to Table 2) to enable the construction of a calibration curve.
3. Prepare appropriate dilutions (total volume 200 μl) of the sample(s) for analysis.

Table 2
Standards for S^{2-} determination

S^2-solution (μl)	Water (μl)	[S^{2-}] μM
0	200	0
5	195	52.0
10	190	104.1
15	185	156.0
20	180	208.2
25	175	260.2

Solutions prepared using a stock solution of 2.08 mM sulfide

4. To each tube (standards and samples), immediately add 0.6 ml of freshly prepared 1 % (w/v) zinc acetate, followed by 50 μl of 12 % (w/v) NaOH. Cap the tubes, invert to mix, and incubate at ambient temperature for 15 min.
5. After incubation, centrifuge the standards/samples at 900 × *g* in a benchtop centrifuge for 60 s.
6. To each tube add 0.15 ml of freshly prepared 0.1 % (w/v) *N*,*N*-dimethyl-*p*-phenylenediamine dihydrochloride dissolved in 5 M HCl. Immediately add 0.15 ml of freshly prepared 10 mM $FeCl_3$ dissolved in 1 M HCl. Vortex mix for 30 s.
7. Incubate the samples at ambient temperature for 30 min (*see* **Note 23**).
8. Record the absorbance of each standard/sample at 670 nm.
9. A calibration curve is constructed by plotting the concentration of sulfide (*x* axis) versus A_{670nm} (*y* axis). The concentration of sulfide present in the protein sample is determined by reference to this calibration curve, taking into account any dilutions made to the sample.

3.8 Determination of Dissolved O_2 Content: The Winkler Assay [21]

1. Fill a 100 ml volumetric flask to the mark with the sample for analysis.
2. Seal the flask with a Suba-Seal and briefly purge the head space of the flask with O_2-free nitrogen gas.
3. Inject 1 ml of solution 1, followed by 1 ml of solution 2 (*see* Subheading 2.4). Mix by inversion.
4. Incubate the sample for 15 min at ambient temperature. During this time a precipitate should form (*see* **Note 24**).
5. Inject 2 ml of 85 % (v/v) phosphoric acid and mix by inversion (*see* **Notes 25, 26**).
6. Transfer the sample to a 250 ml Erlenmeyer flask.

7. Using a burette, immediately titrate the sample with solution 3 (*see* Subheading 2.4).
8. Near the end point of the titration, add a few drops of the indicator solution. Continue to titrate until the blue/black color of the indicator disappears. Record the volume of solution 3 used.
9. During the assay, 1 mol of O_2 indirectly consumes 4 mol of $Na_2S_2O_3$ (*see* **Note 27**).

3.9 Determination of the Reactivity of Iron–Sulfur Cluster Proteins with O_2 (or Nitric Oxide)

1. FNR can act as a sensor for O_2 (or nitric oxide) via its $[4Fe\text{-}4S]^{2+}$ cluster [22–25]. When reacted with O_2 (or nitric oxide) changes in the visible region of the optical spectrum occur because of alterations to the [4Fe-4S] cluster. These changes can be observed via absorbance and/or circular dichroism spectroscopy. The reactions of FNR with O_2 (or nitric oxide) are relatively rapid, typically being completed within 5 min at 20 °C. In contrast, the reaction of WhiD with O_2 takes ≥120 min to reach completion. Thus, it is essential to determine the reaction time course before conducting a titration (*see* **Note 28**).
2. Prior to starting a titration it is essential to baseline the spectrometer with the cuvette(s) and buffer being used during the titration. We typically use 260–900 and 280–700 nm for absorbance and circular dichroism spectroscopy, respectively.
3. Combine an aliquot of sample and buffer (2 ml final volume) in an anaerobic cuvette to give a final concentration of ~30 μM iron sulfur cluster and record the optical spectrum. Use the extinction coefficient for the protein being studied (for FNR, $\varepsilon_{406nm} = 16{,}220\ M^{-1}\ cm^{-1}$) to determine the exact concentration of iron sulfur protein present.
4. Using a gas tight syringe, add an aliquot of buffer containing dissolved atmospheric O_2 (or NONOate). Typically, we add 0.2 molecules of O_2 (or nitric oxide) per iron sulfur cluster per injection (*see* Table 3).
5. Incubate at ambient temperature for the amount of time determined in **step 1**, and then record the optical spectrum.
6. Repeat **steps 4** and **5** until no further changes in the optical spectrum are observed.
7. For O_2 titrations, plots of the ratio $[O_2]$:[FeS] (*x* axis) versus A_{406nm} or A_{420nm} (*y* axis) are particularly useful (A_{360nm} for nitric oxide). Similarly, for CD titrations, the ratio of the introduced analyte can be plotted versus specific wavelengths, *see* Fig. 3. In general absorbance and CD observations provide complementary information [22, 26, 27].

Table 3
Particulars for a typical O_2 titration

Add (μl)	O_2 (μM)	[4Fe-4S] FNR (μM)	Ratio [O_2]:[4Fe-4S]
0	0	30.0	0
50	5.6	29.3	0.2
55	11.4	28.5	0.4
50	16.5	27.8	0.6
50	21.3	27.2	0.8
50	25.9	26.6	1.0
60	31.1	25.9	1.2
60	36.1	25.3	1.4
50	40.1	24.7	1.6
50	43.9	24.2	1.8
50	47.6	23.8	2.0
75	52.8	23.1	2.3
80	58.0	22.4	2.6
80	63.0	21.7	2.9
150	71.5	20.6	3.5
150	79.2	19.6	4.0

Buffer (25 mM HEPES, 2.5 mM $CaCl_2$,100 mM NaCl, 100 mM $NaNO_3$, pH 7.5) containing 228.7 μM dissolved atmospheric O_2 at 20 °C was titrated, by injection, using gas tight syringes (*see* **Note 29**)

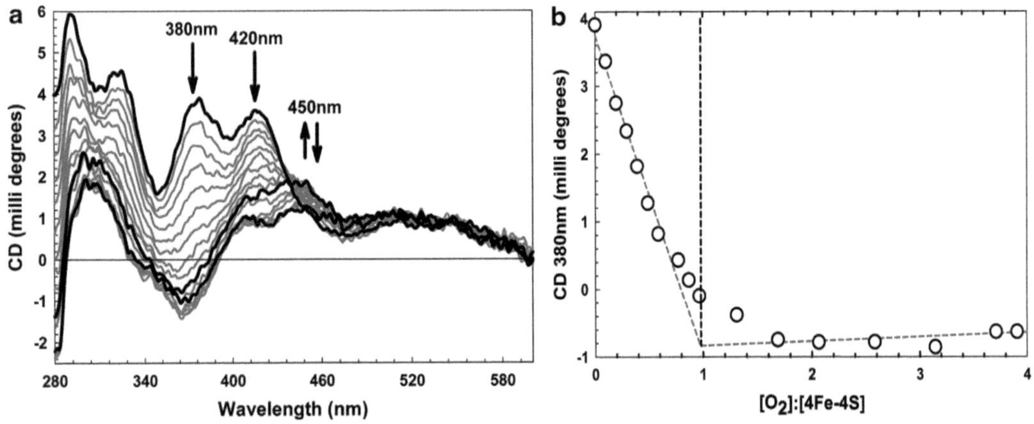

Fig. 3 O_2 titration of [4Fe-4S] FNR. (**a**) As described in Subheading 3.9, air saturated buffer (25 mM HEPES, 2.5 mM $CaCl_2$, 100 mM NaCl, 100 mM $NaNO_3$, pH 7.5; 232 μM O_2, 19 °C) was titrated into FNR (29.7 μM [4Fe-4S]). The upper spectrum corresponds to [4Fe-4S] FNR in the absence of O_2 and the lowest spectrum corresponds to FNR in the presence of 77.4 μM O_2. Spectra recorded at [O_2]:[4Fe-4S] ratios of approximately 0, 1, and 4 are shown in black, intervening spectra are shown in grey. Arrows indicate the direction of movement of spectral features. (**b**) CD intensity at 380 nm plotted versus the [O_2]:[4Fe-4S] cluster ratio. Taking the initial slope and the asymptote at high O_2 levels a binding stoichiometry of ~1 O_2 per [4Fe-4S] cluster is obtained

4 Notes

1. Rinse all Erlenmeyer flasks and other glassware with 1 M HNO_3, followed by Chelex-100 treated water. This will remove residual metal ions (e.g., $Fe^{2+/3+}$) from the surface of the glass.
2. A 2 % (w/v) solution of Fluka casein hydrolysate (for Microbiology) was found to contain 2.2 μM iron.
3. Sealable, polytetrafluoroethylene (PTFE) plastic containers are ideal for this. The container should be pre-treated as described in **Note 1**. Suitable glassware may also be used during this step.
4. To dissolve, warm the solution in a microwave.
5. Proli-NONOate (Cayman Chemicals) spontaneously dissociates in a pH dependent, first order process with a half-life of 1.8 s at pH 7.4, 37 °C, to liberate 2 mol of NO per mole of parent compound. Other NONOates may differ, check the product literature.
6. Alternatively, remove from the anaerobic cabinet, flash-freeze in liquid nitrogen or on dry ice and store at −80 °C. When required, and to reduce the likelihood of O_2-induced damage, return the frozen sample(s) to the anaerobic cabinet and allow to them thaw.
7. For routine work, it is not necessary to remove trace metal ions from the growth medium.
8. This protocol also works well with pGEX vectors used to express GST-FNR fusion proteins.
9. The reader's attention is drawn to the metabolic deficiencies of the *E. coli* BL21 λDE3 strain. An alternative strain of *E. coli* may be required to successfully express other particularly O_2-sensitive iron–sulfur proteins under anaerobic conditions [28]. The plasmid pGS1859 contains the *E. coli fnr* gene under the control of the T7 promoter in pET11d.
10. Apo-proteins can usually be obtained by omitting the addition of excess iron to the growth medium during expression, maintaining good culture aeration (200 rpm) and purifying the protein under aerobic conditions on the bench. If the protein still contains traces of an iron–sulfur cluster post purification, it may be advisable to follow the generic growth conditions in Subheading 3.1, followed by anaerobic protein purification.
11. Recombinant cysteine desulfurase from *Azotobacter vinelandii* (NifS) is purified according to Zheng et al. [29], except for the following: 1 mM IPTG was added when $A_{600nm} = 0.6$; MEM vitamins (Sigma Aldrich; product number M6895) were added 10 min after the addition of IPTG; Streptomycin sulfate was

used at 1.5 g per 100 ml; The sample was loaded onto a 10 ml HiTrap Q Sepharose column (GE Healthcare) and eluted using a 65 ml linear gradient, after which, NifS was ≥95 % pure, as judged by SDS-PAGE.

12. Reconstitution can also be carried out at 30 °C or at ambient temperature. A lower temperature increases the time required to reach desired levels of cluster incorporation.
13. Weak cation exchange resins, such as heparin, are particularly useful for transcriptional regulators. Usually, the 0.2 μm syringe filter is connected directly to the top of the ion exchange column.
14. Alternatively, remove from the anaerobic cabinet, sonicate on ice, and return immediately to the anaerobic cabinet. This may be suitable for less O_2-sensitive iron–sulfur proteins. In general, we do not use dithiothreitol (DTT) or sodium dithionite during purification because of the possibility of chelation of the cluster by DTT or the generation of damaging low pH by the reaction of dithionite with adventitious O_2.
15. The purification can be paused here if necessary. Pool the FNR containing fractions and freeze under anaerobic conditions (*see* **Note 6**).
16. To concentrate dilute material pooled from ion exchange chromatography, dilute the pooled sample tenfold with suitable buffer lacking salt (e.g., 50 mM Tris, 5 % glycerol, pH 7.5) prior to performing this procedure.
17. For large volumes (≥25 ml), a peristaltic pump can be used.
18. It is important to wash on-column reconstituted proteins with copious amounts of buffer to ensure low molecular weight reactants are removed prior to elution. This step can also be used to perform buffer exchange prior to elution in the new buffer.
19. The first 0.5 ml of eluent from a 1 ml column typically does not contain highly concentrated protein. This can be collected in one fraction.
20. If assaying the ^{57}Fe solution from Subheading 2.3, prepare a 400 μM iron solution. Follow the dilutions in Table 1. The actual concentration will be twice that shown in Table 1.
21. Typical dilution factors are as follows: For the ^{57}Fe solution, prepared in Subheading 2.3, the starting dilution should be 1 part sample to 400 parts water. For concentrated proteins the starting dilution should be 1 part sample to 50 parts water.
22. For additional practical considerations, the reader is referred to the method of Beinert [30].
23. In some cases it may be necessary to incubate at 42 °C.

24. $4Mn^{(II)}(OH)_2 + O_2 + 2H_2O \rightarrow 4Mn^{(III)}(OH)_3$. The manganese hydroxides are sparingly soluble and form white Mn(II) or brown Mn(III) precipitates.
25. *Be careful!!* 85 % (v/v) phosphoric acid is very corrosive.
26. $4Mn^{(III)}(OH)_3 + 4I^- + 12H^+ \rightarrow 4Mn^{2+} + 2I_2 + 12H_2O$. Acidification dissolves the brown Mn(III) precipitate oxidizing iodide to iodine.
27. $4S_2O_3{}^{2-} + 2I_2 \rightarrow 2S_4O_6{}^{2-} + 4I^-$. Free iodine is reduced to iodide by thiosulfate. Hence, 4 mol of thiosulfate correspond to 1 mol of dissolved O_2 (*see* **Notes 24–27**).
28. To determine the reaction time course, place 2 ml of aerobic buffer (~220 μM) in an anaerobic cuvette, zero the spectrometer at 420 nm. Inject an aliquot of sample (~30 μM final concentration iron–sulfur cluster), mix by inversion, and monitor decrease in A_{420nm} versus time for over 10–120 min. If A_{420nm} increases at any point, check the sample for precipitation.
29. NONOates are highly soluble, enabling highly concentrated solutions to be obtained. For a 1.15 mM NONOate solution (*see* **Note 5**) the additions would be tenfold less, i.e., 5 μl, resulting in a 0–402 μM NO concentration; covering [NO]:[4Fe-4S] ratios from 0 to 16.

Acknowledgements

This work has been supported by a series of grants from the BBSRC, held jointly between UEA and the University of Sheffield, over several years.

References

1. Balk J, Pilon M (2011) Ancient and essential: the assembly of iron-sulfur clusters in plants. Trends Plant Sci 16:218–226
2. Sheftel A, Stehling O, Lill R (2010) Iron-sulfur proteins in health and disease. Trends Endocrinol Metab 21:302–314
3. Beinert H (2000) Iron-sulfur proteins: ancient structures, still full of surprises. J Biol Inorg Chem 5:2–15
4. Johnson MK (1998) Iron-sulfur proteins: new roles for old clusters. Curr Opin Chem Biol 2:173–181
5. Beinert H, Holm RH, Munck E (1997) Iron-sulfur clusters: nature's modular, multipurpose structures. Science 277:653–659
6. Fleischhacker AS, Kiley PJ (2011) Iron-containing transcription factors and their roles as sensors. Curr Opin Chem Biol 15:335–341
7. Crack JC, Green J, Thomson AJ, Le Brun NE (2012) Iron-sulfur cluster sensor-regulators. Curr Opin Chem Biol 16:35–44
8. Crack JC, Green J, Hutchings MI, Thomson AJ, Le Brun NE (2012) Bacterial iron-sulfur regulatory proteins as biological sensor-switches. Antioxid Redox Signal 17:1215–1231
9. Carter KR, Rawlings J, Orme-Johnson WH, Becker RR, Evans HJ (1980) Purification and characterization of a ferredoxin from *Rhizobium japonicum* bacteroids. J Biol Chem 255:4213–4223

10. Averill BA, Bale JR, Ormejohnson WH (1978) Displacement of iron-sulfur clusters from ferredoxins and other iron-sulfur proteins. J Am Chem Soc 100:3034–3043
11. Macomber L, Imlay JA (2009) The iron-sulfur clusters of dehydratases are primary intracellular targets of copper toxicity. Proc Natl Acad Sci U S A 106:8344–8349
12. Cruz-Ramos H, Crack J, Wu GG, Hughes MN, Scott C, Thomson AJ, Green J, Poole RK (2002) NO sensing by FNR: regulation of the *Escherichia coli* NO-detoxifying flavohaemoglobin, Hmp. EMBO J 21:3235–3244
13. Crack JC, den Hengst CD, Jakimowicz P, Subramanian S, Johnson MK, Buttner MJ, Thomson AJ, Le Brun NE (2009) Characterization of [4Fe-4S]-containing and cluster-free forms of Streptomyces WhiD. Biochemistry 48:12252–12264
14. Crack JC, Gaskell AA, Green J, Cheesman MR, Le Brun NE, Thomson AJ (2008) Influence of the environment on the $[4Fe\text{-}4S]^{2+}$ to $[2Fe\text{-}2S]^{2+}$ cluster switch in the transcriptional regulator FNR. J Am Chem Soc 130:1749–1758
15. Sutton VR, Kiley PJ (2003) Techniques for studying the oxygen-sensitive transcription factor FNR from *Escherichia coli*. Meth Enzymol 370:300–312
16. Yan A, Kiley PJ (2009) Techniques to isolate O_2-sensitive proteins: [4Fe-4S]-FNR as an example. Meth Enzymol 463:787–805
17. Kuchenreuther JM, Grady-Smith CS, Bingham AS, George SJ, Cramer SP, Swartz JR (2010) High-yield expression of heterologous [FeFe] hydrogenases in *Escherichia coli*. PLoS One 5: e15491
18. Stehling O, Smith PM, Biederbick A, Balk J, Lill R, Muhlenhoff U (2007) Investigation of iron-sulfur protein maturation in eukaryotes. Meth Mol Biol 372:325–342
19. Orme-Johnson WH, Holm RH (1978) Identification of iron-sulfur clusters in proteins. Meth Enzymol 53:268–274
20. Lazazzera BA, Beinert H, Khoroshilova N, Kennedy MC, Kiley PJ (1996) DNA binding and dimerization of the Fe-S-containing FNR protein from *Escherichia coli* are regulated by oxygen. J Biol Chem 271:2762–2768
21. Winkler LW (1888) Die Bestimmung des im Wasser gelösten Sauerstoffes (the determination of oxygen dissolved in water). Berichte der deutschen chemischen Gesellschaft 21: 2843–2854
22. Crack JC, Stapleton MR, Green J, Thomson AJ, Le Brun NE (2013) Mechanism of $[4Fe\text{-}4S](Cys)_4$ cluster nitrosylation is conserved amongst NO-responsive regulators. J Biol Chem 288:11492–11502
23. Crack JC, Green J, Cheesman MR, Le Brun NE, Thomson AJ (2007) Superoxide-mediated amplification of the oxygen-induced switch from [4Fe-4S] to [2Fe-2S] clusters in the transcriptional regulator FNR. Proc Natl Acad Sci U S A 104:2092–2097
24. Crack JC, Jervis AJ, Gaskell AA, White GF, Green J, Thomson AJ, Le Brun NE (2008) Signal perception by FNR: the role of the iron-sulfur cluster. Biochem Soc Trans 36:1144–1148
25. Crack JC, Le Brun NE, Thomson AJ, Green J, Jervis AJ (2008) Reactions of nitric oxide and oxygen with the regulator of fumarate and nitrate reduction, a global transcriptional regulator, during anaerobic growth of *Escherichia coli*. Meth Enzymol 437:191–209
26. Stephens PJ, Thomson AJ, Dunn JB, Keiderling TA, Rawlings J, Rao KK, Hall DO (1978) Circular dichroism and magnetic circular dichroism of iron-sulfur proteins. Biochemistry 17:4770–4778
27. Crack JC, Smith LJ, Stapleton MR, Peck J, Watmough NJ, Buttner MJ, Buxton RS, Green J, Oganesyan VS, Thomson AJ, Le Brun NE (2011) Mechanistic insight into the nitrosylation of the [4Fe-4S] cluster of WhiB-like proteins. J Am Chem Soc 133:1112–1121
28. Pinske C, Bonn M, Kruger S, Lindenstrauss U, Sawers RG (2011) Metabolic deficiences revealed in the biotechnologically important model bacterium *Escherichia coli* BL21(DE3). PLoS One 6:e22830
29. Zheng L, White RH, Cash VL, Jack RF, Dean DR (1993) Cysteine desulfurase activity indicates a role for NIFS in metallocluster biosynthesis. Proc Natl Acad Sci U S A 90: 2754–2758
30. Beinert H (1983) Semi-micro methods for analysis of labile sulfide and of labile sulfide plus sulfane sulfur in unusually stable iron-sulfur proteins. Anal Biochem 131:373–378

Chapter 5

Cell-Free Synthesis of the H-Cluster: A Model for the In Vitro Assembly of Metalloprotein Metal Centers

Jon M. Kuchenreuther, Stacey A. Shiigi, and James R. Swartz

Abstract

Many organometallic cofactors are highly complex and require multiple accessory proteins for both their assembly and transfer to a target protein. A cell-free system in which the biosynthetic pathway for a prosthetic group has been fully or even partially reconstructed enables investigations of the reaction sequence as well as the cofactor itself. As a model for the in vitro assembly of protein-bound metal centers, we describe a procedure for the cell-free synthesis of the H-cluster in the context of producing purified and active [FeFe] hydrogenase samples for spectroscopic studies. In general terms, this in vitro system is a combination of non-purified accessory proteins, exogenous substrates, and purified hydrogenase apoprotein. We also describe methods for making the required components used in the cell-free system. Specifically, these procedures include anaerobic expression of heterologous metalloproteins in *Escherichia coli*, anaerobic cell lysate production, and anaerobic metalloprotein purification using *Strep*-Tactin® chromatography.

Key words Cell-free, In vitro, Synthetic biology, Expression, Purification, Cofactor, Accessory protein, Hydrogenase, H-cluster

1 Introduction

Biological cofactors are low-molecular-weight compounds that proteins require for enzyme activity. Proteins without or with their cofactors are called apoenzymes or holoenzymes, respectively. These non-protein compounds can be organic, inorganic, or organometallic in nature, and they typically bind to conserved sequence motifs. Some cofactors are considered coenzymes that loosely bind to a particular protein, while others are prosthetic groups that are more tightly associated with the polypeptide chain. Many prosthetic groups are highly complex and thus require specialized helper proteins and unique reaction sequences for both their assembly and transfer to the active site of a target apoenzyme. For example, nine gene products are involved in assembling the iron–molybdenum nitrogenase cofactor (FeMo-co) [1, 2], and seven gene products

Juan C. Fontecilla-Camps and Yvain Nicolet (eds.), *Metalloproteins: Methods and Protocols*,
Methods in Molecular Biology, vol. 1122, DOI 10.1007/978-1-62703-794-5_5,

are involved in synthesizing the [NiFe] hydrogenase active site cluster [1]. Three proteins with [FeS] clusters—the HydE, HydF, and HydG maturases—are required for producing the [FeFe] hydrogenase holoenzyme by assembling and transferring the catalytic subunit of the H-cluster (i.e., the $[2Fe]_H$ subcluster) [3–5]. Oftentimes, however, the roles of accessory proteins and the precise mechanisms by which they synthesize cofactors are not fully understood. This, along with their typical sensitivity to O_2, often makes metalloenzymes difficult to produce and thus limits the available options for studying them and their active site cofactors.

In vitro biosynthesis systems allow researchers to work with proteins and study biochemical pathways without the need to sustain microbial growth and metabolism [6]. The open access to a cell-free environment enables the precise control of substrate and catalyst concentrations as well as rapid sampling. In vitro studies have proven quite useful for studying enzyme kinetics, identifying substrates, characterizing protein–substrate interactions, and producing modified proteins for spectroscopic analysis [7]. However, reconstructing a multicomponent reaction sequence such as the synthesis of an organometallic prosthetic group can be quite challenging. The necessary enzymes and substrates are not always known, and a cell-free system may initially consist of undefined mixtures of constituents [8]. Nonetheless, if the required components have been elucidated, in vitro approaches for synthesizing complex cofactors can be developed that consist of a defined set of purified proteins and substrates [9, 10].

Regardless of whether cell-free systems are completely or partially defined, they can have several useful purposes. For example, if a biosynthetic pathway for an organometallic cofactor can be reconstituted in vitro, the system can be used as a platform to study reaction sequences as well as the concerted functionality of accessory proteins [4, 10]. With the open access of an in vitro system, active site metal centers can be readily labeled using isotopologs, which provides intriguing possibilities for making unique cofactors and protein samples that can be probed and characterized using various analytical methods and spectroscopic techniques. The use of non-purified accessory proteins can benefit such isotopic enrichment studies by enabling facile post-reaction purification of the target holoenzyme containing the labeled cofactor [11, 12]. One must still consider that many metal centers require unique biosynthetic pathways with specific accessory proteins and substrates, and the cell-free systems encompassing these pathways would also be unique. Nevertheless, several aspects of the methods that we present here should extend to various other in vitro metalloprotein systems.

In this chapter, we describe procedures for reconstructing the pathway for H-cluster biosynthesis and [FeFe] hydrogenase maturation in a cell-free environment. The pathway is reconstituted by combining non-purified Hyd maturases, extrinsic low-molecular-weight substrates, and purified hydrogenase apoprotein devoid of the catalytic $[2Fe]_H$ subcluster [12]. Specifically, we detail a 50 mL reaction that typically generates active hydrogenase quantities (5–10 mg) sufficient for multiple samples for advanced spectroscopic analysis. This highly effective and scalable cell-free system depends on the anaerobic expression and protein preparation methods that we also developed [13], and which would likely benefit the engineering of other in vitro metalloprotein systems as well. Therefore, prior to describing the cell-free hydrogenase activation procedure, we provide general methods for readily producing large quantities and high concentrations of either non-purified or purified metalloproteins that typically contain stoichiometric amounts of metal centers without requiring in vitro reconstitution. Activity assays, protein characterization assays, and variations of the cell-free maturation protocol for using smaller reaction volumes (~100 μL) and purified maturases are beyond the scope of this chapter, and the details for these methods can be found elsewhere [4, 12, 14, 15].

2 Materials

The O_2 sensitivity of various metalloenzymes generally requires that studies with these proteins be carried out under strictly anaerobic conditions. A vinyl anaerobic chamber with latex gloves and an attached airlock (Coy Laboratories) offers a convenient approach to create an anaerobic working environment. The glove box should be maintained at 2–4 % H_2 and 96–98 % N_2, and a palladium catalyst (Coy Laboratories) within the chamber uses the H_2 to react with O_2 to form H_2O, keeping the O_2 levels below 1 ppm. The procedures outlined in this chapter require some materials that we suggest be permanently stored within the anaerobic chamber (*see* Table 1). For the use of a stirred cell concentrator as described, an external gas line (100 % N_2 or Ar with regulated pressure) should be inserted through one of the glove box's ports, and a valve and connector should be provided within the chamber. We recommend transferring consumables and plastic components to the chamber at least 1 day before their use. Aerobic liquids exceeding 100 mL should remain in the chamber a minimum of 3 days, or alternatively, they can be flushed with 100 % N_2 for ~30 min before transferring them to the chamber. Ultrapure deionized water (18 MΩ) should be used for all solutions. Anaerobic water is readily prepared by transferring hot autoclaved water to the glove box immediately following sterilization; the bottle of hot water should be tightly sealed when transferring it through the airlock. If reagents or proteins must be kept cold to prevent degradation or denaturation, we recommend using –20 °C bench top cold blocks

Table 1
Recommended supplies for the anaerobic chamber

Item	Details	Procedure(s)
Anaerobic water	1 L	Multiple
Liquid waste container	Capacity: 1–2 L	Multiple
Automatic pipettor	e.g., Pipet-Aid® (BD Biosciences)	Multiple
Serological pipets	10 mL, 25 mL (BD Biosciences)	Multiple
Magnetic stir plate		Multiple
Electric balance		Cell lysate preparation
pH probe and pH meter		Multiple
Ring stand and clamp		Protein purification
Vortex mixer		Cell-free reaction
Microcentrifuge	e.g., MiniSpin® (Eppendrof)	Protein concentration

(Nalgene) rather than transferring ice into the anaerobic chamber. We find that −20 °C cold blocks are more effective and introduce less O_2 into the glove box. Standard materials for SDS-PAGE are needed to analyze protein solutions during cell lysate preparations and protein purifications.

2.1 Anaerobic Protein Expression

1. Antibiotic(s): Experiment dependent.
2. LB agar plates containing the appropriate antibiotic(s).
3. LB growth medium: LB Broth (Miller), 25 g/L, autoclaved.
4. 1.1× LB growth medium: LB Broth (Miller), 27.5 g/L, autoclaved.
5. Glucose: 25 % (w/v) D-glucose [250 mL basis]: While stirring, heat ~125 mL water to ~60 °C. Add 62.5 g D-glucose. After the solid dissolves, add water to a final volume of 250 mL. Sterile filter and store at 23 °C.
6. MOPS/KOH: 1 M 3-(N-morpholino)-propanesulfonic (MOPS) acid, pH 7.4: Prepare a 1 M solution of MOPS acid. Add potassium hydroxide (KOH) pellets until the solution is pH 7.4. Sterile filter and store at 23 °C.

7. IPTG: 300 mM Isopropyl β-D-1-thiogalactopyranoside (IPTG). Sterile filter, flash-freeze using liquid N_2, and store at −80 °C.
8. SBC growth medium: Supplemented and buffered complex growth medium [1 L basis]: Prepare and autoclave 900 mL of 1.1× LB growth medium. Add 10 mL of glucose solution and 100 mL of MOPS/KOH buffer using sterile techniques. Store at 23 °C (*see* **Note 1**).
9. Fe^{3+} solution: Ferric ammonium citrate, 25 g/L. Sterile filter and store at 4 °C (*see* **Note 2**).
10. L-cysteine: Solid chemical (*see* **Note 2**).

2.2 Cell Lysate Production

The following supplies should be transferred to the anaerobic chamber at least 1 day before preparing a cell lysate:

1. Beaker with a magnetic stir bar—50 mL capacity for every 1 L of culture, 1 per lysate.
2. Metal spatula.
3. Centrifuge tube—40–50 mL capacity, 1 per L of culture per lysate (*see* **Note 3**).
4. Glass vials and rubber stoppers.
5. Metal caps and crimper.
6. 10× BugBuster® Protein Extraction Reagent (Novagen).
7. Benzonase® Nuclease (Novagen).
8. Chicken egg white lysozyme—Solid chemical.
9. 20× anaerobic HEPES: 4-(2-hydroxyethyl)-1-piperazineethanesulfonic (HEPES) acid, 1 M, pH 8.2: Prepare a 1 M solution of HEPES free acid. Add KOH pellets until the solution is pH 8.2. Sterile filter and transfer to the anaerobic chamber.
10. 20× anaerobic KCl: Potassium chloride (KCl), 1 M. Sterile filter and transfer to the anaerobic chamber.
11. Anaerobic lysis solution: Anaerobic and buffered BugBuster® lysis solution [50 mL basis]: Add 5 mL of 10× BugBuster® Protein Extraction Reagent to 40 mL of anaerobic water. Add 5 mL of 20× anaerobic HEPES (this results in a 2× final concentration; i.e., 100 mM). Add 50 mg of chicken egg white lysozyme. Add Benzonase® Nuclease to a final concentration of 25–50 U/mL (*see* **Note 4**).

2.3 Metalloprotein Purification and Concentration

The following supplies should be moved into the anaerobic chamber at least 1 day before carrying out the protein purification:

1. *Strep*-Tactin® Superflow® high capacity (HC) resin (IBA GmbH)—Amount is experiment-dependent (*see* **Note 5**).

2. Gravity flow chromatography column (Bio-Rad) (*see* **Notes 6** and **7**).
3. Ring stand and clamp for holding the purification column.
4. Automatic pipettor and serological pipets—10 and 25 mL capacity.
5. Liquid waste container.
6. Amicon stirred cell concentrator (Millipore).
7. Ultracel regenerated cellulose ultrafiltration membrane (Millipore) (*see* **Note 8**).
8. Amicon centrifugal filters—0.5 mL capacity (*see* **Note 9**).
9. Microcentrifuge (*see* **Note 10**).
10. 20× anaerobic HEPES: *See* Subheading 2.2.
11. 20× anaerobic KCl: *See* Subheading 2.2.
12. 2 N KOH: 2 M Potassium hydroxide (KOH). Store at 23 °C.
13. DTB solution: 25 mM d-Desthiobiotin (DTB), pH 8.0. Prepare a solution of 25 mM DTB in water. Adjust the pH to 8.0 using 2 N KOH. Sterile filter, transfer to the anaerobic chamber, and store at 23 °C.
14. HABA solution: 20 mM 2-(4-Hydroxyphenylazo)benzoic acid (HABA), pH 8.0: Prepare a solution of 20 mM HABA in water. Adjust the pH to 8.0 using 2 N KOH. Sterile filter, transfer to the anaerobic chamber, and store at 23 °C.
15. Anaerobic wash buffer: Anaerobic HEPES buffer [1 L basis]: Add 50 mL of 20× anaerobic HEPES and 50 mL of 20× anaerobic KCl to 900 mL of anaerobic water. For each purification, prepare 50× the volume of resin to be used.
16. Anaerobic elution buffer: Anaerobic wash buffer with 2.5 mM DTB [50 mL basis]: Add 5 mL of the DTB solution to 45 mL of anaerobic wash buffer. For each purification, prepare 5× the volume of resin to be used.
17. Regeneration buffer: Anaerobic wash buffer with 1 mM HABA [100 mL basis]: Add 5 mL of the HABA solution to 95 mL of anaerobic wash buffer. For each purification, prepare 10× the volume of resin to be used (*see* **Note 11**).

2.4 Cell-Free Synthesis of the $[2Fe]_H$ Subcluster

The in vitro [FeFe] hydrogenase maturation reactions are a mixture of proteins and extrinsic low-molecular-weight compounds. The accessory protein constituents are provided by three cell lysates—each containing one of the Hyd maturases (HydE, HydF, or HydG)—and are prepared using the methods described in Subheadings 3.1–3.2. Purified and buffer-exchanged CpI apoprotein is prepared using the procedures in Subheadings 3.1–3.3.

The following supplies should be transferred to the anaerobic chamber at least 1 day before carrying out the cell-free reaction:

1. PD-10 desalting columns containing 8.3 mL of Sephadex™ G-25 Medium (GE Healthcare)—6–8 columns per 50 mL reaction.
2. 100 mL beaker with a magnetic stir bar.
3. 1.5 mL microcentrifuge tubes.
4. Two 40–50 mL centrifuge tubes (with screw caps)—rated for 20,000 × *g*.

Chemicals for the standard maturation reaction are commercially available. Protocols for preparing the solutions of the substrates are provided below. We recommend that some solutions be prepared on the day of the experiment within the anaerobic chamber and using anaerobic water. Other solutions can be prepared beforehand and either stored within the glove box (23 °C) or flash-frozen using liquid N_2 and stored in a freezer (–80 °C). Also, we have demonstrated the use of isotopically enriched analogs such as ^{57}Fe [11] and various L-tyrosine isotopologs [12] with this cell-free system. When using analogs, solutions should be prepared similarly to the standard compounds, if possible.

1. 1 N H_2SO_4: 0.5 M Sulfuric acid (H_2SO_4). Store at 23 °C.
2. 1 N HCl: 1 M Hydrochloric acid (HCl). Store at 23 °C.
3. 2 N KOH: *See* Subheading 2.3.
4. 20× anaerobic HEPES: *See* Subheading 2.2.
5. Anaerobic Fe^{2+}: Prepare 100 mM $Fe(NH_4)_2(SO_4)_2$ with anaerobic water. Add 1 N HCl to a final concentration of 0.05 N and store in the anaerobic chamber (*see* **Note 12**).
6. Anaerobic S^{2-}: Prepare 50 mM Na_2S with anaerobic water and store in the anaerobic chamber (*see* **Note 13**).
7. PLP: Prepare 250 mM pyridoxal 5′-phosphate (PLP) in water. Adjust the pH to 6.5–7.5 using 2 N KOH. Flash-freeze with liquid N_2 and store at –80 °C.
8. SAM: Prepare 250 mM *S*-adenosyl-L-methionine (SAM) in water. Adjust the pH to 3.0–4.0 using 1 N H_2SO_4. Flash-freeze with liquid N_2 and store at –80 °C (*see* **Note 14**).
9. GTP: Prepare 500 mM guanosine-5′-triphosphate (GTP) in water. Adjust the pH to 6.5–7.5 using 2 N KOH. Flash-freeze with liquid N_2 and store at –80 °C (*see* **Note 15**).
10. Anaerobic DTT: Prepare 250 mM dithiothreitol (DTT) with anaerobic water on the day of the experiment.
11. Anaerobic cysteine: Prepare 100 mM L-cysteine with anaerobic water on the day of the experiment.

12. Anaerobic DTH: Prepare 250 mM sodium dithionite (DTH) with anaerobic water on the day of the experiment.
13. Tyrosine: Solid chemical (*see* **Note 16**).

3 Methods

3.1 High Yielding Anaerobic Expression of Heterologous Proteins in *Escherichia coli*

Traditionally, *E. coli*-based recombinant expression of metalloproteins has involved a few different approaches. One general approach consists of aerobically expressing the O_2-sensitive protein for long periods of time and at low temperatures so that protein translation is extremely slowed down [16, 17]. A second method also involves aerobic induction of protein production, albeit at higher temperatures and for shorter periods [18]. With this latter approach, the brief aerobic induction phase is sometimes followed by a long anaerobic incubation phase, during which no cell growth occurs but O_2-sensitive metal centers can be assembled in vivo [19]. Nevertheless, most of the metalloprotein is translated under aerobic conditions when using any of these methods, and they typically yield small amounts of soluble protein. The resulting metalloprotein samples also commonly have substoichiometric amounts of metal clusters, which therefore require reconstitution in vitro. We find that these methods are not sufficient for the high concentrations of functional metalloproteins necessary for cell-free H-cluster synthesis and for readily preparing samples for spectroscopic analysis.

In this section, we present a general protocol for expressing an O_2-sensitive protein under strictly anaerobic growth and mixed acid fermentation [13]. Our methods are capable of producing ~100 mg of soluble, heterologous protein per liter of culture. Moreover, metalloproteins produced with these methods typically contain stoichiometric amounts of metal clusters without in vitro reconstitution [4]. Herein, we describe a 1 L cell culture as the basis for the protocol.

1. Using standard molecular biology procedures, clone the gene for the target protein into an expression vector between the T7 promoter and terminator sequences.
2. Introduce the expression plasmid into *E. coli* strain BL21(DE3) by transformation using standard procedures, and select for transformants on LB plates containing the appropriate antibiotic(s) (*see* **Note 17**).
3. Pick a single colony and inoculate a 5 mL overnight starter culture of LB growth medium containing the appropriate antibiotic(s).
4. Incubate the overnight starter culture at 37 °C, and shake at 250 rpm.

5. Subculture 2 mL of the starter culture (OD_{600} ~2.5) into 1 L of SBC growth medium contained in a 2 L baffled shake flask (*see* **Note 18**).
6. Aerobically incubate the 1 L culture at 30 °C and shake at 250 rpm.
7. At an OD_{600} of 0.4–0.6, transfer the 1 L culture along with a sterile magnetic stir bar (2–3 cm) and a 1 L centrifuge bottle with a screw cap to the anaerobic chamber (*see* **Note 19**).
8. Pour the culture into the centrifuge bottle (*see* **Note 20**).
9. Place the sterile magnetic stir bar into the culture, and continuously stir the culture during anaerobic growth and mixed acid fermentation.
10. Add IPTG to a final concentration of 300 μM to induce strictly anaerobic expression of the target protein (*see* **Note 21**).
11. Screw the cap onto the bottle and incubate at 23 °C for 8–10 h (*see* **Notes 22–23**).

3.2 Production of E. coli-Based Cell Lysates Containing Anaerobically Expressed Heterologous Proteins

Here, we describe a straightforward procedure for anaerobically preparing a clarified cell lysate using a commercial lysis buffer. This method does not require special equipment within the anaerobic chamber and also minimizes the number of steps that may be detrimental to an O_2-sensitive protein. Cell lysates produced using this method can be used for in vitro studies (*see* Subheading 3.4), or an anaerobically expressed protein can be purified from the cell extract (*see* Subheading 3.3).

1. Carry out the strictly anoxic expression of a target O_2-sensitive protein as described in Subheading 3.1.
2. Following protein expression (*see* Subheading 3.1, **step 11**), remove the bottle containing the culture from the anaerobic chamber, and centrifuge the culture at 4,000 × *g* for 20 min (*see* **Note 24**).
3. Transfer the sealed centrifuge bottle back into the anaerobic chamber.
4. Decant the supernatant into a waste container (*see* **Note 25**).
5. Weigh the empty beaker.
6. Using a metal spatula, transfer the wet cell paste to the beaker.
7. Weigh the total amount of wet cell paste recovered (*see* **Note 26**).
8. For every 1 g of wet cell paste, prepare 4 mL of anaerobic lysis solution.
9. Add the lysis solution to the beaker and resuspend the cells.
10. Add a magnetic stir bar to the solution and stir for 30–60 min (*see* **Note 27**).

11. Slowly add 20× anaerobic KCl to the stirred crude cell lysate to a final concentration of 50 mM (*see* **Note 28**).
12. If the pH of the lysate is <7.5, adjust the pH to 7.5–7.8 using 20× anaerobic HEPES. This solution is the "crude cell lysate" (*see* **Note 29**).
13. Save 20–50 μL of the crude cell lysate for SDS-PAGE analysis.
14. Transfer the crude cell lysate to 40 mL centrifuge tubes with screw caps, and tightly seal the tubes.
15. Remove the centrifuge tubes from the glove box.
16. Clarify the lysate via centrifugation at 20,000 × *g* for 20 min (*see* **Notes 24** and **30**).
17. Transfer the tubes containing the lysate back into the glove box.
18. Using a pipette and without disturbing the pellet, carefully transfer the supernatant to a clean beaker or a plastic conical tube. This solution is the "clarified cell lysate."
19. Save 20–50 μL aliquots of the clarified lysate for SDS-PAGE analysis.
20. Aliquot the clarified cell lysate to glass vials and anaerobically seal the vials with a rubber stopper and crimped metal cap.
21. Remove the vials as well as the 20–50 μL aliquots from the glove box.
22. Flash-freeze the lysate in the glass vials using liquid N_2 and store at –80 °C.
23. Carry out SDS-PAGE analysis using standard procedures to characterize the heterologous protein content in the crude and clarified cell lysates (*see* Fig. 1).

3.3 Anaerobic Purification of Metalloproteins Using Strep-Tactin® Affinity Chromatography

Combining the high yielding anaerobic expression and cell lysate preparation methods described in Subheadings 3.1 and 3.2 generally results in cell extracts with heterologous metalloprotein concentrations of 5–50 μM. Such concentrations are sufficient for using the *Strep*-Tactin® affinity chromatography (K_d ~1 μM), which offers multiple advantages for isolating metalloproteins. Unlike metal affinity chromatography (e.g., Ni-NTA resin for purifying hexahistidine tagged proteins), the *Strep*-Tactin® approach does not require a buffer-exchange step to remove high salt concentrations in the eluate, and proteins are not susceptible to potential leaching of their metal clusters due to interactions with the resin. In this section, we present a protocol for readily isolating and concentrating metalloproteins that contain a *Strep*-tag® II affinity tag by using an in-house prepared *Strep*-Tactin® Superflow® HC gravity flow column.

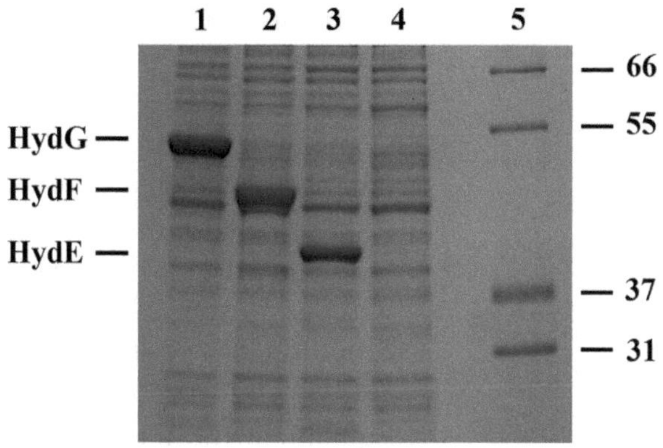

Fig. 1 Example of a Coomassie-stained gel from SDS-PAGE of clarified *E. coli* cell lysates. The image of the gel shows soluble proteins from clarified cell extract (0.2 μL) produced using *E. coli* strain BL21(DE3) Δ*iscR::kan* and the methods detailed in Subheading 3.2. *Lanes 1–3* show the non-purified *Shewanella oneidensis* HydG (54 kDa), HydF (45 kDa), and HydE (42 kDa) maturases, respectively. Based on band intensities and comparison to a bovine serum albumin standard, we estimate that each lysate has 5–20 mg/mL (10–40 μM) of the Hyd maturase. *Lane 4* shows the proteins from a control Δ*iscR* lysate without any Hyd maturases. *Lane 5* is the Mark12™ protein standard

1. Anaerobically express a target protein with an N-terminal or C-terminal *Strep*-tag® II affinity tag, as described in Subheading 3.1.
2. Produce a clarified cell lysate containing the target protein by following **steps 1–19** in Subheading 3.2 (*see* **Note 31**).
3. Transfer the *Strep*-Tactin® Superflow® HC resin and a –20 °C bench top cold block (optional) to the anaerobic chamber.
4. Attach the empty gravity flow column to the ring stand.
5. Load the column with a suspension of the *Strep*-Tactin® resin and allow it to settle.
6. Equilibrate the resin with five column volumes of anaerobic wash buffer (*see* **Note 32**).
7. Using a pipette, transfer the clarified cell lysate to the gravity flow column, taking care not to resuspend the resin. Begin collecting flow through fractions in plastic conical tubes (*see* **Note 33**).
8. Wash the resin with three column volumes of anaerobic wash buffer to remove the *E. coli* lysate.
9. Elute the tagged target protein with three column volumes of anaerobic elution buffer.
10. Remove 20–50 μL of each fraction from the glove box for SDS-PAGE analysis (*see* **Note 34**).

11. Carry out SDS-PAGE analysis using standard procedures to determine which fractions contain the purified target protein (*see* **Note 35**).
12. Combine fractions with the purified protein. This solution is the "as-isolated target protein."
13. Concentrate the protein solution within the anaerobic chamber (go to **step 14**), or store the purified protein in the as-isolated state (go to **step 27**). If the as-isolated target protein must be concentrated to volumes below the capability of the stirred cell, use the 0.5 mL centrifugal filters instead (go to **step 24**).
14. Equilibrate the ultrafiltration membrane for the stirred cell concentrator by soaking it in 50 mL of anaerobic wash buffer for 10 min.
15. According to the manufacturer's instructions, assemble the stirred cell concentrator (*see* **Note 36**).
16. Transfer the as-isolated target protein into the assembled stirred cell.
17. Tightly seal the concentrator and connect it to the external gas supply (*see* **Note 37**).
18. Adjust the regulator setting to pressurize the concentrator at 50–75 % the maximum operating pressure for the selected membrane.
19. Pressurize the concentrator by opening the external gas valve within the glove box.
20. Immediately turn on the magnetic stir plate to begin mixing the protein solution.
21. Concentrate the protein until the solution is ~2× the minimal process volume of the stirred cell.
22. Stop the concentration step by turning off the stir plate, closing the external gas valve, and slowly opening the relief valve on the stirred cell (*see* **Note 38**).
23. Transfer the protein solution from the stirred cell to an appropriately sized plastic conical tube. If this solution must be concentrated to volumes below the minimal process volume of the stirred cell, proceed to **step 24**. If the concentration process is complete, proceed to **step 27**.
24. Concentrate the purified target protein using 0.5 mL Amicon centrifugal filters as follows:

    Equilibrate the membrane for each filter device with 0.4 mL of anaerobic wash buffer by centrifuging at 5,000 × *g* for 3 min using a microcentrifuge within the glove box. Discard both the flow through and the retentate. Transfer the target protein solution (up to 0.4 mL final volume) to each filter device and centrifuge at 5,000 × *g* for 3 min (*see* **Note 39**). Discard the flow through (*see* **Note 40**). Using a micropipette, slowly resuspend the protein to create a well-mixed solution (*see* **Note 38**).

25. To collect the concentrated protein, invert each filter device into a new 1.5 mL microcentrifuge tube.
26. Centrifuge the tubes at 2,500 × *g* for 1 min.
27. Aliquot the purified target protein to glass vials, and anaerobically seal the vials with a rubber septum and crimped metal cap (*see* **Note 41**).
28. Remove the sealed vials from the anaerobic chamber and flash-freeze the purified target protein using liquid N_2. Store at –80 °C.
29. Wash the *Strep*-Tactin® resin with one column volume of anaerobic wash buffer (*see* **Note 42**).
30. Regenerate the resin using regeneration buffer.
31. Remove the gravity flow column with the *Strep*-Tactin® resin from the glove box.
32. Store the *Strep*-Tactin® resin at 4 °C.

3.4 Preparation of Reaction Mixtures for the Cell-Free Synthesis of the [2Fe]$_H$ Subcluster and Post-translational [FeFe] Hydrogenase Activation

In this section, we describe a protocol for the in vitro maturation of the CpI hydrogenase from *Clostridium pasteurianum* as a model procedure for the cell-free synthesis of active site cofactors. Reaction mixtures are a combination of exogenous substrates, cell lysates that contain accessory proteins without an affinity tag, and inactive hydrogenase apoprotein that is the target for the cell-free synthesized cofactor. As shown in Fig. 2, the mixtures are prepared by combining groups of substituents in the following order: (1) solutions of proteins excluding the hydrogenase apoprotein, (2) exogenous substrates, and (3) hydrogenase apoprotein. Here, we provide specific details for activating ~15 mg of CpI hydrogenase and subsequently purifying the holoenzyme from a 50 mL reaction that uses non-purified HydE, HydF, and HydG maturases (*see* Table 2). Such reactions typically generate active hydrogenase quantities (5–10 mg of re-purified and active enzyme) sufficient for multiple advanced spectroscopy samples. Due to the reddish brown color of the CpI hydrogenase, we also provide images that illustrate how a metalloprotein may appear throughout this procedure (*see* Fig. 3).

1. Using *E. coli* strain BL21(DE3) Δ*iscR::kan* and the methods described in Subheading 3.1, anaerobically express the HydE maturase for 6–8 h in 1 L of culture supplemented with Fe^{3+} solution and L-cysteine at final concentrations of 250 mg/L and 1 mM, respectively.
2. Using the procedure in Subheading 3.2, prepare 10–20 mL of clarified cell lysate from the *E. coli* cells produced in **step 1**. This solution is the "HydE^{lysate}" (*see* **Note 43**).

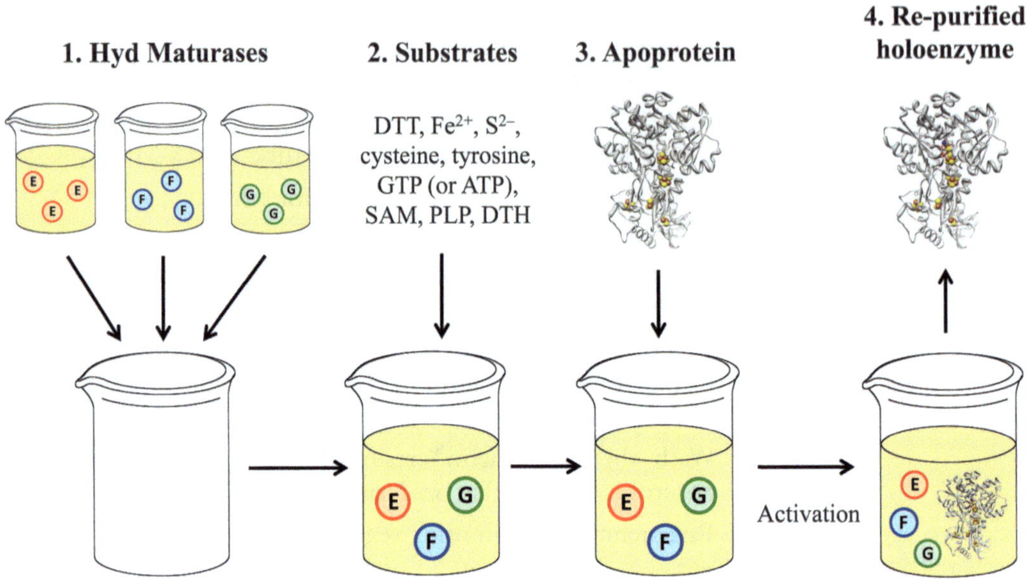

Fig. 2 Schematic for preparing cell-free mixtures for H-cluster synthesis and [FeFe] hydrogenase maturation. (1) The reaction mixture is prepared by first combining three *E. coli* lysates that each contain one of the HydE, HydF, and HydG maturases. (2) Next, low-molecular-weight substrates are added to the mixture, (3) followed by the addition of purified and desalted hydrogenase apoprotein. (4) After the mixture is incubated under anaerobic conditions for 12–18 h, the active [FeFe] hydrogenase can be readily re-purified from the reaction mixture and subsequently characterized

Table 2
Composition of hydrogenase maturation reaction mixtures

Group	Substituent	Initial concentration	Final concentration	Amount added[a]
1	$HydE^{lysate}$	1×	0.05×	2.5 mL
1	$HydF^{lysate}$	1×	0.10×	5 mL
1	$HydG^{lysate}$	1×	0.50×	25 mL
2	Anaerobic DTT	250 mM	1 mM	0.2 mL
2	Anaerobic Fe^{2+}	100 mM	1 mM	0.5 mL
2	Anaerobic S^{2-}	50 mM	0.5 mM	0.5 mL
2	Anaerobic cysteine	100 mM	2 mM	1 mL
2	PLP	250 mM	1 mM	0.2 mL
2	SAM	250 mM	2 mM	0.4 mL
2	GTP (or ATP)	500 mM	20 mM	2 mL
2	Tyrosine, solid	–	2 mM	18 mg
2	Anaerobic DTH	250 mM	2 mM	0.4 mL
2	20× anaerobic HEPES	1,000 mM	50 mM	2.5 mL
3	Desalted $apoCpl^{purified}$	25 μM	5 μM	10 mL

[a]Basis: 50 mL total reaction volume

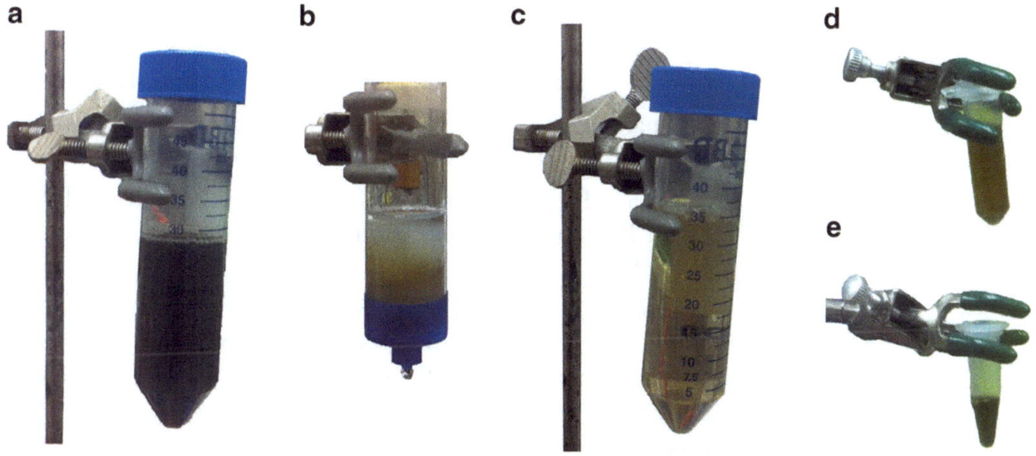

Fig. 3 Images showing the appearance of various protein fractions during the purification and concentration of in vitro activated CpI hydrogenase. (**a**) The clarified cell-free reaction mixture; dark brown/black solution. (**b**) Bound CpI holoenzyme (brown band) during the elution step. (**c**) Eluate containing holoCpI^{purified} (~5 μM); brown transparent solution. (**d**) HoloCpI^{purified} (~50 μM) after concentration with a stirred cell concentrator; brown solution. (**e**) HoloCpI^{purified} (~500 μM) after concentration with Amicon centrifugal filters; dark brown opaque solution

3. Repeat **steps 1–2** for expression of the HydF maturase in 1 L of culture and the preparation of 10–20 mL of the "HydF^{lysate}."
4. Repeat **steps 1–2** for the expression of the HydG maturase in 2 L of culture and the preparation of 20–40 mL of the "HydG^{lysate}."
5. Repeat **steps 1–2** for the expression of the CpI–*Strep*-tag II apoprotein in 2 L of culture and the preparation of 20–40 mL of the "CpI–*Strep*-tag II^{lysate}."
6. Using the protocol in Subheading 3.3, purify the CpI apoprotein from the CpI–*Strep*-tag II^{lysate} and concentrate the protein to 20–40 μM.
7. For every 2.5 mL of concentrated CpI apoprotein, use one PD-10 desalting column per the manufacturer's instructions to buffer-exchange the CpI–*Strep*-tag II hydrogenase into anaerobic wash buffer (*see* **Note 44**).
8. Concentrate the desalted CpI–*Strep*-tag II to ~25 μM, as described in **steps 14–23** in Subheading 3.3. This solution is the "desalted apoCpI^{purified}."
9. Transfer the following to the anaerobic chamber: solutions of 2 N KOH, PLP, SAM, and either GTP or ATP; solid forms of DTT (8 mg/reaction), DTH (22 mg/reaction), L-cysteine (12 mg/reaction), and L-tyrosine (18 mg/reaction); and protein solutions produced in **steps 1–8** (*see* **Note 45**).

10. If previously made and stored at –80 °C, allow the $HydE^{lysate}$, $HydF^{lysate}$, $HydG^{lysate}$, and desalted $apoCpI^{purified}$ to thaw within the glove box and reach ~23 °C (*see* **Note 46**).
11. In separate 1.5 mL microcentrifuge tubes, prepare the anaerobic DTT, anaerobic cysteine, and anaerobic DTH solutions.
12. In a 15 mL plastic conical tube, combine 2 mL of GTP (or ATP), 1 mL of anaerobic cysteine, 0.4 mL of SAM, and 0.2 mL of PLP.
13. Combine 2.5 mL $HydE^{lysate}$, 5.0 mL $HydF^{lysate}$, and 25.0 mL $HydG^{lysate}$ in a 100 mL glass beaker, and stir the protein mixture using a magnetic stir plate and stir bar (*see* **Notes 47–48**).
14. Add 0.2 mL of anaerobic DTT and 0.5 mL of anaerobic Fe^{2+} to the stirred reaction mixture. Wait 5–10 min.
15. Add 0.5 mL of anaerobic S^{2-} to the reaction mixture.
16. Add the 3.6 mL mixture of GTP/cysteine/SAM/PLP to the reaction mixture.
17. Add 18 mg of L-tyrosine to the reaction mixture (*see* **Note 16**).
18. Add 0.4 mL of anaerobic DTH to the reaction mixture.
19. If the pH of the reaction mixture is lower than 7.4, adjust the pH to 7.4–7.8 by adding 20× anaerobic HEPES buffer. The required amount varies.
20. Add 10 mL of 25 μM desalted $apoCpI^{purified}$ to the reaction mixture (*see* **Note 49**).
21. Transfer 25 mL of the reaction mixture to two 40–50 mL centrifuge tubes (*see* **Note 50**).
22. Seal the centrifuge tubes with screw caps and incubate the hydrogenase maturation reaction mixture at 23 °C for 12–18 h.
23. When the reaction is finished, remove the anaerobically sealed centrifuged tubes from the glove box.
24. Clarify the reaction mixture by centrifuging at 20,000 × *g* for 10 min (*see* Fig. 3a and **Note 24**).
25. Transfer the centrifuge tubes back into the glove box.
26. Using a pipette and without disturbing the pellet, carefully transfer the clarified reaction mixture to an empty beaker with a magnetic stir bar (*see* **Note 51**).
27. If the pH of the mixture is lower than 7.4, adjust the pH to 7.4–7.8 by slowly adding 20× anaerobic HEPES buffer while stirring (*see* **Note 52**).
28. As described in Subheading 3.3, purify and concentrate the active CpI holoenzyme ($holoCpI^{purified}$; *see* Fig. 3b–e and **Notes 53–55**).

4 Notes

1. We find that glucose-supplemented LB broth (Miller's modification) growth medium works best for strictly anoxic protein expression. Terrific broth growth medium with glucose or glycerol also leads to high yields of heterologous protein, although in some cases, we observe decreased levels of both soluble and active [FeS] protein. In a previous report, we added sodium fumarate with the LB growth medium to serve as an electron acceptor; however, we have since observed that this substrate is neither required nor beneficial. We also find that a minimal medium or a defined medium with glucose [20] do not work with this protein expression protocol, mainly because anaerobic growth and mixed acid fermentation do not occur unless commercial yeast extract or tryptone is also included.
2. The Fe^{3+} solution and L-cysteine are applicable to the methods described in this chapter that pertain to the [FeS] protein-related examples.
3. The centrifuge tubes should be rated for 20,000 × *g* and compatible with an appropriately rated centrifuge rotor.
4. We find that the crude cell lysate is at pH 6.0–7.0 when using an anaerobic lysis solution with 50 mM HEPES buffer (pH 8.2). Therefore, we suggest using a lysis solution with 100 mM HEPES to obtain a lysate with a pH of ~7.5.
5. *Strep*-Tactin® Superflow® HC resin is rated to isolate 3–15 mg protein per mL of sedimented resin, although we find that ~1 mg of protein is purified per mL of resin for some metalloproteins.
6. It may be challenging to use a fast protein liquid chromatography (FPLC) unit such as an AKTA™ Purifier (GE Healthcare) while controlling the temperature during purifications within the anaerobic chamber. We find that gravity flow chromatography at ~25 °C is adequate for our studies with various [FeS] proteins.
7. The gravity flow chromatography column should be 0.7–2.5 cm in diameter, and its total volumetric capacity should be 3–4 times the amount of resin to be used for the purification.
8. The nominal molecular weight limit (NMWL) should be at least 50 % smaller than the molecular weight of the target protein.
9. We find that the Amicon Ultra 0.5 mL Centrifugal Filters are effective for concentrating [FeS] proteins such as [FeFe] hydrogenases without affecting the integrity of the protein sample, and we have achieved concentrations of ~3 mM (200 mg/mL)

of active CpI using these ultrafiltration devices with a 30 kDa molecular weight cutoff.

10. The microcentrifuge should have a maximum relative centrifugal force (rcf) higher than $12{,}000 \times g$.
11. Regenerating the Strep-Tactin® resin with anaerobic buffer and/or under anaerobic conditions is optional and may be more convenient than regenerating the resin outside of the glove box using aerobic buffers.
12. HCl prevents the oxidation of Fe^{2+} to Fe^{3+}, which we observe occurs over time even within the anaerobic chamber. The addition of HCl is not necessary if anaerobic Fe^{2+} is prepared on the day of the experiment.
13. Oxidized solid Na_2S hydrate can also be used. In order to remove the layer of oxidized Na_2S, add a large chunk of solid (0.5–5.0 g) to hot water for 5–10 s. Quickly remove the crystal using a pair of tweezers, weigh the crystal, and immediately transfer it to the anaerobic chamber. Prepare the S^{2-} solution using anaerobic water.
14. Some investigators synthesize SAM to produce a highly pure reagent for their studies. However, we find that commercial products from Sigma (SAM chloride and SAM *p*-toluenesulfonate salt) are sufficient for cell-free H-cluster synthesis, and SAM solutions remain effective for at least 6 months when stored at –80 °C. We have also observed that SAM-e (or SAM tosylate) is effective for in vitro hydrogenase maturation, and relatively inexpensive "sample" quantities of 10–100 g can be requested from NutriScience.
15. We find that adenosine-5′-triphosphate (ATP, which is substantially less expensive than GTP) can substitute for GTP for cell-free H-cluster synthesis and hydrogenase maturation. It is noteworthy that both the precise role of GTP/ATP as well as the requirement for high concentrations of these NTPs are currently not understood.
16. The solubility of L-tyrosine in water is ~2.5 mM at 25 °C. We find that adding the solid chemical is appropriate and effective for large volume reaction mixtures (~0.5 mg tyrosine/mL). For smaller volume reactions, a slurry of 250 mM tyrosine can be used. Dissolved tyrosine can also be used by preparing 500 mM tyrosine with anaerobic water, and then adding an equal volume of 1 N KOH to dissolve the amino acid (250 mM final concentration) immediately before adding it to the cell-free reaction mixture.
17. Variants of the *E. coli* strain BL21(DE3) can also be used. For example, we find that the engineered Δ*iscR::kan* strain is useful for producing [FeS] proteins [21].

18. When expressing [FeS] proteins, we add the Fe^{3+} solution to the growth medium (250 mg/L final concentration) during the aerobic growth phase.
19. We recommend carrying out anaerobic growth in a 1 L centrifuge bottle, which facilitates centrifuging the culture the following day. Alternative sealable containers such as plastic 2 L shake flasks are also sufficient.
20. We find that the cell culture does not need to be de-gassed to displace dissolved O_2. By the time IPTG is added to induce protein expression, the remaining O_2 has likely been consumed by the cells.
21. Additional substrates that improve the production of a functional target protein should be added with the IPTG. With this procedure, addition of L-cysteine (1–5 mM final concentration) is useful for the anaerobic expression of proteins with [FeS] clusters. The solution may turn black during anoxic growth, presumably from the formation of iron sulfide. Cysteine concentrations greater than 1 mM or the addition of iron concurrent with cysteine increases the likelihood that the black substance develops.
22. We find the optimal length of time for anaerobic induction and induction strength is protein dependent and can range from 4 to 20 h and 0.2–1 mM IPTG, respectively. Also, the final OD_{600} is typically 1.5–2.5, depending on the length of the anaerobic growth phase.
23. While SBC growth medium contains 100 mM MOPS buffer (pH 7.6), the pH of the culture still decreases to 5.5–6.0 during anaerobic growth. Controlling the pH during mixed acid fermentation is optional, although in our studies, we find that this is not necessary for the production of [FeS] proteins.
24. Based on our active [FeFe] hydrogenase production studies [13], we find that air does not appear to diffuse through the bottles and negatively affect the O_2-sensitive metalloproteins during the centrifugation step outside of the glove box.
25. Washing the cell pellet before the lysis step is optional, although we do not find this necessary for the purposes of our studies.
26. We find that 3–4 g of wet cell paste is typically generated from a 1 L culture when using the anaerobic expression methods described in Subheading 3.1.
27. The rpm setting on the stir plate depends on the viscosity of the crude cell extract. At the onset of lysis, the solution will become highly viscous due to the chromosomal DNA, and a high rpm setting should be used. The viscosity of the solution will decrease as the endonuclease digests the DNA, at which point

the rpm setting should be decreased to prevent the extract from splashing out of the beaker.

28. Endonuclease is commonly inhibited by >150 mM monovalent cation concentrations. The solution of 20× anaerobic KCl is added following cell lysis in order to achieve optimal endonuclease activity.
29. Even though a buffered lysis solution (pH ~8.0) is used to extract the proteins, we find that the crude cell lysate typically has an initial pH of 6.0–7.0. We believe this is due to the absence of a cell wash step along with the high concentrations of acetate produced in *E. coli* during mixed acid fermentation.
30. We find that the crude cell extract may become black due to insoluble material that accumulates from the addition of iron and cysteine for producing [FeS] proteins. The combination of using a commercial lysis buffer and centrifugation at 20,000 × *g* results in the separation of the black constituents from the clarified lysate. We also find that when using other lysis techniques such as homogenization or sonication, the black constituents remain a part of the clarified lysate and often negatively affect subsequent purifications such as *Strep*-Tactin® affinity chromatography.
31. Before purifying the target metalloprotein, flash-freezing and storing the cell lysate from Subheading 3.2 is optional. Both the lysate preparation and the purification procedures can be completed in 1 day.
32. We find that at least 200 μM DTH in anaerobic buffers can negatively affect the performance of the *Strep*-Tactin® Superflow® HC resin, and we advise excluding strong reducing agents from the purification solutions.
33. Throughout the purification, we suggest collecting fractions that are 10–20 % of the total volume of the cell lysate. For example, if 50 mL of lysate is used for the purification, then one should collect 5–10 mL fractions.
34. During SDS-PAGE analysis (**step 11**), the protein fractions can be kept at 23 °C, in a –20 °C cold block, or anaerobically sealed and flash-frozen using liquid N_2.
35. A relatively quick activity assay (if applicable) can also be used to establish which fractions contain the tagged target enzyme, and some metalloproteins may have a distinct color that make them easily identifiable.
36. Prior to loading and concentrating the target protein, the membrane can be tested for holes by using anaerobic wash buffer and following **steps 16–20**.

37. Before pressurizing the concentrator, make sure the tube for the permeate is inserted into a collection reservoir such as a beaker.
38. At this point, additional protein or other solutions such as a glassing agent (e.g., glycerol) can be added. Repeat **steps 16–22** for the stirred cell, and **steps 24–26** for the centrifugal filters.
39. While higher speeds and longer times will further concentrate the protein, we choose these parameters in order to prevent concentration polarization of the protein at the membrane interface, and to minimize the likelihood of the protein denaturing and precipitating.
40. Alternatively, the flow through can be kept to verify that none of the target protein transferred through the membrane.
41. If using the purified target protein to make samples for advanced spectroscopic analysis, proceed with the appropriate protocol for preparing the samples for the particular spectroscopy.
42. *Strep*-Tactin® Superflow® HC resin is presently expensive relative to the metal affinity resins typically used for purifying proteins with hexahistidine tags. We recommend regenerating the resin immediately following a purification experiment and then storing it at 4 °C. We find it is crucial that DTH is not present during the regeneration steps, and a wash step ensures that no strong reducing agent is present when the regeneration buffer is added.
43. We find that 3–4 g of wet cell paste is produced per liter of culture, from which 10–20 mL of clarified cell lysate can be made.
44. This step is essential to remove the d-desthiobiotin in order to re-purify the CpI holoenzyme from the in vitro hydrogenase maturation mixture.
45. If the anaerobic Fe^{2+} and S^{2-} solutions are not being stored in the glove box, transfer solid forms of $Fe(NH_4)_2(SO_4)_2$, and Na_2S along with the other substrates and protein mixtures to the anaerobic chamber.
46. The $HydE^{lysate}$, $HydF^{lysate}$, and $HydG^{lysate}$ may appear cloudy due to insoluble material that accumulates during the freeze/thaw process. We speculate this substance develops from the use of the BugBuster® lysis solution. The precipitate does not appear to affect hydrogenase maturation, and clarification of the lysates to remove the insoluble material before the cell-free reaction is not required.
47. We find that cell-free H-cluster synthesis and hydrogenase activation is more effective (i.e., higher final concentrations of

active [FeFe] hydrogenase are obtained) when adding a higher ratio of HydG^{lysate} to HydE^{lysate} and HydF^{lysate} rather than a 1:1:1 volumetric ratio. Small-scale reactions (e.g., 50–100 μL) with the substrate and apoprotein concentrations shown in Table 2 can be used beforehand to determine the ratio of the maturase lysates that results in the highest yield of active hydrogenase [4].

48. We observe that the addition of a cell lysate from *E. coli* strain BL21(DE3) Δ*iscR::kan* that does not contain heterologous proteins (i.e., Δ*iscR*^{lysate}, final concentration of 0.10×) improves in vitro hydrogenase maturation, although the lysate is not essential. In general, the Δ*iscR*^{lysate} is made using the protocol in Subheading 3.2 (albeit under aerobic conditions) from cells grown aerobically to an OD_{600} of 2–3 using LB growth medium supplemented with 50 mg/L kanamycin.
49. If the active CpI is to be purified following in vitro maturation, we do not advise final concentrations less than 5 μM CpI due to the difficulty in re-purifying the holoenzyme when using Strep-Tactin® Superflow® HC resin (K_d ~1 μM).
50. We observe the accumulation of insoluble material during the maturation reaction, even when using clarified lysates. We advise that the insoluble material should not be resuspended prior to clarification.
51. The clarified reaction mixture appears dark brown or black (*see* Fig. 3a), likely due to both the reddish brown color of the [FeS] proteins as well as the accumulation of FeS during the reaction. We find that the presence of the brown/black components does not affect the purification of the CpI holoenzyme.
52. We observe that the pH decreases during the reaction. A pH of 8 is recommended for *Strep*-Tactin® chromatography, and we have observed ineffective purification of active [FeFe] hydrogenase if the pH of the reaction mixture is not adjusted before loading it onto the resin.
53. If a particular cell-free cofactor synthesis system cannot achieve holoenzyme concentrations greater than 1 μM, alternative chromatography methods should be considered.
54. During the elution step, the target protein removed from the column may have a distinct color associated with it, as shown for the CpI holoenzyme in Fig. 3b. We find that for a 50 mL reaction with 5 μM of CpI apoprotein added (~15 mg), roughly 40–60 % of the enzyme is isolated into 25–50 mL of elution buffer. Thus, the as-isolated CpI holoenzyme eluate fraction (which is yellowish brown in color as shown in Fig. 3c) is ~5 μM (0.3 mg/mL) of active CpI.

55. The amount of active [FeFe] hydrogenase in the reaction mixture as well as the different purification fractions can be readily quantified within the glove box by using a colorimetric methyl viologen reduction assay with a UV-visible spectrophotometer. Since this assay is specific to hydrogenases, describing it is beyond the scope of this chapter, and details for the assay can be obtained from our previously published study [12].

Acknowledgements

This work was supported by the United States Department of Energy, Office of Basic Energy Sciences, Division of Materials Sciences and Engineering under Award #DE-FG02-09ER46632 to JRS.

References

1. Fontecilla-Camps JC, Amara P, Cavazza C et al (2009) Structure-function relationships of anaerobic gas-processing metalloenzymes. Nature 460(7257):814–822
2. Rubio LM, Ludden PW (2008) Biosynthesis of the iron-molybdenum cofactor of nitrogenase. Annu Rev Microbiol 62:93–111
3. Peters JW, Broderick JB (2012) Emerging paradigms for complex iron-sulfur cofactor assembly and insertion. Annu Rev Biochem 81:429–450
4. Kuchenreuther JM, Britt RD, Swartz JR (2012) New insights into [FeFe] hydrogenase activation and maturase function. PLoS One 7 (9):e45850
5. Posewitz MC, King PW, Smolinski SL et al (2004) Discovery of two novel radical S-adenosylmethionine proteins required for the assembly of an active [Fe] hydrogenase. J Biol Chem 279(24):25711–25720
6. Hodgman CE, Jewett MC (2012) Cell-free synthetic biology: thinking outside the cell. Metab Eng 14(3):261–269
7. Reissmann S, Hochleitner E, Wang H et al (2003) Taming of a poison: biosynthesis of the NiFe-hydrogenase cyanide ligands. Science 299(5609):1067–1070
8. Kuchenreuther JM, Stapleton JA, Swartz JR (2009) Tyrosine, cysteine, and S-adenosyl methionine stimulate in vitro [FeFe] hydrogenase activation. PLoS One 4(10):e7565
9. Curatti L, Hernandez JA, Igarashi RY et al (2007) In vitro synthesis of the iron-molybdenum cofactor of nitrogenase from iron, sulfur, molybdenum, and homocitrate using purified proteins. Proc Natl Acad Sci U S A 104(45):17626–17631
10. Curatti L, Ludden PW, Rubio LM (2006) NifB-dependent in vitro synthesis of the iron-molybdenum cofactor of nitrogenase. Proc Natl Acad Sci U S A 103(14):5297–5301
11. Kuchenreuther JM, Guo Y, Wang H et al (2013) Nuclear resonance vibrational spectroscopy and electron paramagnetic resonance spectroscopy of ^{57}Fe-enriched [FeFe] hydrogenase indicate stepwise assembly of the H-cluster. Biochemistry 52(5):818–826
12. Kuchenreuther JM, George SJ, Grady-Smith CS et al (2011) Cell-free H-cluster synthesis and [FeFe] hydrogenase activation: all five CO and CN^- ligands derive from tyrosine. PLoS One 6(5):e20346
13. Kuchenreuther JM, Grady-Smith CS, Bingham AS et al (2010) High-yield expression of heterologous [FeFe] hydrogenases in Escherichia coli. PLoS One 5(11):e15491
14. Fish WW (1988) Rapid colorimetric micromethod for the quantitation of complexed iron in biological samples. Meth Enzymol 158:357–364
15. Bradford MM (1976) A rapid and sensitive method for the quantitation of microgram quantities of protein utilizing the principle of protein-dye binding. Anal Biochem 72:248–254
16. Lanz N, Grove T, Gogonea C et al (2012) RlmN and AtsB as models for the overproduction and characterization of radical SAM proteins. In: Hopwood DA (ed) Methods in enzymology. Elsevier Inc, Berlin, pp 125–152
17. Rubach JK, Brazzolotto X, Gaillard J et al (2005) Biochemical characterization of the HydE and HydG iron-only hydrogenase maturation enzymes from Thermatoga maritima. FEBS Lett 579(22):5055–5060

18. Ugulava NB, Gibney BR, Jarrett JT (2000) Iron-sulfur cluster interconversions in biotin synthase: dissociation and reassociation of iron during conversion of [2Fe-2S] to [4Fe-4S] clusters. Biochemistry 39(17):5206–5214
19. King PW, Posewitz MC, Ghirardi ML et al (2006) Functional studies of [FeFe] hydrogenase maturation in an Escherichia coli biosynthetic system. J Bacteriol 188(6):2163–2172
20. Kuchenreuther JM (2011) Engineering in vivo and in vitro systems for expression and activation of [FeFe] hydrogenases. Dissertation. Stanford University, Stanford, pp 1–164
21. Akhtar MK, Jones PR (2008) Deletion of iscR stimulates recombinant clostridial Fe–Fe hydrogenase activity and H2-accumulation in Escherichia coli BL21(DE3). Appl Microbiol Biotechnol 78(5):853–862

Chapter 6

Electrochemistry of Metalloproteins: Protein Film Electrochemistry for the Study of *E. coli* [NiFe]-Hydrogenase-1

Rhiannon M. Evans and Fraser A. Armstrong

Abstract

Protein film electrochemistry is a technique which allows the direct control of redox-active enzymes, providing particularly detailed information on their catalytic properties. The enzyme is deposited onto a working electrode tip, and through control of the applied potential the enzyme activity is monitored as electrical current, allowing for direct study of inherent activity as electrons are transferred to and from the enzyme redox center(s). No mediators are used. Because the only enzyme present in the experiment is bound at the electrode surface, gaseous and liquid phase inhibitors can be introduced and removed whilst the enzyme remains in situ. Potential control means that kinetics and thermodynamics are explored simultaneously; the kinetics of a reaction can be studied as a function of potential. Steady-state catalytic rates are observed directly as current (for a given potential) and non-steady-state rates (such as interconversions between different forms of the enzyme) are observed from the change in current with time. The more active the enzyme, the higher the current and the better the signal-to-noise. In this chapter we outline the practical aspects of PFE for studying electroactive enzymes, using the *Escherichia coli* [NiFe]-hydrogenase 1 (Hyd-1) as an example.

Key words Protein film electrochemistry, Hydrogenase, Hydrogen

1 Introduction

In recent years, the study of redox active enzymes through protein film electrochemistry (PFE) has flourished. Electroactive metalloproteins can be directly adsorbed (immobilized) onto various electrodes, but particularly pyrolytic graphite "edge" (see later), which has a rough surface [1–3] and is electrochemically "silent" across a wide potential range in water. Examples of enzymes and their reactions studied by PFE include H^+ reduction and/or H_2 oxidation by [FeFe]- and [NiFe]-hydrogenases, [4–7] CO_2 reduction and/or CO oxidation by carbon monoxide dehydrogenase (CODH), [8, 9] nitrate reduction by nitrate reductase, [10] succinate oxidation

Juan C. Fontecilla-Camps and Yvain Nicolet (eds.), *Metalloproteins: Methods and Protocols*, Methods in Molecular Biology, vol. 1122, DOI 10.1007/978-1-62703-794-5_6,

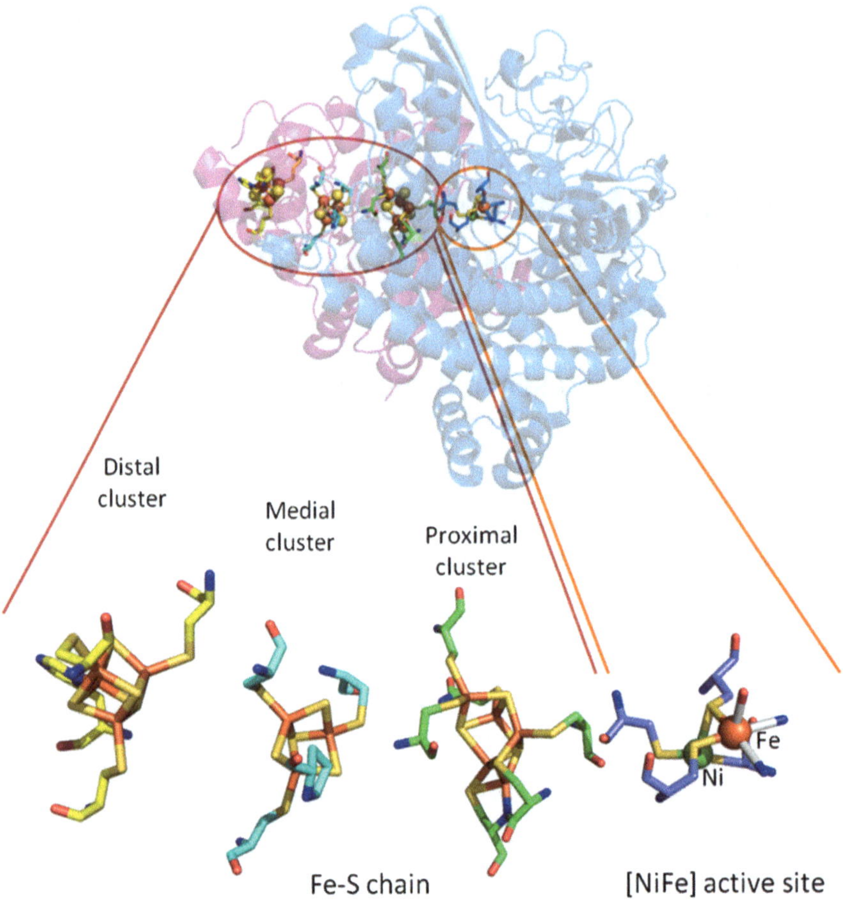

Fig. 1 The structure of the O_2 tolerant [NiFe]-hydrogenase Hyd-1 from *Escherichia coli* (3USE.pdb). The active site Ni (*green sphere*) and Fe (*orange sphere*) are shown with the three ligating cysteine residues in purple (two terminal and two bridging). The CO and two CN^- ligands which ligate the Fe atom are shown as sticks. The three Fe-S clusters, named according to their position relative to the active site, are shown ("distal" ligated by yellow carbon side chains, "medial" ligated by cyan carbon side chains and "proximal" ligated by green carbon side chains). For clarity only one heterodimer of Hyd-1 is shown

and/or fumarate reduction by fumarate reductase, [3] (see also Armstrong and Hirst, [11]). In the work described herein, we use pyrolytic graphite edge (PGE) electrodes for the adsorption of a [NiFe]-hydrogenase from *Escherichia coli* ("Hyd-1," see below) to exemplify the types of experiments that can be performed using PFE to interrogate enzyme kinetics and thermodynamics.

Hydrogenases catalyze the interconversion of molecular hydrogen ($H_2 \rightleftharpoons 2H^+ + 2e^-$) in microbes. The two major classes are known as [FeFe]- and [NiFe]-hydrogenases according to the first-row transition metal content of their active site [12]. In the case of [NiFe]-hydrogenases (Fig. 1), the active site Fe atom has one CO and two CN^- ligands and is buried in the protein matrix, being ligated to the protein backbone by four cysteine –S donors (two terminal, i.e., ligated to the Ni, and two bridging the Ni and Fe). The active site is electronically linked by a chain of Fe–S clusters to its natural redox partner (a cytochrome contained in a subunit that is embedded in the bacterial membrane and linked to the quinone pool, not shown) [13]. A series of water and gas channels provide routes for substrate and inhibitor access and product release [14–16]. *E. coli* produces two periplasmic facing membrane bound [NiFe]-hydrogenases; hydrogenase-1 and hydrogenase-2 ("Hyd-1" and "Hyd-2") that are further classified as "O_2 tolerant" and "O_2 sensitive," respectively [6].

The term "O_2 tolerance" refers to the ability of [NiFe]-hydrogenases to perform H_2 oxidation in the presence of O_2 [17]. Hydrogen, CO and O_2 are expected to bind at transition metals using π-back-donation [18–21]. Whereas CO is usually a reversible inhibitor, the so-called "O_2 sensitive" [NiFe]-hydrogenases are inactivated rapidly by O_2 in a poorly reversible process [4, 6, 7]. Recently there has been much work to elucidate the role of the Fe–S clusters in "O_2 tolerant" Hyd-1 [4, 6, 7, 22]. A major theme is the role of the Fe–S clusters to ensure rapid delivery of sufficient electrons to the active site to reduce O_2 completely to harmless water. For a more detailed discussion of this mechanism of O_2 tolerance, see references [4, 17].

Protein film electrochemistry is an excellent tool for studying the inhibition of enzymes by various substances such as the inhibition of [NiFe]-hydrogenases with O_2 or CO [4, 7], the inhibition of CODH by various inhibitors such as cyanide, cyanate, and hydrogen sulfide [8] or the inhibition of [FeFe]-hydrogenases with formaldehyde and CO [5]. Inhibitors can be introduced into the electrochemical cell in the gaseous headspace (using mass flow controllers to create precise gas mixtures), or injected as a saturated solution (Fig. 4 and section 3.7). The immobilization of miniscule amounts of enzyme on the electrode (*see* Fig. 3 and section 3.2) allows for these inhibitory substances to be added during the experiment whilst the enzyme response is monitored as a change in current. The inhibitor can then be removed and the enzyme's recovery monitored; gaseous substances are flushed from the cell by the constant flow of gases through the cell headspace, the equilibration time of which is dependent on the volume of buffer in the working cell and the gas flow rate. Nonvolatile inhibitors are removed from the experiment by buffer exchange [5], during which time the electrical connection is maintained.

In order to study the inherent activity of the enzyme in response to changes in applied potential, presence or absence of inhibitors etc., the electron transfer to and from the enzyme (interfacial electron transfer, IET) and the mass transport of substances to and from the enzyme must not be limiting. These possible sources of rate limitation can be represented as a series of resistors (Fig. 3) and can be overcome provided the enzyme is well oriented for fast IET, and by rotating the working electrode at sufficiently high speed to overcome mass transport limitation (*see* section 3.6).

In this work we describe the basic electrochemical setup for studying *Escherichia coli* [NiFe]-hydrogenase-1 (Hyd-1), and the limitations and practical aspects that must be taken into consideration for the accurate measurement of catalytic activity when various parameters are altered, for example, rotation rate of the working electrode, substrate (H_2) concentration, introduction and removal of inhibitors, and the applied potential.

2 Materials

All solutions are prepared using ultra pure water (Millipore, 18 MΩ cm) and are purged with N_2 for at least 3 h to remove dissolved O_2 before taking into the glove box. All gases (e.g., BOC gases) are of high purity/precision grade.

2.1 Casting in House Pyrolytic Graphite "Edge" Working Electrode

1. Aradur hardener and Araldite resin (Robnor Resins) stored at room temperature (*see* **Note 1**).
2. Pyrolytic Graphite plate (PG, 2″ × 2″ × 0.125″, Momentive Performance Materials) (*see* **Note 2**).
3. Silver loaded epoxy adhesive (RS Components Ltd)—stored at room temperature and prepared according to manufacturer's instructions.
4. Sandpaper grades used in this work include Tufbak Durite grades P60, P120, and P400.

2.2 Electrochemistry Setup, Equipment, and Reagents

1. All electrochemistry described is carried out an anaerobic glove box (Vacuum Atmospheres or MBraun) containing a N_2 atmosphere ($O_2 < 2$ ppm).
2. Protein film electrochemistry experiments are carried out in a sealed, glass electrochemistry cell (*see* Fig. 4) made in-house (section 3.1).
3. The working electrode (WE, *see* Fig. 4) is a pyrolytic graphite "edge" rotating disk electrode (RDE) of geometric surface area 0.03 cm^2 (section 3.1). The working electrode is rotated at a constant speed (100–4,000 rpm) depending on experiment to avoid mass transport limitation (Fig. 3 and section 3.6) controlled by an electrode rotator (EcoChemie or EG&G, **Note 4**).
4. The reference electrode (RE, *see* Fig. 4) used is a saturated calomel electrode (SCE, *see* **Note 6**).
5. The counter electrode (CE, *see* Fig. 4) used is a platinum wire (*see* **Note 3**).

6. Protein film electrochemistry measurements are made using an electrochemical analyzer (Autolab potentiostat PGSTAT128N, Metrohm Autolab B.V.) and electrochemical software (Nova, Metrohm Autolab B.V.).
7. A circulating water bath with refrigeration capabilities (VWR, *see* **Note 5**) is connected to the water-jacketed working compartment of the electrochemical cell. The side arm containing the reference electrode is always maintained at room temperature (*see* **Note 6**).
8. Sodium chloride (NaCl, 0.01 M) is used to fill the side arm (*see* **Note 7**) for electrolytic connection of the reference SCE to the working compartment of the electrochemical cell (Fig. 4).
9. A mixed buffer solution [18] is used to carry out experiments over a wide pH range without the need for change in buffer composition and comprises: 1.23 g/L (15 mM) Sodium acetate ($pK_a = 4.76$ [23]), 3.11 g/L (15 mM) CHES (2-(cyclohexylamino)ethanesulfonic acid, $pK_a = 9.30$ [23]), 3.57 g/L (15 mM) HEPES (*N*-(2-hydroxyethyl)piperazine-*N'*-(2-ethanesulfonic acid, pK_a 7.55 [23]), 2.93 g/L (15 mM) MES (2-(*N*-morpholino)ethanesulfonic acid, $pK_a = 6.10$ [23]), 3.65 g/L (15 mM) TAPS, (*N*-[Tris(hydroxymethyl)methyl]-3-aminopropanesulfonic acid, $pK_a = 8.40$ [23]), and 5.84 g/L (100 mM) sodium chloride, adjusted to the desired pH using HCl and NaOH at the working temperature of the intended experiment. Buffer is stored at room temperature in the glove box (short term, *see* **Note 8**). Long-term storage is at 4 °C.
10. Precise gas mixtures are created using mass flow controllers (Sierra Instruments), introduced into the electrochemical cell headspace via a needle (Fig. 4, *see* **Note 9**).

3 Methods

3.1 Fabrication of the Pyrolytic Graphite "Edge" Working Electrode

The pyrolytic graphite "edge" working electrode (WE) is housed in an insulating PTFE casing (*see* **Note 10**). A stainless steel core runs through the center of the casing and the screw thread at the top is used to attach to the rotating spindle (Fig. 2), which sits inside the rotator head and makes a connection to the potentiostat (Fig. 4). The tip of the electrode casing has a crevice of diameter ~0.2 cm into which the cylindrical graphite rod (surface area = ~0.03 cm^2 cut from the graphite plate, *see* **Note 2**) is slotted (*see* **Note 11**). Electrical connection between the steel rod and the graphite is made by application of silver epoxy resin (*see* **Note 12**). The graphite shaft is protected from solution with an insulating araldite Epoxy resin (*see* **Note 13**). The joint between the PTFE insulating casing

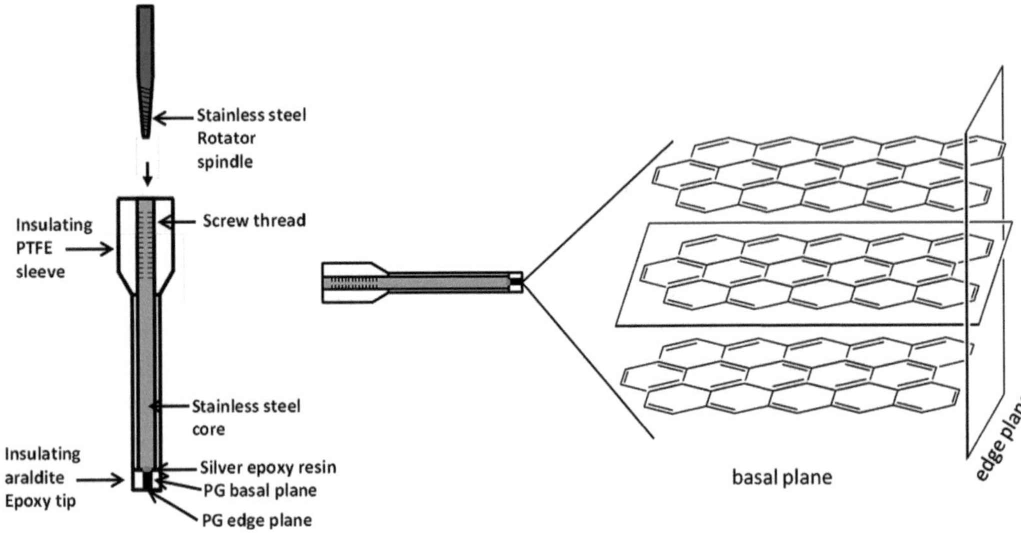

Fig. 2 The components of an "in-house" pyrolytic graphite (PG) edge working electrode (*left*) and the orientation of the planes of the PG tip relative to the PTFE electrode shaft (*right*)

and the araldite epoxy tip case is sealed with clear nail varnish. The connectivity of the electrode can be tested with a digital multimeter, with a resistance of less than 10 Ω, around 4 or 5 Ω being optimal.

3.2 Making or "Growing" an Enzyme Film on a Working Electrode

A "blank" control experiment in cyclic voltammetry mode (*see* section 3.4) should first be performed to ensure that the only significant current observed in later experiments results from the Faradaic current (*see* section 3.4) reporting on the intrinsic catalytic activity of the adsorbed enzyme. The orientation of the enzyme is crucial for proper electron transfer between the enzyme and the electrode (interfacial electron transfer, IET) because an internal electron relay system only approaches a limited region of the protein surface. A rough surface such as PGE greatly increases the contact area, analogous to stabilizing eggs in an egg box. For measurements to report on the intrinsic nature of the enzyme mechanisms, IET should not be rate limiting. Similarly, mass transport of substrate to and products from the electrode surface cannot be rate limiting either. We can represent the possible rate limiting steps for electron transport and enzyme turnover (measured directly as current) as a series of resistors (Fig. 3).

There are numerous procedures for adsorbing an enzyme onto an electrode surface. The simplest method, used in this work, is to "spot" a solution of enzyme [24] onto a freshly polished surface (*see* **Note 11** and ref. [2] for a detailed description of the surface of PGE electrodes polished with sandpaper and alumina slurry) as detailed below. Immersion of various electrode surfaces into an

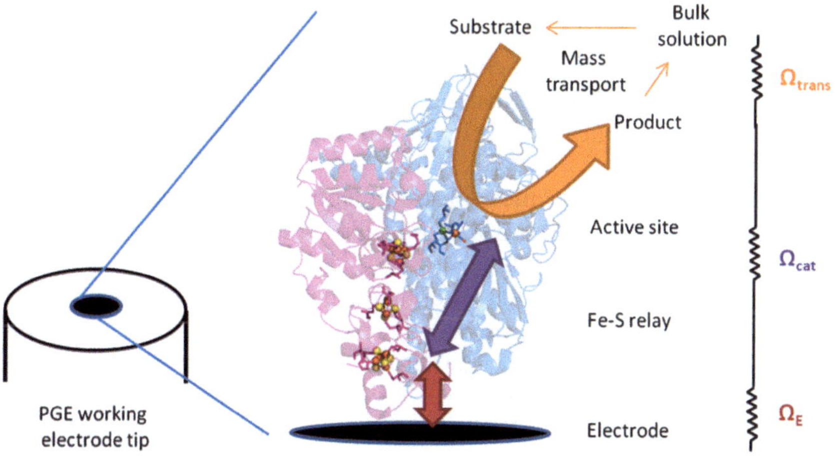

Fig. 3 The orientation of enzyme relative to the surface of the working electrode and the bulk solution is important. The possible rate limitations are represented as a series of resistors, where ΩE = interfacial electron transport, Ωcat = intramolecular electron transport for inherent catalytic activity and Ωtrans = mass transport of substrate and product(s). Figure adapted from Vincent et al. [24]

enzyme solution (the "forming solution"), with or without the cycling of potential ("cyclic voltammetry," *see* section 3.4) to monitor film growth is also successful. The addition of a co-adsorbate (e.g., cationic compounds such as polymyxin B sulfate or poly-L-lysine) to the solution of enzyme often improves the adsorption (through noncovalent interactions) and stability of the enzyme film on the electrode surface, and tests should always be carried out to optimize the voltammetry [24]. Immobilization and stabilization can also be improved substantially by covalent attachment which can be particularly beneficial if the site of linkage optimizes electron transfer between the electrode and enzyme (IET, *see* Fig. 3 and section 3.4). Strategies for the covalent attachment of hydrogenases to graphite electrodes were reported by Rudiger et al. who coupled surface-exposed glutamate side chains to the carbon surface via amide linkages [25]. More recently Hyd-1 was covalently attached via surface lysine residues to multiwalled carbon nanotubes coated on to PGE [26]. Neither of these methods allowed for site-specific binding of the enzyme since more than one side chain is surface exposed in each case.

In this work, Hyd-1 (isolated aerobically as previously described [4] and stored flash frozen in liquid N_2) is applied to the working electrode PGE tip (Fig. 2) using the "spotting" procedure [24] to create or "grow" an enzyme "film" where Hyd-1 adsorbs directly to the surface of the edge surface of the graphite tip (*see* **Note 11**): The PGE electrode is abraded using P400 Tufbak Durite sandpaper and wiped with cotton wool to remove any excess carbon. The surface of the electrode is then immediately "spotted" with enzyme (~0.5 – 1 μL at 10 – 100 μM) and allowed to soak for ~30 s before holding under a stream of ultra pure water (Millipore, 18 MΩ cm) to remove excess enzyme. While speed is desirable, longer soaking times may be required, or application and reapplication of the enzyme solution, and this should be checked.

3.3 Electrochemical Setup (Three Electrode System)

Electrochemical measurements are performed using the electrochemical setup shown in Fig. 4, and using the materials described in section 2.2.

In the PFE experiments described herein, steady-state enzyme activity is measured directly as current (i) as a function of the change in potential difference between the working electrode and the solution interface. A voltage, E, is applied, and the current flow between the WE and CE is measured according to:

$$E = \Delta\varnothing_{e/s}{}^{(WE)} + iR + \Delta\varnothing_{e/s}{}^{(RE)}$$

where $\Delta\varnothing_{e/s}{}^{(WE)}$ is the potential difference between the working electrode (WE) and the bulk solution, i.e., at the electrode/solution (e/s) interface, and $\Delta\varnothing_{e/s}{}^{(RE)}$ is the potential difference between the reference electrode (RE) and the bulk solution, i.e., at the e/s interface. The potentiostat fixes the potential between the WE and the RE. The applied potential only reflects $\Delta\varnothing_{e/s}{}^{(WE)}$

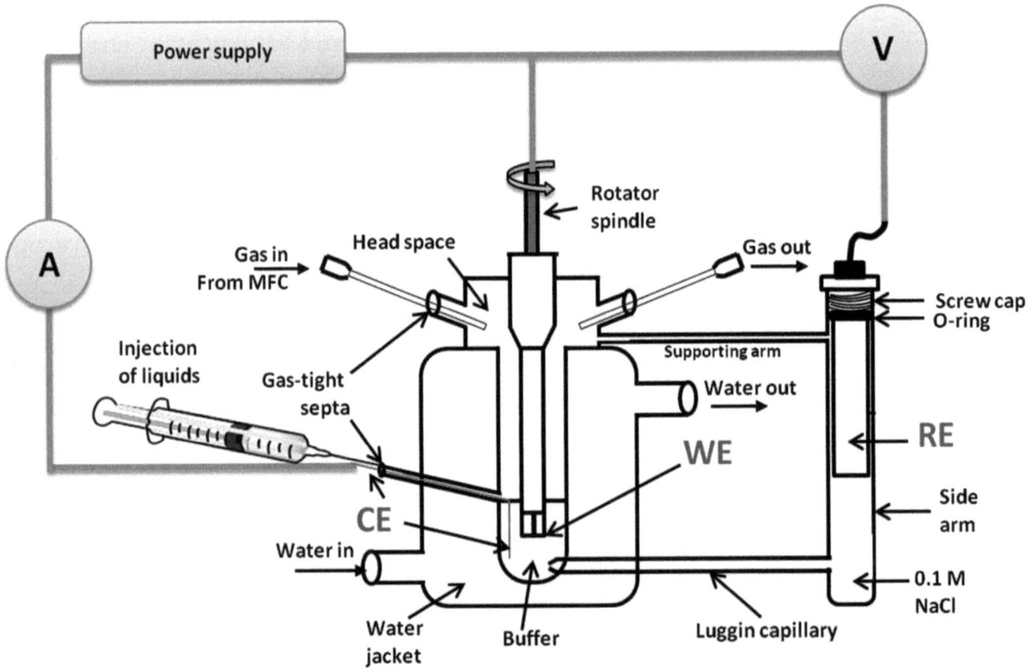

Fig. 4 A typical electrochemical cell setup. The three-electrode system, where *RE* reference electrode, *WE* working electrode, *CE* counter electrode, *V* voltmeter, and *A* ammeter and the circuit imposed by a potentiostat are shown

when $\Delta\varnothing_{e/s}^{(RE)}$ is constant. The term iR describes the bulk solution resistance to the flow of current between the WE and RE. When iR is significant (>1–2 mV, for example in all cases apart from when WE is a microelectrode), a counter electrode (CE) must be included in the system so that no significant current is drawn through the RE and $\Delta\varnothing_{e/s}^{(RE)}$ thus remains at a constant, precise value. This three-electrode system is shown as a schematic in Fig. 4 in relation to the electrochemical apparatus used in this work.

3.4 Protein Film Electrochemistry Techniques

A number of complementary techniques are used in PFE. The applied potential can be varied in cyclic voltammetry, to give insight into thermodynamic parameters and how the current varies with potential: alternatively, the potential can be held constant while a gas or inhibitor is added, or stepped to more positive or more negative values in chronoamperometry, *see* later to record the kinetics of reactions at a defined potential.

1. Cyclic voltammetry: In cyclic voltammetry, the applied potential, $\Delta\varnothing_{e/s}^{(WE)}$, is swept linearly between two limits (E_1 and E_2). The rate of change in potential with time is defined as the scan rate (mV/s) which is a parameter set by the operator. In Fig. 5 for example, the applied potential is swept from −0.636 V (E_1) to +0.241 V (E_2) in the "oxidative scan" and back to −0.636 V in the "reductive scan." A scan rate of 30 mV/s was used.

   The current (i) observed in a cyclic voltammetry experiment is composed of Faradaic (i_f) and non-Faradaic (i_c) currents according to:

   $$i = i_c + i_f = C_d\frac{\delta E}{\delta t} + i_f = \nu C_d + i_f$$

   where C_d describes the storage of charge or capacitance of the e/s interface in response to the change in applied potential over time, $\frac{\delta E}{\delta t}$ which depends on scan rate, ν.

2. Non-Faradaic currents: The current response shown in Fig. 5 is an example of non-Faradaic current which arises when a "blank" electrode (i.e., without adsorbed enzyme or "electroactive species") is used in cyclic voltammetry. Non-Faradaic current (capacitance current) arises from the ionic interactions between the charged WE and the ions in bulk solution which form an electrical "double layer." The net charge of the e/s interface remains neutral; as the electrode charge alters when the potential is swept, the charge at the solution interface changes in an equal but opposite manner, altering the double-layer composition through redistribution of ions at the interface and producing the observed effect (akin to a capacitor). The size of the non-Faradaic current depends on the surface area of the electrode, the applied potential (or change in potential for a

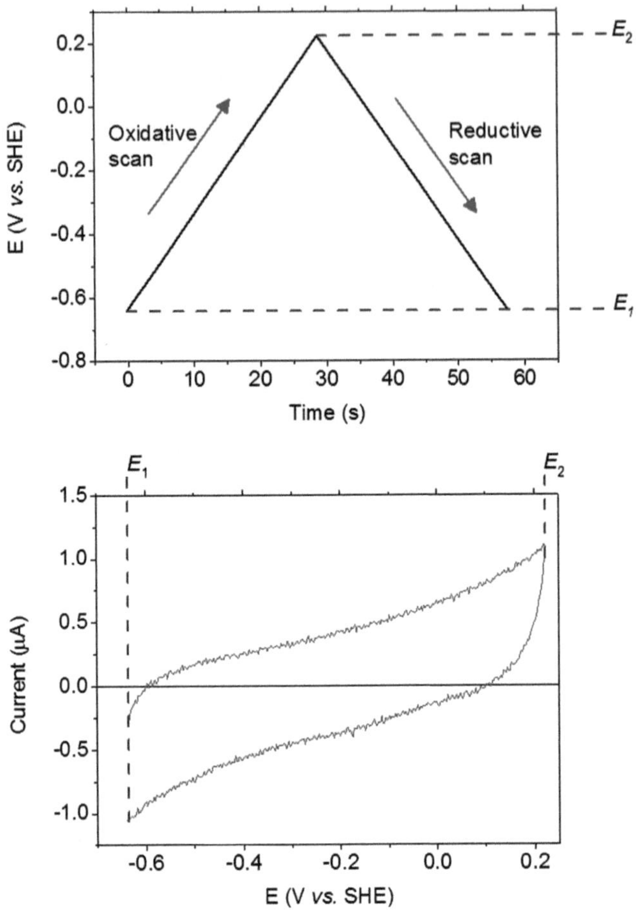

Fig. 5 An example of a cyclic voltammetry experiment with a "blank" electrode producing a non-Faradaic current

potential step experiment in chronoamperometry, *see* later), the scan rate and the composition of the bulk solution. A "blank" CV experiment should always be performed prior to growing an enzyme film (*see* section 3.2) as a control experiment.

3. Faradaic current: Faradaic current arises from oxidation and/or reduction processes creating a flow of electrons between a redox-active species (e.g., hydrogenase) adsorbed on the electrode surface and the electrode itself. The Faradaic or "sample" current overlays the non-Faradaic current, as exemplified in Fig. 6. Proteinic processes (originating in the enzyme) can be divided into catalytic and non-catalytic turnover signals. Non-turnover signals reflect reversible reduction and oxidation of redox centers in the adsorbed enzyme without the transport of electrons being linked to the turnover of substrate (*see* Vincent et al. [24], for discussion).

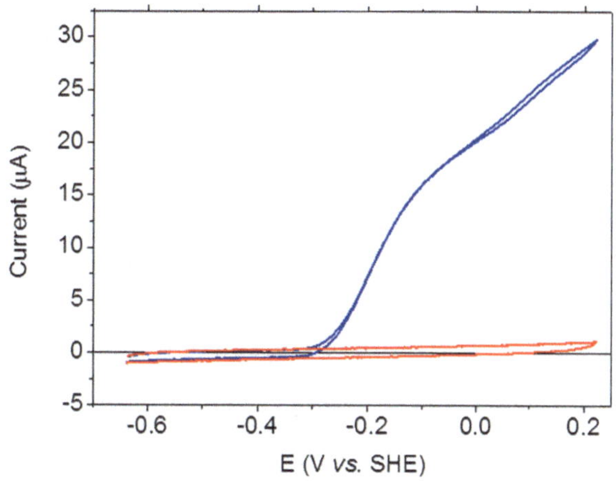

Fig. 6 Cyclic voltammetry overlay of "blank" non-Faradaic current (*red*) and redox-active Faradaic current produced by Hyd-1 (*blue*)

For steady-state catalysis, the current is directly proportional to rate via the expression:

$$i = k_{(\mathrm{E})} n A \Gamma F$$

where $k_{(\mathrm{E})}$ is the potential-dependent turnover frequency of the enzyme, n is the number of electrons involved in the reaction, A is the electrode surface area (usually in cm^2), Γ is the electroactive enzyme coverage per unit area and F is the Faraday constant (96,484 C/mol). As a rough guide, for an electroactive coverage of 1^{-13} $\mathrm{mol/cm}^2$, (about an order of magnitude lower than that expected to yield faint "non-turnover" signals due to electrons passing in and out of the enzyme's sites), a current of 1 μA measured at a typical electrode with a surface area of 0.03 cm^2 would correspond to a turnover frequency of 1700 s^{-1} for a two-electron reaction (i.e., H_2 oxidation). In principle, it is usually very difficult to ascertain the electroactive coverage as non-turnover signals are usually absent (see above). The Faradaic current should ideally reflect only the rate of the redox processes occurring within the enzyme adsorbed on the electrode surface.

However, as discussed previously (*see* section 3.2 and Fig. 3) a number of other important processes can limit the current observed, namely, IET and mass transport limitations. These processes must be taken into consideration when designing a PFE experiment and their contributions minimized as much as possible so that the observed current reflects the kinetics and/or thermodynamics of enzyme processes only.

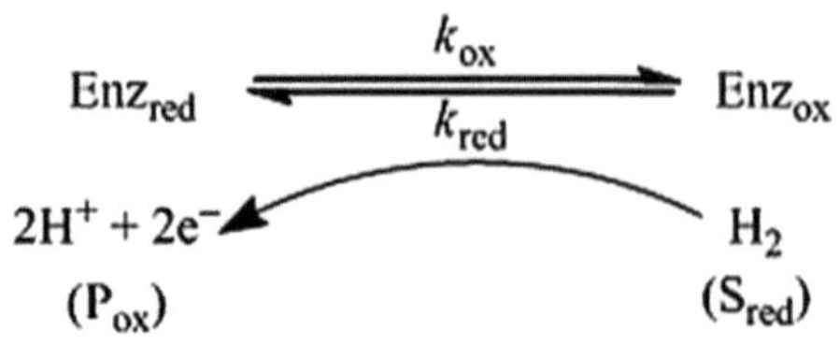

Fig. 7 A simple scheme illustrating the change in redox state of the enzyme as it is reduced and product is oxidized

Catalytic conversion of H_2 (reduced substrate) to $2H^+$ and $2e^-$ is outlined in Fig. 7. Intramolecular electron transfer in enzymes is generally not rate limiting: reorganization energies for buried redox centers are usually small and edge-to-edge distances between adjacent redox centers are <14 Å [27]. The current–potential relationship is (ideally) a simple sigmoidal waveform.

The current, measured for a solution containing a mixture of oxidized and reduced forms of the enzyme substrate depends upon E (the applied potential). The potential measured for an aqueous solution in equilibrium with H_2 gas depends on the pH of the solution (ox) and the partial pressure of H_2 (red) according to the Nernst equation:

$$E_{eq} = E^{\theta} + \frac{RT}{nF} \ln \frac{[ox]}{[red]}$$

Here, E_{eq} is the equilibrium potential (i.e., the applied potential where there is no net flow of electrons, -0.41 V at pH 7, under 1 bar H_2 at 25 °C), E^{θ} is the standard reduction potential for the mixture of redox couples (0 V at pH 0, under 1 bar H_2 at 25 °C), R is the gas constant, and T is the absolute temperature.

For a current to flow, a potential must be applied. The difference between applied potential and the equilibrium potential ($E - E_{eq}$) is known as the overpotential (η) [11]. The observed current increases with applied potential. The current should reach a maximum value at high potential ($E \gg E_{eq}$) providing there is no mass transport limitation of substrate and product to and from the electrode surface and should reflect the maximum turnover rate of substrate at the active site (k_{cat}). In order to determine k_{cat} in this way experimentally, however, the electro-active coverage of the electrode surface must be known (see above).

The sigmoidal response is rarely seen, due to non-idealities; instead a residual slope is seen, even at high potentials. This effect is attributed to there being a number of different orientations of the hydrogenase molecules on the surface of the electrode [28] which gives rise to a dispersion (a spread) in the rate of IET rate constants. A protein molecule which can orientate in such a way that a single redox center (e.g., the distal Fe–S cluster of Hyd-1, Fig. 3) lies uniformly close to the electrode surface will undergo rapid and efficient interfacial electron transfer with minimal dispersion. Interfacial electron transfer is usually described using the Butler–Volmer equation which relates how the observed current varies as a exponential function of the applied potential (see refs. [24] and [29] for more details).

4. Activating Hyd-1: The O_2-tolerant [NiFe]-hydrogenase known as Hyd-1 is isolated aerobically (no H_2 is required for growth) and it exists in more than one oxidized inactive state (for a discussion of the "as isolated" states of Hyd-1 see refs. [4, 7]). Prior to each experiment, the enzyme must be reductively "activated" by poising the potential at −0.56 V for 300 s then monitoring H_2 oxidation activity at −0.06 V for 100 s. These potential "steps" are repeated until H_2 oxidation activity stabilizes at −0.06 V (except for slow film loss, see later). This procedure is an example of chronoamperometry.
5. Chronoamperometry: Chronoamperometry is a PFE technique which deconvolutes the potential and time domains seen in cyclic voltammetry. In chronoamperometry, the applied potential is held constant and the current response is monitored as a function of time following a rapid perturbation such as injection of inhibitor or a potential step. Since the current itself is directly related to rate, the current response after the perturbation is equivalent to monitoring the "rate of change of rate," or the rate at which a situation at one steady state adjusts to a new steady state.

Figure 8 shows an example of a chronoamperometry experiment where, after first fully activating the enzyme (see above) and then poising the electrode at −0.65 V for 600 s (not shown), anaerobic inactivation is induced at +0.39 V for 10,000 s in 100 % H_2 to form the state known as Ni–B (also known as "Ready"; see Evans et al. [4]) which is a Ni(III)-OH species. The potential is then stepped to a less positive value in the range +0.235 → +0.035 V, and the rapid reactivation of Ni–B is monitored through the increase in current.

The potential dependence of the rate of reactivation is determined by fitting the data for reactivation as previously described [4] for a range of different potential values. When stepping the potential from one value to another, an interfering charging "spike" is observed due to the redistribution of ions at the e/s interface as described above. The duration of the spike must be taken into account when fitting data to obtain kinetic information of an enzymatic process and can be viewed as a "dead time" which will mask

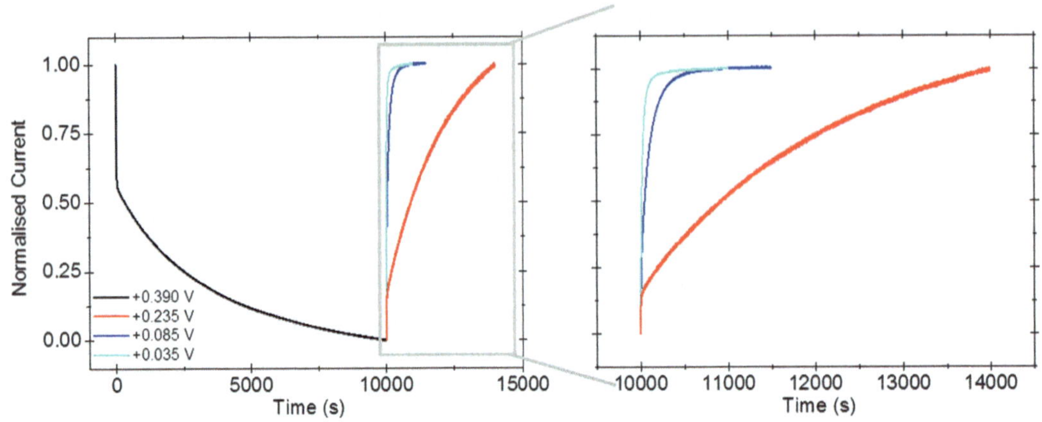

Fig. 8 An example of a chronoamperometry experiment where the current response to change in potential is shown for Hyd-1 at pH 6.0 and 5 °C, 100 % H_2. The enzyme is anaerobically inactivated by poising at high potential (+0.39 V, black) for 10,000 s in 100 % (shown only once for ease of comparison). Inactivation was then immediately followed by stepping to a lower potential (e.g., +0.235 V, *red*, +0.085 V, *blue* and +0.035 V, *cyan*) to monitor the potential dependence of reactivation. All currents have been normalized for ease of comparison. For more information regarding the fitting of data to determine rate of inactivation and reactivation, and the potential dependence of the rate of reactivation of anaerobically generated states in the range +0.235 to +0.035 V *see* Evans et al. [4]

reactions occurring on the same time scale as the charging [4, 30]. One way to overcome this is to carry out the experiments at lower temperatures so that the reactivation kinetics are slowed down enough that a significant proportion of the half-life of the reaction is not obscured by the charging spike [4].

3.5 Cyclic Voltammetry for Measuring Film Stability and K_m

Cyclic voltammetry provides a fingerprint wave shape of the potential dependence of the catalytic activity of an enzyme, providing IET and mass transport between the electrode surface and bulk solution are not limiting. If the rate of catalysis is always at a steady state for each potential and the film is stable, the forward and reverse scans overlay (when corrected for capacitance). The wave shape observed in cyclic voltammetry for a hydrogenase deviates from the sigmoidal waveform predicted by the Nernst equation (*see* Fig. 9) due not only to dispersion effects as the potential is swept higher (*see* section 3.4), but also due to non-steady state effects. Upon reversal of the scan direction we observe "hysteresis" whereby the wave shape does not retrace on the sweep to low potential. As the potential is increased above a certain value, active enzyme converts slowly to Ni–B; the rate of this reaction is largely potential-independent [4] as it is controlled by a chemical step (the ligation of OH^-). The slow interconversion persists when the scan direction is reversed, but close to a value known as E_{switch}, the current rises again as Ni–B converts back to active enzyme (see Fig. 8). The reactivation process is much faster and the rate depends on potential [4]. The term

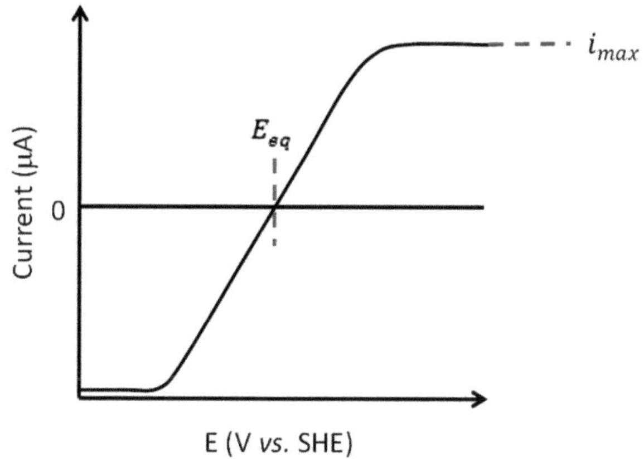

Fig. 9 Sigmoidal waveform of current response as a function of potential where the maximum current (i_{max}) reflects the maximum turnover of the substrate at the active site providing there is no mass transport limitation. The equilibrium potential (E_{eq}) reflects the potential where there is no net flow of electrons and therefore current = 0

E_{switch} is conveniently estimated from the minimum of the first derivative (di/dE) of the reductive sweep of a cyclic voltammetry scan [24] and depends on many experimental parameters such as scan rate and extent of inactivation to Ni-B, as well as temperature and pH. Whatever the exact origin of E_{switch} it serves as a very useful guide as to the ease of reactivation of Ni-B when the experimental conditions used are identical for each enzyme. As the scan continues to lower potentials, the driving force for H_2 oxidation is lowered and the current begins to decrease.

"Film loss" refers to the exponential and irreversible loss of current over time as enzyme desorbs from the electrode surface, unfolds or loses cofactor. The level of film loss is dependent on a number of factors such as the conditions of the experiment and the potential the electrode is poised at. One observation that we have made consistently is that enzymes perform best when the electrode is held at a constant potential: changes in potential, and particularly large *steps* in potential appear destabilizing. The rate/extent of film loss can be determined in chronoamperometry experiments by measuring the slow loss of current in a given set of conditions for 600–1,200 s prior to a change in experimental conditions. At the end of the experiment the initial conditions are restored to compare the final current to that initially measured. Film loss can also depend upon the way in which the film is prepared (e.g., by first abrading the electrode surface with sand paper or alumina [2]). Film loss can thus differ from film to film, and will indeed appear to be contributing less when using an older film.

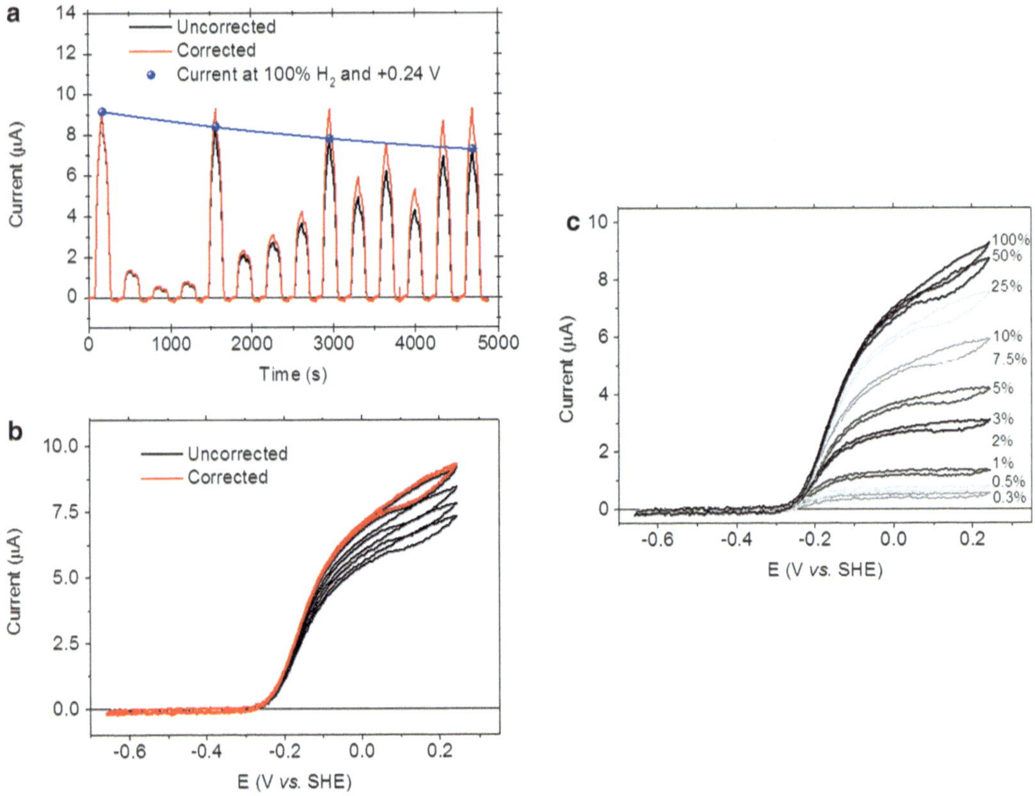

Fig. 10 An example of a typical cyclic voltammetry experiment in order to measure the current response at different levels of H_2 at pH 6.0 and 30 °C, $\omega = 4{,}500$. (**a**) Every fourth scan was recorded under 100 % H_2 in order to monitor the extent of film loss at each potential. For demonstrative purposes the film loss is shown at +0.24 V (*blue spheres*), fitted to an exponential decay (*blue line*). All uncorrected data (*black*) are fitted to this exponential decay function and the percentage film loss is calculated relative to the current at $t = 0$. All current is then corrected for the extent of film loss (*red*). (**b**) The uncorrected and corrected current response at 100 % H_2 shows an excellent overlay of the scans performed under identical conditions. (**c**) The corrected scans at each percentage H_2 show the actual current response to change in H_2 as the applied potential is swept

When a large number of cyclic voltammetry experiments are to be performed, for example *see* Fig. 10 for measuring the potential dependence of K_m, film loss can be monitored by always returning to 100% H_2 every third or fourth scan. For ease of demonstration, the film loss is shown at +0.24 V only.

Ideally, film loss would be minimized by covalent attachment of the enzyme to the electrode surface, and film stability has indeed been shown to improve with such methods [26]. However, it is not always scientifically sound to correct for film loss, especially when the loss of current can be attributed to the formation of inactive (and unrecoverable) states. It is worthwhile noting here that in the experiments performed in Fig. 10 a rotation rate of 4,500 rpm was required at low H_2 concentrations to avoid mass transport limitations (*see* section 3.6).

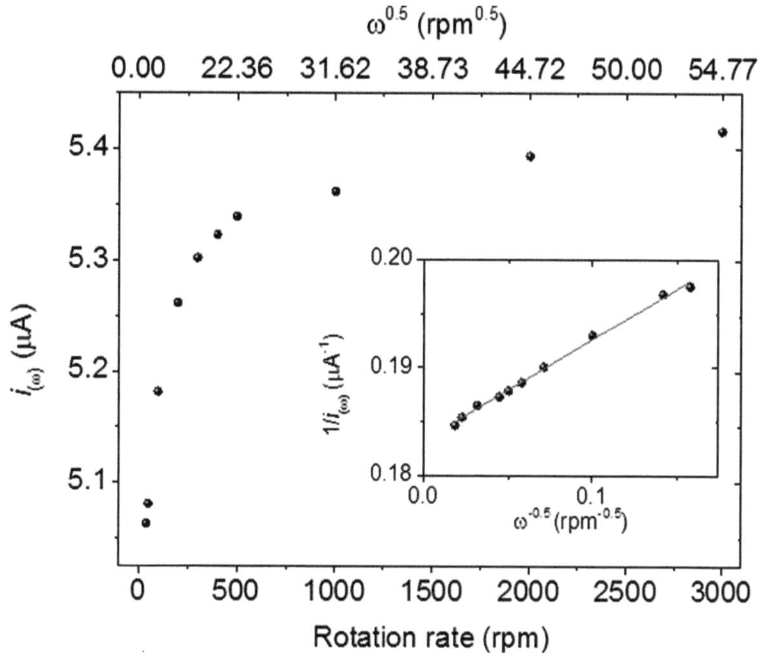

Fig. 11 A plot of the current response to change in rotation rate and a Koutecky–Levich plot (*inset*) of the reciprocal of the limiting current at each rotation rate against the reciprocal rotation rate

3.6 Rotation Rate Dependence of Observed Current

As described earlier, to ensure that the current reflects the inherent activity of the enzyme and is not limited by the diffusion of substrate and products to and from the electrode surface, the WE should be rotated at a sufficiently high speed to overcome mass transport limitation. In the first place, the Levich equation (below) shows that the current increases in accordance with the square root of the rotation rate when a reaction is diffusion controlled:

$$i_{(\omega)} = 0.62nFAD^{2/3}\nu^{-1/6}\omega^{1/2}C$$

At a given potential, $i_{(\omega)}$ is the current at rotation rate ω, D is the diffusion coefficient of the substrate, ν is the kinematic viscosity of the solvent, and C is the concentration of the analyte.

For an enzyme reaction, we expect the current to approach a limit as the rotation rate is increased because the rate will ultimately be controlled by the turnover frequency of the enzyme. Fig. 11 shows the limiting current can be extracted from a double reciprocal plot, i.e., $1/i_{(\omega)}$ versus $1/\omega$. The intercept gives the reciprocal of the maximum limiting current ($i_{\lim}$) at infinite rotation rate according to the Koutecky–Levich equation:

$$\frac{1}{i_{(\omega)}} = \frac{1}{i_{\lim}} + \frac{1}{0.62nFAD^{2/3}\nu^{-1/6}\omega^{1/2}C}$$

This analysis can be made for any potential value and for any substrate concentration. In most cases, particularly at high concentrations of substrate, an obvious limiting value is reached at quite low rotation rates. In the case of Fig. 11, the electrode was poised at −0.15 V and the limiting current at each rotation rate ($i_{(\omega)}$) was measured in 100 % H_2.

The intercept in this plot gives an i_{lim} value of 1/0.1833 = 5.45 μA. The curve is however, obvious, and a rotation rate of at least 1,000 rpm was used for subsequent experiments using this enzyme film at −0.15 Vand 100 % H_2. When cyclic voltammetry experiments are going to be performed, in order to determine the correct rotation rate for the experiment in question, a Levich plot should be performed at the potential where the current (and hence activity) is greatest. Likewise, a Koutecky–Levich plot should be performed using the lowest concentration of substrate to be used in the experiment. It may prove impractical to study a very active enzyme, acting on a low concentration of substrate, under completely diffusion-controlled conditions, since even the highest rotation rates available will not provide fresh substrate at sufficiently high flux.

3.7 Inhibition Studies

As mentioned earlier, [NiFe]-hydrogenases can be sub-classified according to their O_2 tolerance, i.e., their ability to oxidize H_2 in the presence of O_2. Protein film electrochemistry is an excellent tool for studying enzyme inhibition due to its ability to probe isolated potential-dependent states directly by poising the electrode potential. The enzyme remains in situ on the electrode surface whilst the inhibitor, whether added as a solution or gas, can be varied and quickly removed when required, by flushing out or solution exchange [4, 5].

Addition of gaseous reagents, for example O_2, is achieved in two ways [4, 7]. The first way is to inject a known volume of O_2-saturated buffer into a known buffer volume in the working compartment of the electrochemical cell. The O_2 concentration is given by Henry's Law. The mixing time of the injected solution depends on the rotation rate and the volumes involved. Exposure of the enzyme to the O_2 will be transient as the O_2 will immediately begin to be flushed from the system as the total buffer volume equilibrates with the headspace gas. A second way of introducing a gas such as O_2 is to change its composition in the supply to the headspace, using the mass flow controllers: this method gives a steady level but initial equilibration time is longer. The two methods can be combined, and injection can be made under a constant flow of the gas through the headspace.

The fact that the enzyme is immobilized has important implications because even nonvolatile inhibitors can be added then removed to check for reversibility. A good example is the inhibition of [FeFe]-hydrogenases by formaldehyde, which has a very high Henry's constant and is normally regarded as causing irreversible damage to an enzyme. When studied by PFE, the H^+ reduction activity of the hydrogenase is immediately lowered by formaldehyde injection but is restored to normal after the electrolyte solution is exchanged to remove formaldehyde [5].

4 Notes

1. Immediately prior to use, mix 3 parts resin by mass to 1 part hardener by mass. Roughly, 3 g resin to 1 g hardener makes 3 working electrodes. Cut the end off a 1 mL pipette tip and use to mix resin and hardener together in a weighing boat, stirring well. Avoid making bubbles. If it is difficult to mix, warm slightly.
2. The graphite plate is mounted and waxed onto a glass plate. Holding in place on a vacuum chuck, the graphite is cut into strips measuring 3 mm × 3 mm using a diamond saw. The graphite strips are then removed from the wax and degreased before turning in order to create cylindrical rods of diameter ~2 mm, and snapped into lengths of ~1 cm.
3. Take a Precision seal® 5–6 mm rubber septum (Sigma-Aldrich) and thread a needle through the center. Carefully thread the Pt wire part way into the needle's end and pull the needle back through the septum to pull the Pt wire through the septum. Discard the needle appropriately. The septum is then placed in the CE sidearm and held in place with parafilm.
4. The electrode rotator should fit snugly against the electrochemical cell to ensure gas tightness of experiments.
5. To ensure a good flow of water through the water jacket of the electrochemical cell (Fig. 4), the water inlet should be hooked up to the lower water sidearm and the outlet to the upper water sidearm.
6. The SCE reference electrode is stored in saturated KCl at room temperature and rinsed in ultra pure water (Millipore, 18 MΩ cm) immediately prior to use. During PFE experiments, the RE is housed in a non-isothermal side arm containing 0.10 M NaCl and connected to the main cell compartment by a Luggin capillary (*see* Fig. 4). Potentials (E) are converted with respect to the standard hydrogen electrode (SHE) using the correction $E_{SHE} = E_{SCE} + 241$ mV at 25 °C [29].
7. Fill the side arm halfway with NaCl. Tip the cell to flow the solution along the Luggin capillary. Make sure that there are no bubbles in the Luggin capillary as this will interrupt the electrolytic connection between the working compartment and the RE. Once all visible air is removed, fill the side arm to the top. Place a gloved finger over the top of the side arm and gently push to force solution through the end of the Luggin capillary.

This ensures there are no air bubbles through the entire length of the Luggin capillary. Insert the RE and screw in gently. This will push solution through the Luggin into the working compartment. Wash the working compartment with ultra pure water before adding buffer.

8. A plastic bottle is safer to use when transferring via vacuum/gas purge systems of glove boxes.
9. Precision seal® 5–6 mm rubber septa (Sigma-Aldrich) are used to seal the end of the gas inlet and outlet side arms of the electrochemical cell and held in place using parafilm (*see* Fig. 4). Needles are used to pierce the septa for delivery of gas into the cell headspace. Gases are humidified by flow through a bubbler prior to entrance into the headspace. An outlet bubbler is always used to monitor the gas-tightness of the cell setup.
10. A shaft is drilled into the top of the PTFE casing. The PTFE sleeve is heated in an oven to 150 °C and using the expansion, the stainless steel core slides into the PTFE shaft. The steel core is made 0.05 mm larger than the diameter of the PTFE shaft and, on cooling, the PTFE shrinks to give a tight fit around the steel core.
11. Test the fitting of the PG rods in the crevice of the electrode tip. Gently slide the PG rod into the crevice, making sure not to snap the rod. The PG rod is mounted into the casing such that the basal, aromatic, plane lies vertically in line with the length of the electrode casing. The "edge" of the PG rod is 90° to this such that the edges of the aromatic layers are projected towards the electrolyte (*see* Fig. 2). Polishing the edge surface of the PG tip provides a rough surface with carbon atoms of unsatisfied valences which react with water or O_2 to give polar functionalities such as C = O, C–OH and COOH that bind the enzyme applied (*see* section 3.2) through non-covalent bonding to functional groups on the protein surface [1, 2, 24, 31].
12. Silver will give a peak in the voltammetry. Use a scalpel to scrape off any silver glue but be careful not to damage the graphite.
13. A mold to cast the insulating tip is created using ~1.5 turns of Scotch magic tape around the edge of the PTFE casing ensuring the edge of the tape lies above the end of the graphite. This is filled with mixed Aradur hardener and Araldite resin (*see* **Note 1**) by a gentle flow of the mixture down the inside of the tape until the level covers the whole tip of the graphite. This method should avoid making bubbles. Do not overfill the mold—dripping onto the casing causes the electrode weight to be unevenly distributed, making rotation at high speeds

nonuniform and it also makes removal of the Scotch tape difficult. Allow electrodes to cure at room temperature overnight. Polish with coarse sandpaper (e.g., P60 or P120), until the graphite appears through the araldite epoxy resin, followed by P400 until the electrode tip is flat.

Acknowledgements

This research was supported by the Biological and Biotechnological Sciences Research Council (Grants BB/H003878-1 and BB/I022309-1). We thank Elena Nomerotskaia for help with protein purification.

References

1. Davies TJ, Hyde ME, Compton RG (2005) Nanotrench arrays reveal insight into graphite electrochemistry. Angew Chem 117: 5251–5256
2. Blanford C, Armstrong F (2006) The pyrolytic graphite surface as an enzyme substrate: microscopic and spectroscopic studies. J Solid State Electrochem 10:826–832
3. Léger C, Elliott SJ, Hoke KR, Jeuken LJC, Jones AK, Armstrong FA (2003) Enzyme electrokinetics: using protein film voltammetry to investigate redox enzymes and their mechanisms. Biochemistry 42:8653–8662
4. Evans RM, Parkin A, Roessler MM, Murphy BJ, Adamson H, Lukey MJ, Sargent F, Volbeda A, Fontecilla-Camps JC, Armstrong FA (2013) The principles of sustained enzymatic hydrogen oxidation in the presence of oxygen—the crucial influence of high potential Fe-S clusters in the electron relay of [NiFe]-hydrogenases. J Am Chem Soc 135:2694–2707
5. Foster CE, Krämer T, Wait AF, Parkin A, Jennings DP, Happe T, McGrady JE, Armstrong FA (2012) Inhibition of [FeFe]-hydrogenases by formaldehyde and wider mechanistic implications for biohydrogen activation. J Am Chem Soc 134:7553–7557
6. Lukey MJ, Parkin A, Roessler MM, Murphy BJ, Harmer J, Palmer T, Sargent F, Armstrong FA (2010) How Escherichia coli is equipped to oxidize hydrogen under different redox conditions. J Biol Chem 285:3928–3938
7. Lukey MJ, Roessler MM, Parkin A, Evans RM, Davies RA, Lenz O, Friedrich B, Sargent F, Armstrong FA (2011) Oxygen-tolerant [NiFe]-hydrogenases: the individual and collective importance of supernumerary cysteines at the proximal Fe-S cluster. J Am Chem Soc 133:16881–16892
8. Wang VCC, Can M, Pierce E, Ragsdale SW, Armstrong FA (2013) A unified electrocatalytic description of the action of inhibitors of nickel carbon monoxide dehydrogenase. J Am Chem Soc 135:2198–2206
9. Parkin A, Seravalli J, Vincent KA, Ragsdale SW, Armstrong FA (2007) Rapid and efficient electrocatalytic CO_2/CO interconversions by carboxydothermus hydrogenoformans CO dehydrogenase I on an electrode. J Am Chem Soc 129:10328–10329
10. Anderson LJ, Richardson DJ, Butt JN (2001) Catalytic protein film voltammetry from a respiratory nitrate reductase provides evidence for complex electrochemical modulation of enzyme activity. Biochemistry 40: 11294–11307
11. Armstrong FA, Hirst J (2011) Reversibility and efficiency in electrocatalytic energy conversion and lessons from enzymes. Proc Natl Acad Sci USA 108:14049–14054
12. Vignais PM, Billoud B, Meyer J (2001) Classification and phylogeny of hydrogenases1. FEMS Microbiol Rev 25:455–501
13. Volbeda A, Darnault C, Parkin A, Sargent F, Armstrong FA, Fontecilla-Camps JC (2013) Crystal structure of the O_2-tolerant membrane-bound hydrogenase 1 from Escherichia coli in complex with its cognate cytochrome b. Structure 21:184–190
14. Volbeda A, Amara P, Darnault C, Mouesca J-M, Parkin A, Roessler MM, Armstrong FA, Fontecilla-Camps JC (2012) X-ray crystallographic and computational studies of the O_2-tolerant [NiFe]-hydrogenase 1 from

Escherichia coli. Proc Natl Acad Sci USA 109:5305–5310
15. Fritsch J, Scheerer P, Frielingsdorf S, Kroschinsky S, Friedrich B, Lenz O, Spahn CMT (2011) The crystal structure of an oxygen-tolerant hydrogenase uncovers a novel iron-sulphur centre. Nature 479:249–252
16. Shomura Y, Yoon K-S, Nishihara H, Higuchi Y (2011) Structural basis for a [4Fe-3S] cluster in the oxygen-tolerant membrane-bound [NiFe]-hydrogenase. Nature 479:253–256
17. Cracknell JA, Wait AF, Lenz O, Friedrich B, Armstrong FA (2009) A kinetic and thermodynamic understanding of O_2 tolerance in [NiFe]-hydrogenases. Proc Natl Acad Sci USA 106:20681–20686
18. Jones AK, Lamle SE, Pershad HR, Vincent KA, Albracht SPJ, Armstrong FA (2003) Enzyme electrokinetics: electrochemical studies of the anaerobic interconversions between active and inactive states of *Allochromatium vinosum* [NiFe]-hydrogenase. J Am Chem Soc 125:8505–8514
19. Lamle SE, Albracht SPJ, Armstrong FA (2004) Electrochemical potential-step investigations of the aerobic interconversions of [NiFe]-hydrogenase from *Allochromatium vinosum*: insights into the puzzling difference between unready and ready oxidized inactive states. J Am Chem Soc 126:14899–14909
20. Rousset M, Montet Y, Guigliarelli B, Forget N, Asso M, Bertrand P, Fontecilla-Camps JC, Hatchikian EC (1998) [3Fe-4S] to [4Fe-4S] cluster conversion in *Desulfovibrio fructosovorans* [NiFe] hydrogenase by site-directed mutagenesis. Proc Natl Acad Sci USA 95:11625–11630
21. Dementin S, Belle V, Bertrand P, Guigliarelli B, Adryanczyk-Perrier G, De LAL, Fernandez VM, Rousset M, Leger C (2006) Changing the ligation of the distal [4Fe4S] cluster in NiFe hydrogenase impairs inter- and intramolecular electron transfers. J Am Chem Soc 128:5209–5218
22. Roessler MM, Evans RM, Davies RA, Harmer J, Armstrong FA (2012) EPR spectroscopic studies of the Fe-S clusters in the O_2-tolerant [NiFe]-hydrogenase Hyd-1 from *E. coli*, and characterization of the unique [4Fe-3S] cluster by HYSCORE. J Am Chem Soc 134: 15581–15594
23. Sambrook J, Russell DW (2001) Molecular cloning a laboratory manual, 3rd edn. Cold Spring Harbor Laboratory Press, Cold Spring Harbor, NY
24. Vincent KA, Parkin A, Armstrong FA (2007) Investigating and exploiting the electrocatalytic properties of hydrogenases. Chem Rev 107:4366–4413
25. Rudiger O, Abad JM, Hatchikian EC, Fernandez VM, De Lacey AL (2005) Oriented immobilization of *Desulfovibrio gigas* hydrogenase onto carbon electrodes by covalent bonds for nonmediated oxidation of H_2. J Am Chem Soc 127:16008–16009
26. Krishnan S, Armstrong FA (2012) Order-of-magnitude enhancement of an enzymatic hydrogen-air fuel cell based on pyrenyl carbon nanostructures. Chem Sci 3:1015–1023
27. Page CC, Moser CC, Chen X, Dutton PL (1999) Natural engineering principles of electron tunnelling in biological oxidation-reduction. Nature 402:47–52
28. Léger C, Jones AK, Albracht SPJ, Armstrong FA (2002) Effect of a dispersion of interfacial electron transfer rates on steady state catalytic electron transport in [NiFe]-hydrogenase and other enzymes. J Phys Chem B 106:13058–13063
29. Bard AJ, Faulkner LR (2001) Electrochemical methods. Wiley, New York
30. Hirst J, Armstrong FA (1998) Fast-scan cyclic voltammetry of protein films on pyrolytic graphite edge electrodes: characteristics of electron exchange. Anal Chem 70:5062–5071
31. Armstrong FA, Anthony Cox P, Hill HAO, Lowe VJ, Nigel Oliver B (1987) Metal ions and complexes as modulators of protein-interfacial electron transport at graphite electrodes. J Electroanal Chem Inter Electrochem 217:331–366

Chapter 7

FTIR Spectroscopy of Metalloproteins

Oscar Gutiérrez-Sanz, Olaf Rüdiger, and Antonio L. De Lacey

Abstract

Absorption of infrared radiation by proteins gives important information about their structure and function. The most intense infrared bands correspond to the overlap of all the peptide bond absorption. Additionally, in many metalloproteins their prosthetic groups have intrinsic ligands or bind substrates/ inhibitors that absorb intensively in the infrared. Here, we describe thoroughly several Fourier transform infrared methods for studying structure–function relationships in metalloproteins, using hydrogenases as an example.

Key words Metalloenzymes, Infrared, Spectroelectrochemistry, SEIRA, Hydrogenase

1 Introduction

Absorption of infrared radiation by proteins gives important information about their structure and function. Functional groups of proteins are detected in the infrared spectra when they are changes of the dipole moment of one or more of their vibrational modes [1]. The most intense infrared bands of a protein correspond to the amide I and amide II ones, which correspond to the overlap of all the peptide bond absorption. The amide I band is mostly due to the carbonyl stretching vibrations, which have frequencies in the range of 1,630–1,680 cm^{-1} dependent on the protein secondary structure [2]. The amide II band is mostly due to the bending mode of N-H in peptide bonds, their frequency being around 1,550 cm^{-1} [2]. Some functional groups of amino acid residues can also be detected by infrared absorption, such as thiols, carboxylic acids, and aromatic rings, although their band intensities are very weak. A further complication for infrared spectroscopy of proteins in solution is the strong absorption by water, which leaves only a narrow window of the spectrum for analysis [2]. This problem can be tackled in different ways. Dissolving the protein in deuterated water allows shifting the operational spectral window

Juan C. Fontecilla-Camps and Yvain Nicolet (eds.), *Metalloproteins: Methods and Protocols*, Methods in Molecular Biology, vol. 1122, DOI 10.1007/978-1-62703-794-5_7,

so that the amide I and amide II bands can be studied [3]. Besides, by measuring difference infrared spectra of the protein sample between two stable functional states, the molecular changes taking place in the protein structure can be indentified from the background absorption of the rest of the protein and the solvent [4]. In this context, another useful technique is surface-enhanced infrared absorption (SEIRA), which requires immobilization of the studied protein on metals such as Au, Ag, and Cu. SEIRA is only selective to molecular absorption at a range of 0–10 nm from the metal surface, thus eliminating all contribution from the solution [5].

In some metalloproteins, such as hydrogenases or CO dehydrogenases, the prosthetic groups have intrinsic ligands (carbonyls or cyanides) that absorb intensively in the infrared [6]. More frequently, several substrates or inhibitors (nitric oxide, carbon monoxide, azide, etc.) bound to the metallic sites of proteins can also be detected by infrared spectroscopy [7–10]. Therefore, in these cases structure–function relationships in metalloproteins can be studied using infrared techniques. As many metalloproteins have different redox states for their metallic ions, Fourier transform infrared (FTIR) spectroelectrochemical studies are very useful for this purpose [11].

Here, we thoroughly describe several FTIR methods for studying structure–function relationships in metalloproteins, using hydrogenases as a representative example.

2 Materials

2.1 Chemicals and Solutions for FTIR Measurements of Hydrogenases

All aqueous solutions were prepared with Milli Q (18.2 MΩ cm) water, and all reagents were of analytical grade.

1. Hydrogenase sample in H_2O: Concentrate purified hydrogenase in pH 5–8 buffered water solution to a concentration ≥80 μM by ultrafiltration in Centricon-30 (Amicon) for measurements in the transmission liquid cell or ≥800 μM for measurements in the spectroelectrochemical cell or for SEIRA.
2. Hydrogenase sample in D_2O solution: Prepare deuterated pH 5–8 buffer solutions in 99.9 % D_2O with Tris(hydroxy-D-methyl)amino-d_2-methane or anhydrous sodium acetate. Add deuterium chloride to adjust the pD (*see* **Note 1**). Exchange the hydrogenase sample with the deuterated buffer by repeated ultrafiltration with Centricon-30 inside a dry-box (MBraun). Leave overnight the deuterated hydrogenase sample in the dry-box to allow exchange of all solvent-accessible protons of the protein.
3. 0.1 M $Na_2S_2O_4$ (sodium dithionite) solution in 20 mM Tris–HCl pH 8.0: In a glass vial with two tubes in angle, add 3.5 mg of sodium dithionite in one tube, and fill the other one

with 2 mL of 20 mM Tris–HCl pH 8.0 buffer. Close the vial with a suba-seal septum stopper (Sigma-Aldrich), and eliminate the air inside the vial using a vacuum/Ar manifold (*see* **Note 2**). Then, dissolve the sodium dithionite in the buffer solution by agitating the vial. Keep the vial in an ice bath until its use.

4. 40 % NH_4F (w/v): Dissolve 40 g of NH_4F in 100 mL of H_2O.
5. 2 % HF (v/v): Add 4.17 mL of concentrated HF to 100 mL H_2O to form (*see* **Note 3**).
6. Plating solution: Dissolve 3.7812 g of Na_2SO_4 (0.15 M), 2.4817 g of $Na_2S_2O_3$ $5H_2O$ (0.05 M), and 0.5349 g of NH_4Cl (0.05 M) in 100 mL H_2O.
7. 1 mM 4-aminothiophenol: Dissolve 1.3 mg in 10 mL ethanol.
8. Gases: H_2 and Ar 99.999 % pure (Air Liquide). Gas lines end in needles (0.50 × 90 mm Becton-Dickinson Spinal needle), and the residual O_2 was removed with O_2-trap filters (Varian) placed before them.
9. Redox mediator cocktail, 0.5 mM of each: Weigh 18.37 mg of indigo-tetrasulfonate (Sigma-Aldrich), 11.6 mg of indigo-carmine (Sigma-Aldrich), 9.22 mg of anthraquinone-1,5-disulfonic acid (ICN Pharmaceuticals), 8.22 mg of anthra-quinone-2-sulfonate (Serva), 10.23 mg of benzyl viologen (Sigma-Aldrich), and 6.43 mg of methyl viologen (Sigma-Aldrich), and dissolve them in 5 mL of 2 M KCl solution. Dilute 1:10 in water for use.

2.2 Instrumentation

1. Glass gas-type vials: 1 and 3 mL glass vials with conical inset (Análisis vínicos S.L) with suba-seal septa stoppers.
2. Gastight 100 μL microsyringe.
3. Glass vacuum/Ar manifold with firestone valve and ports ended in needles (0.50 × 90 mm Becton Dickinson Spinal needle).
4. Thermostatized transmission cell (82 μm pathlength) with CaF_2 windows of 4 mm thickness.
5. Spectroelectrochemical cell: The cell was designed by Moss et al. [11]. A 6 μm thick gold mesh (200 gold mesh, Buckbee-Mears) is used as the working electrode. The working electrode is placed between two CaF_2 disks (20 mm × 4 mm, Apollo Scientific Limited) that are glued to methacrylate rings as support. A Pt foil is placed at the side of one of the CaF_2 windows as counter-electrode. Two rubber rings surround the CaF_2 for fixing the working and counter-electrodes and for sealing the cell. Around the periphery of the windows there is a space for the electrolyte solution, allowing electric contact between the working and counter-electrodes and with a Ag/AgCl reference electrode. The two CaF_2 windows are fitted

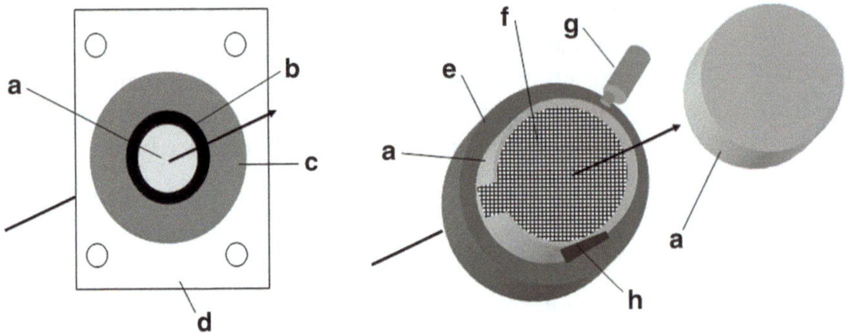

Fig. 1 Diagram of FTIR spectroelectrochemical cell: (**a**) CaF_2 window; (**b**) rubber o-ring; (**c**) teflon body; (**d**) steel plate; (**e**) methacrylate ring; (**f**) gold mesh working electrode; (**g**) Ag/AgCl reference electrode; (**h**) Pt counter-electrode

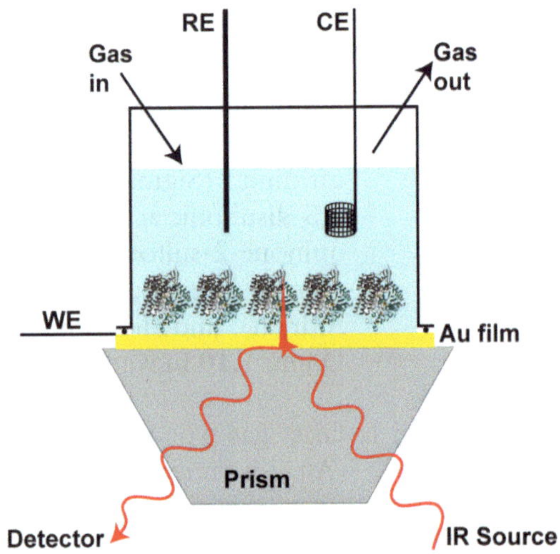

Fig. 2 Schematic representation of the SEIRA setup used. An IR-transparent silicon prism is used as support of the gold thin film that will act as working electrode (WE) and also provide the SEIRA effect. An o-ring seals the bottom of the electrochemical glass cell. The cell should be equipped with a water jacket to control the temperature during the experiment. A Pt mesh is used as counter-electrode (CE) and a Ag/AgCl as reference electrode (RE)

into a Teflon body and pressed together between two steel plates (Fig. 1).

6. FTIR spectrometer with mercury cadmium telluride (MCT) detector and purged atmosphere. SEIRA measurements require building a sample compartment with a Kretschmann-ATR configuration using a trapezium-shaped silicon crystal ($20 \times 25 \times 9$ mm $W \times L \times H$, 60°) (Fig. 2).
7. Potentiostat for electrochemical control of FTIR measurements.

3 Methods

3.1 FTIR Transmission Spectra of Hydrogenase

1. Cool the MCT detector by filling its liquid nitrogen reservoir.
2. Set the parameters for the spectra measurements: spectral range (4,000–1,000 cm^{-1}), number of scans (100–1,000, *see* **Note 4**), resolution (typically 1–2 cm^{-1}), gain (typically 1), aperture (*see* **Note 5**), and interferometer velocity (typically 3–4 cm/s, *see* **Note 6**).
3. Measure a background spectrum with the empty sample holder (*see* **Note 7**).
4. Measure and save a blank spectrum of the cell filled with the buffer solution at the temperature that will be used in the experiment (*see* **Note 8**).
5. Introduce 25–50 μL of hydrogenase sample in the cell with a syringe (*see* **Note 9**).
6. Introduce the cell in the sample compartment, and wait until the background signals stabilize.
7. Measure the spectrum of the sample at the set parameters. Subtract the blank spectrum to the sample one, and do a baseline correction with the spectrometer's software in the wavelength range of interest (2,150–1,800 cm^{-1} for the detection of the CO and CN^{-} ligands of the active site; 1,700–1,500 cm^{-1} for the detection of the peptide bonds) for enhancing the sample's bands (Fig. 3).

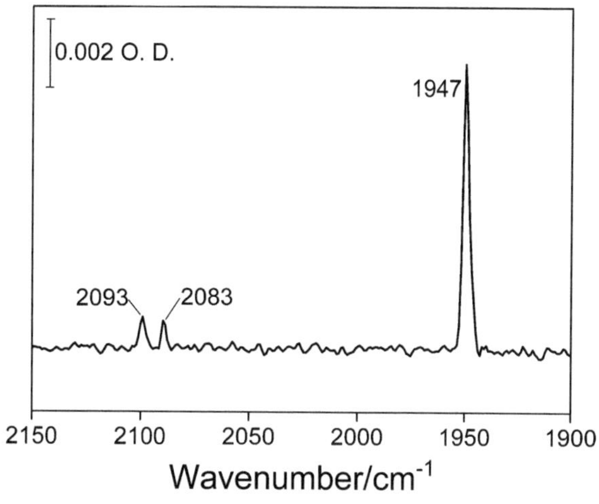

Fig. 3 FTIR transmission spectrum of 0.1 mM *Desulfovibrio gigas* hydrogenase as isolated under aerobic conditions. The number of averaged scans was 1,024 and resolution was 2 cm^{-1}

8. Recover the sample from the cell, and add more hydrogenase solution to a total volume of 50 μL in a glass vial stopped with a septum.
9. Remove the air from the vial in the vacuum/Ar manifold. Add 1 μL of sodium dithionite solution through the septum to remove residual O_2. Replace the Ar of the gas phase by 1 atm H_2, and incubate the hydrogenase sample during 4 h at room temperature for obtaining the completely reduced and activated enzyme (*see* **Note 10**). Fill the cell with the reduced hydrogenase sample with the gastight microsyringe (*see* **Note 11**).
10. Measure the spectrum of the reduced enzyme as done with the oxidized one.

3.2 FTIR Spectroelectrochemistry of Hydrogenases

1. Repeat **steps 1–4** of Subheading 3.1.
2. Mix 1 μL of the redox mediator solution with 10 μL of ≥800 μM hydrogenase solution, and deposit it on the gold mesh of the spectroelectrochemical cell (*see* **Note 9**).
3. Assemble the spectroelectrochemical cell, and fill the enclosed space surrounding the CaF_2 windows and the capillary connection to the reference electrode with 50 mM HEPES pH 8 and 0.2 M KCl (*see* **Note 12**).
4. Screw on the reference electrode, place the spectroelectrochemical cell in the sample compartment, and connect the cables from the potentiostat to the three electrodes of the cell. Set the applied redox potential to the sample with the potentiostat during 10 min to reach equilibrium.
5. Measure the FTIR spectra at different applied potentials for doing redox titrations of the different states of the hydrogenase active site.
6. Difference spectra of the hydrogenase are obtained by subtracting spectra measured at different redox potentials using the spectrometer software (Fig. 4).

3.3 SEIRA Measurements of Hydrogenases

1. A thin gold film is deposited on the surface of the silicon prism by an electroless (chemical) deposition technique [12]. First rinse the prism with ethanol and photographic lens cleaning paper to remove any remaining Au from previous experiments. Polish the surface of the prism with 1 μm alumina for 1 min on a soft cloth. Then rinse the prism with water, and immerse the prism on a 40 wt% NH_4F aqueous solution for 2 min; this will reduce the oxides on the surface of the silicon. Rinse with water, dry the surface of the prism with a N_2 flow, and place the prism in a bath at 65 °C, keeping the surface dry. Wait a couple of minutes for the temperature to equilibrate, and prepare a 1:1:1 mixture with 90 mL of the plating solution, the 2 % (v/v) HF solution and the $NaAuCl_4$ solution.

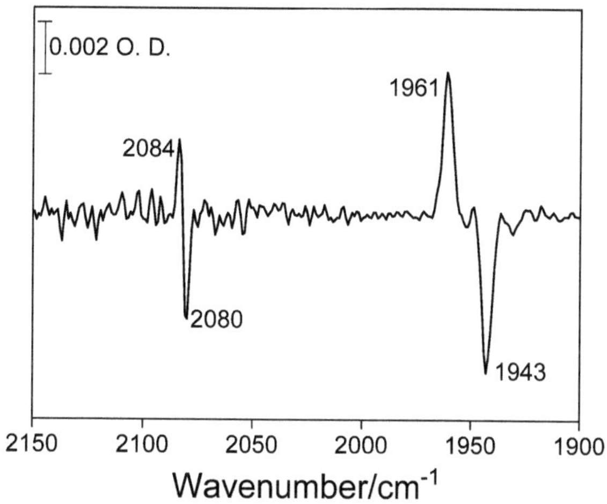

Fig. 4 FTIR difference spectrum of 0.9 mM *Ralstonia eutropha* regulatory hydrogenase between −700 and −350 mV vs. Ag/AgCl (3 M KCl). The number of averaged scans was 1,024 and resolution was 2 cm^{-1}

Mix it and deposit a drop in the center of the silicon prism surface. Leave it to react for 60 s, and stop the reaction rinsing with water. Assemble the electrochemical cell with the prism, and electrochemically clean the Au surface in 0.1 M H_2SO_4 by several oxidation/reduction cycles between 0 and 1.4 V (vs. Ag/AgCl).

2. Rinse the deposited gold surface of the cell with water and ethanol before its chemical modification. Next, cover overnight the Au surface with the 1 mM ethanolic solution of 4-aminothiophenol to form a self-assembled monolayer (SAM). Rinse again the cell with ethanol and 10 mM PBS pH 6.5. Fill the cell with the minimum amount of 10 mM PBS to cover the gold surface, and place it in the spectrometer. Start to collect backgrounds until a constant spectral signature is obtained (the difference spectra between two consecutive spectra are almost flat). Then collect ten spectra and average them; this will be used as a background for the protein immobilization.

3. Add 5 μL of the hydrogenase solution in H_2O, and immediately start to collect spectra. Immobilization can be followed by observing the amide band region (Fig. 5a, amide I at ~1,650 cm^{-1}, amide II at ~1,550 cm^{-1}) [13]. After 30 min add 10 μL of a 0.1 M solution of *N*-hydroxysuccinimide (NHS) solution, collect three spectra, and add 10 μL of a 0.1 M solution of *N*-(3-dimethylaminopropyl)-*N'*-ethylcarbodiimide hydrochloride (EDC) (*see* **Note 13**). Keep recording spectra until no significant increase in the amide band intensities is observed.

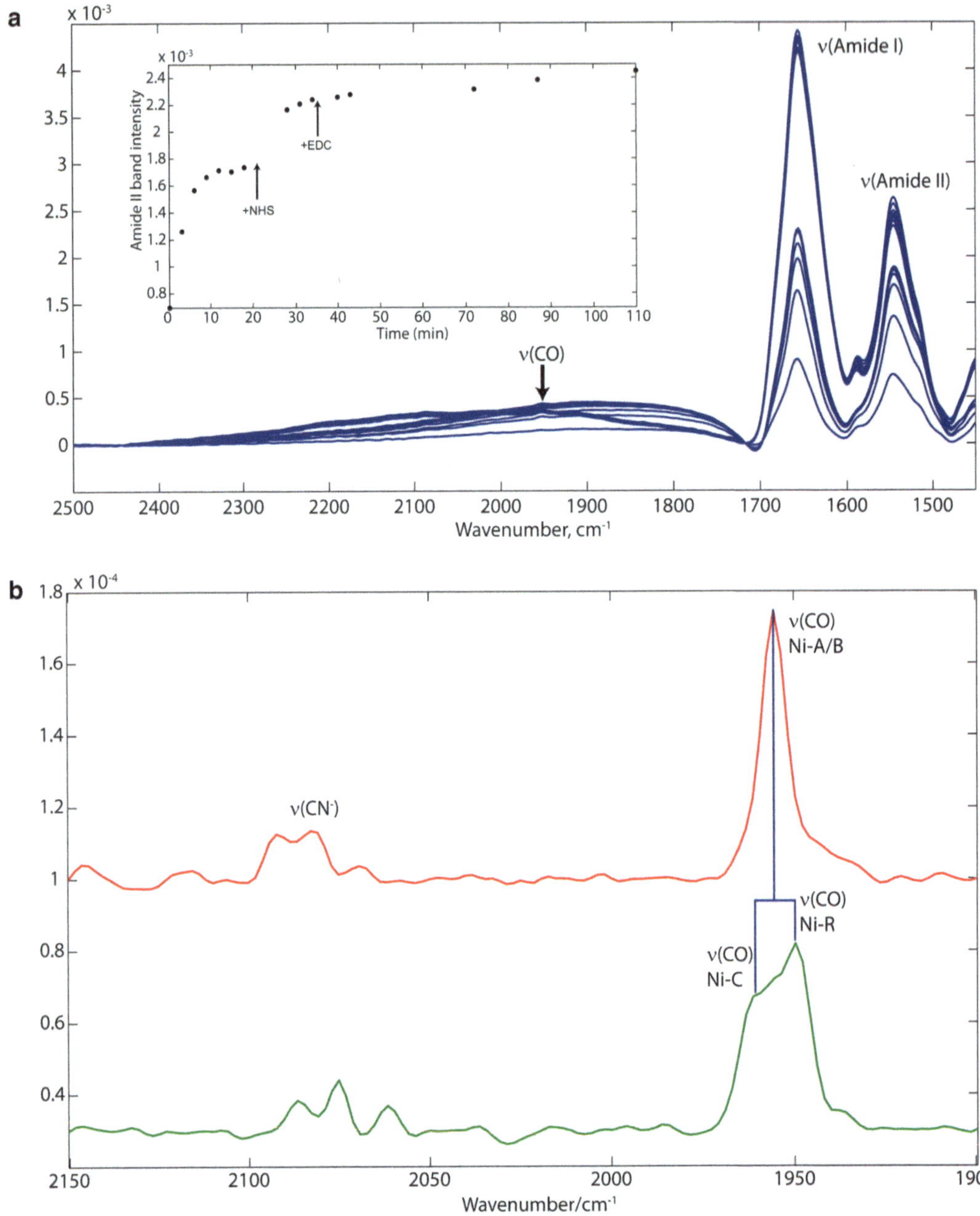

Fig. 5 (**a**) Amide band intensity growth during protein immobilization. The insert shows the evolution of the amide band II intensity over time. It also indicates when NHS and EDC were added. (**b**) Detail of the CO/CN⁻ frequency region showing the bands of the sample as isolated in the oxidized states (*red trace*) and once reduced (*green trace*)

4. Carefully remove the protein solution from the cell and fill it with 0.1 M PBS. Close the electrochemical cell as in Fig. 2, connect all the electrodes, set the desired temperature in the water circulator, and start to flush the cell with N_2. Infrared

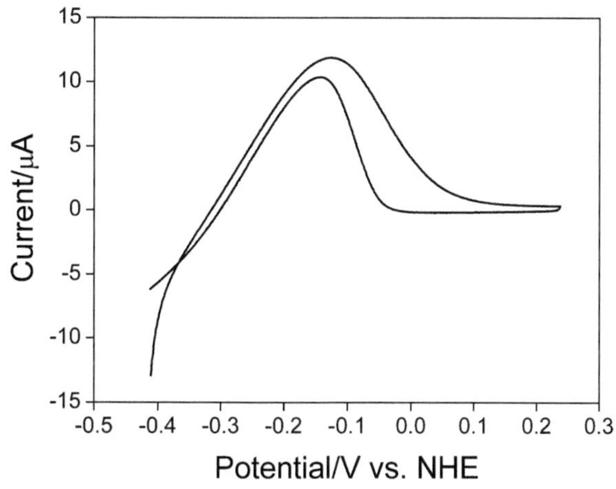

Fig. 6 Cyclic voltammetry of a 4-ATP-modified prism with DvMF hydrogenase covalently attached to the SAM and activated under H_2. Measurement done at 30 °C, pH 6.5, 1 bar H_2, and 20 mV/s

signals at 1,955, 2,081, and 2,090 cm^{-1} arising from the active site of the oxidized hydrogenase (one CO and two CN^-) should be visible at this point. After purging the cell with N_2, the hydrogenase can be reduced by introduction of H_2 into the cell (*see* **Note 14**) or by applying a constant negative potential of −0.6 V vs. Ag/AgCl until the spectrum shows only the reduced state bands (Fig. 5b). A cyclic voltammogram should show catalytic currents of H_2 oxidation (Fig. 6). Spectra of the states participating in the catalytic activity can be recorded at the desired potential and under different substrate conditions (*see* **Note 15**).

4 Notes

1. For the determination of the pD of deuterated buffers it must be taken into account that pD = pH + 0.4.
2. Sodium dithionite reducing power decreases greatly at pH ≤7. The sodium dithionite solutions are very unstable in the presence of oxygen; thus, they have to be prepared under anaerobic conditions.
3. Hydrofluoric acid is extremely toxic. Special protection measures have to be taken to manipulate its concentrated solution. A HF first aid kit is recommended.
4. The signal-to-noise ratio increases with the number of scans averaged for each measurement. Consequently, a large number of scans are required for diluted samples. Conversely, the

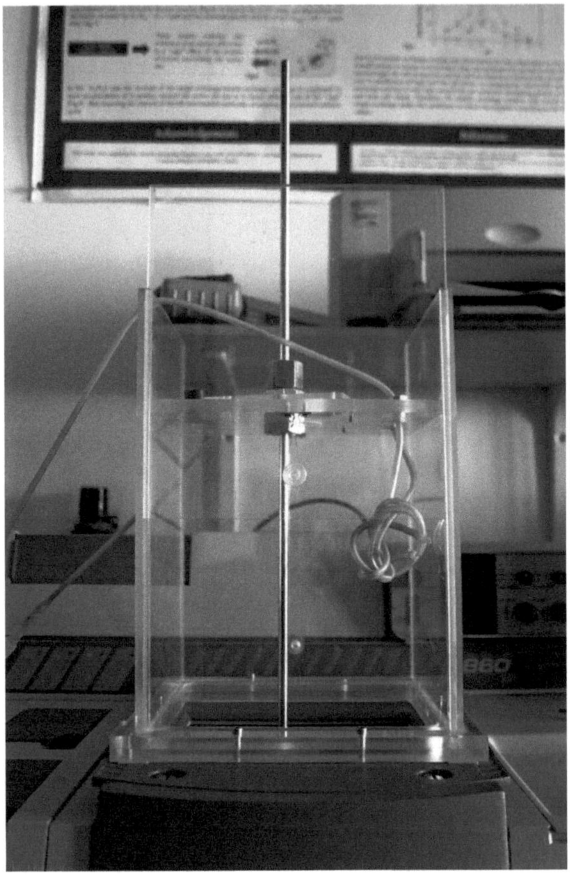

Fig. 7 Photograph of a box that covers the sample compartment and that has a mobile bar screwed to the spectral cell

experiment's time increases with the number of scans performed; thus, fewer scans are preferable for kinetic experiments or unstable samples.

5. The aperture must be low when using an MCT detector in order to decrease the amount of infrared radiation reaching the detector, avoiding its saturation.
6. A faster velocity of the interferometer leads to lower spectral resolution. On the other hand, slower velocity leads to longer measurement time.
7. The CO_2 and H_2O signals in the background spectrum should be low in order to minimize interferences in the sample measurements. Using a box that covers the sample compartment and that has a mobile bar screwed to the spectral cell allows measuring the background before each sample measurement without exposing the compartment to the exterior atmosphere (Fig. 7).

8. If the goal is to analyze the amide bands of the protein the solutions should be prepared with D_2O instead of H_2O.
9. It is important not to form bubbles while filling the cell. If bubbles are in the optical pathlength a sine wave will be obtained in the spectrum.
10. The incubation time depends on the type of hydrogenase and on the temperature. For example, the incubation time for a NiFeSe hydrogenase at room temperature is 10′–15′, whereas if an anaerobic NiFe hydrogenase is incubated at 40 °C the time required is reduced from 4 h to 90 min.
11. It is important to keep anaerobic conditions in this step. The cell can be flushed with H_2 during 10 min before filling the cell with the reduced sample or be filled in a glove-box.
12. Before using this buffer it is necessary to remove the dissolved air using the vacuum/Ar manifold and gas-type vials with septa stoppers.
13. NHS and EDC activate the carboxylic acids from the protein to form covalent attachments with the amine groups on the SAM.
14. Switching to H_2 in the cell will lead to a slow activation of the enzyme, and the open circuit potential will move to a value close to −600 mV vs. Ag/AgCl.
15. Avoid applying potentials more negative than that of thiol SAM desorption potentials, i.e., −0.7 V vs. Ag/AgCl for the 4-ATP at pH 6.5.

References

1. Banwell CD (1983) Infra-red Spectroscopy. In: Banwell CD (ed) Fundamentals of molecular spectroscopy, 3rd edn. McGraw-Hill, London, pp 72–123
2. Wharton CW (1986) Infra-red and Raman spectroscopic studies of enzyme structure and function. Biochem J 233:26–36
3. Thorneley RNF, George SJ (2000) Time-resolved Infra-red Spectroscopy of Functioning Nitrogenase. In: Triplett EW (ed) Prokaryotic nitrogen fixation: a model system for analysis of a biological process. Horizon Scientific Press, Wymondham, pp 81–99
4. Radu I, Schleeger M, Bolwein CA, Heberle J (2009) Time-resolved methods in biophysics. 10. Time-resolved FT-IR difference spectroscopy and the application to membrane-proteins. Photochem Photobiol Sci 8:1517–1528
5. Wang LX, Jiang XE (2012) Bioanalytical applications of surface-enhanced infrared absorption spectroscopy. Chin J Anal Chem 40:975–982
6. De Lacey AL, Rousset M, Cammack R, Fernandez VM (2007) Activation and inactivation of hydrogenase function and the catalytic cycle: spectroelectrochemical studies. Chem Rev 107:4304–4330
7. Sakurai T, Takahasi J, Huang H (1996) FT-IR spectra of the azide-type 3 copper in laccase and ascorbate oxidase. Chem Lett 8:651–652
8. George SJ, Allen JWA, Ferguson SJ, Thorneley RNF (2000) Time-resolved infrared spectroscopy reveals a stable ferric heme-NO intermediate in the reaction of *Paracoccus pantotrophus* cytochrome cd_1 nitrite reductase with nitrite. J Biol Chem 275:33231–33237

9. Hendrich MP, Upadhay AK, Riga J, Arciero DM, Hooper AB (2002) Spectroscopic characterization of the NO adduct of hydroxylamine oxidoreductase. Biochemistry 41:4603–4611
10. Chen J, Huang S, Seravalli J, Gutzman H Jr, Swartz DJ, Ragsdale SW, Baglley KA (2003) Infrared studies of carbon monoxide binding to carbon monoxide dehydrogenase/acetyl-CoA synthase from *Moorella thermoacetica*. Biochemistry 42:14822–14830
11. Moss D, Nabedryk E, Breton J, Mantele W (1990) Redox-linked conformational changes in proteins detected by a combination of infrared spectroscopy and protein electrochemistry. Eur J Biochem 187:565–572
12. Miyake H, Ye S, Osawa M (2002) Electroless deposition of gold thin films on silicon for surface-enhanced infrared spectroelectrochemistry. Electrochem Commun 4:973–977
13. Millo D, Pandelia M-E, Utesch T, Wisitruangsakul N, Mroginski MA, Lubitz W, Hildebrandt P, Zebger I (2009) Spectroelectrochemical study of the [NiFe] hydrogenase from *Desulfovibrio vulgaris* Miyazaki F in solution and immobilized on biocompatible gold. J Phys Chem B 113:15344–15351

Chapter 8

Characterizing Millisecond Intermediates in Hemoproteins Using Rapid-Freeze-Quench Resonance Raman Spectroscopy

Hirotoshi Matsumura and Pierre Moënne-Loccoz

Abstract

The combination of rapid freeze quenching (RFQ) with resonance Raman (RR) spectroscopy represents a unique tool with which to investigate the nature of short-lived intermediates formed during the enzymatic reactions of metalloproteins. Commercially available equipment allows trapping of intermediates within a millisecond to second time scale for low-temperature RR analysis resulting in the direct detection of metal–ligand vibrations and porphyrin skeletal vibrations in hemoproteins. This chapter briefly discusses RFQ-RR studies carried out previously in our laboratory and presents, as a practical example, protocols for the preparation of RFQ samples of the reaction of metmyoglobin with nitric oxide (NO) under anaerobic conditions. Also described are important controls and practical procedures for the analysis of these samples by low-temperature RR spectroscopy.

Key words Rapid-freeze-quench, Cryogenic temperatures, Resonance Raman Spectroscopy, Nitric oxide, Heme, Porphyrin, Iron-nitrosyl, Hemoproteins, Myoglobin

1 Introduction

Elucidating the kinetic rates of formation and decay of structurally defined transient intermediates during catalysis is critical to our understanding of reaction mechanisms in enzymes. The kinetic schemes of biological reactions may be partly revealed by electronic absorption and/or fluorescence stopped-flow analysis, but more advanced spectroscopic techniques are required to define the sub-molecular structure and exact chemical nature of reactive species within the active site of enzymes. Optically active transient species which accumulate during catalysis can be detected by stopped-flow experiments and trapped by rapid freeze quenching (RFQ) techniques, as originally developed for electron paramagnetic resonance (EPR) studies [1, 2]. Briefly, RFQ systems consist of a syringe drive ram system, a rapid mixer leading to turbulent conditions

Juan C. Fontecilla-Camps and Yvain Nicolet (eds.), *Metalloproteins: Methods and Protocols*,
Methods in Molecular Biology, vol. 1122, DOI 10.1007/978-1-62703-794-5_8,

(Reynolds number >2,000) as used in stopped-flow apparatus, a variable aging loop, and a spraying nozzle. Mixed solutions are sprayed into a precooled cryosolvent or onto a solid cold plate to stop the progress of the reaction. The overall aging time of the sample incorporates the mixing time, flow time to the nozzle, time of flight from the nozzle to the freezing agent, and freezing time. Commercially available RFQ apparatus are limited to millisecond time resolution by the relationship between the linear flow rate and the back pressure buildup at the nozzle. With exit nozzle diameters of 200–300 μm and a distance of 5 mm between the nozzle and the cryosolvent, volumetric flow rates greater than 1 ml/s provide fine sprays with short flight times to achieve quenching times below 10 ms and as low as 4 ms at maximum linear flow rates. Reaction times can be increased by inserting calibrated aging loops between the mixer and the spraying nozzle to generate a series of RFQ samples trapped at different time points during a reaction. Characterization of the resulting frozen samples with spectroscopies amenable to cryogenic temperatures can help generate kinetic profiles of transient intermediates whose growth and decay rates can be correlated with observations from stopped-flow experiments. While liquid isopentane at −140 °C remains commonly used in RFQ-EPR, we have found liquid ethane at −140 to −170 °C to be a very practical cryosolvent with comparable heat capacity to isopentane [3, 4]. Specifically, its lower freezing point, higher boiling point, and lower viscosity at temperatures lower than −140 °C allow for relatively easy packing of the sample and for the evaporation of any remaining cryosolvent after incubation at −80 °C to yield solvent-free samples.

Because studies of early events in enzyme catalysis may require submillisecond time resolutions, considerable efforts have been made by several groups to design and build apparatusses with higher operating pressure and flow rates or micro-mixer and rotating cold plates [5–8]. Greater time resolutions are also highly desirable since they lead to greater kinetic homogeneity and can offer opportunities to maximize the concentrations of intermediates by increasing the concentrations of reactants. Nevertheless, conventional stopped-flow and millisecond RFQ experiments remain a logical first step to pre-steady-state kinetic investigations.

Resonance Raman (RR) spectroscopy is widely used to identify and characterize molecular structures associated with catalytic mechanisms, particularly in metalloproteins and in enzymes with chromophoric prosthetic groups. Because RR spectroscopy is a scattering technique, it can be used on frozen samples in a 180° backscattering geometry. In hemoproteins, the main Soret band, with an electronic absorbance at approximately 400 nm and an extinction coefficient typically above 50,000 M^{-1} cm^{-1}, provides strong resonance enhancement of porphyrin skeletal vibrations that are sensitive to the oxidation state, spin state, and coordination number of the central heme iron [9–11]. Ligand-to-metal charge transfer (LMCT) transitions also produce resonance enhancement of metal–ligand vibrations and, in many instances, of intra-ligand vibrations [9–11].

The RFQ-RR methodology was described previously by Hilderbrandt and co-workers using the reaction of metmyoglobin (metMb) with azide [12], and in 2005, we described the potential of microsecond hyperquenching and RR spectroscopy to maximize the buildup of early complexes in rapid reactions [13]. Despite these efforts, only a limited number of RFQ-RR investigations to date have been carried out successfully. Our first success with a true millisecond intermediate was in characterizing a nonheme diiron-peroxo intermediate produced in the ferroxidase activity of frog ferritin [14]. Trapped within 25 ms after mixing apo-ferritin and Fe (II) in oxygenated buffer, a blue intermediate emerged, with a 650-nm charge transfer transition and resonance-enhanced ν(Fe–O)s and ν(O–O) vibrations which supported a μ-1,2 bridging Fe–O–O–Fe geometry [14]. More recently, we have used RFQ-RR to monitor the reaction of nitric oxide (NO) with oxymyoglobin as a preliminary approach to understanding the NO dioxygenase activity catalyzed by flavohemoglobins [15]. This reaction allows aerobic microorganisms to fend off elevated concentrations of NO generated by the immune response of their host, since NO reacts very efficiently with the Fe(III)-superoxo complex to form nitrate [16, 17]. A transient high-spin ferric heme intermediate was initially detected by stopped-flow UV-vis and RFQ-EPR techniques and had been tentatively assigned to an Fe(III)-peroxynitrite species; however, our RFQ-RR data demonstrated that the intermediate is in fact an Fe(III)-nitrato complex [15]. This same RFQ-RR approach has since been used to show that the heme-based gas-sensor DevS of *Mycobacterium tuberculosis* catalyzes the NO dioxygenation reaction as its initial response to NO exposure, before induction of the dormancy regulon [18]. In another study, we used RFQ-EPR and RFQ-RR to delineate the process of heme acquisition by the secreted hemophore HasA from *Pseudomonas aeruginosa* [19].

In this chapter, we provide a step-by-step description of RFQ sample preparation and the procedures we use to acquire and analyze low-temperature RR spectra. We use the reaction of metMb with NO as a model reaction to compare time points in the course of a reaction where changes in porphyrin skeletal modes and metal–ligand vibrations can be easily tracked. Myoglobin is one of the most thoroughly studied heme proteins, and the kinetics of the reaction of metMb with NO are well defined [20–23]. Upon exposure to excess NO, the aqua ligand to the six-coordinate ferric high-spin heme is replaced by NO via a dissociative ligand substitution to form ferric nitrosyl complex (Eq. 1):

$$\text{metMb}(\text{H}_2\text{O}) + \text{NO} \underset{k_{\text{off}}}{\overset{k_{\text{on}}}{\rightleftharpoons}} \text{metMb}(\text{NO}) + \text{H}_2\text{O} \tag{1}$$

The low-temperature RR characterization of RFQ samples from this reaction supports the kinetics of a conversion of the ferric heme from a predominantly high-spin configuration in metMb to a pure low-spin state in the Fe(III)–NO complex. Additional measurements using isotopically labeled NO allow for the unambiguous assignment of vibrations from this $\{FeNO\}^6$ species.

2 Materials

2.1 Protein and Reagents

1. 50 mM phosphate buffer, prepared by adding potassium phosphate dibasic (K_2HPO_4) and potassium phosphate monobasic (KH_2PO_4) in ultrapure water and adjusting the pH to 7.4. Store the phosphate buffer in a glass bottle and purge with high-purity argon for 20 min prior to transfer to anaerobic glovebox containing less than 1 ppm O_2.
2. Lyophilized metMb from equine heart stored at −20 °C (≥95 % purity).
3. Sodium dithionite (85 % purity).
4. Zeba™ desalting spin columns (7K MWCO).
5. ^{14}NO (99.5 %, Airgas) and ^{15}NO gas (>98 % ^{15}N).
6. 1 M NaOH solution: Dissolve sodium hydroxide in 20 mL water in glovebox; seal tightly in an Erlenmeyer flask with a septum and a copper wire.
7. Sodium selenate as a Raman intensity standard.

2.2 Rapid Freeze-Quench

1. A System 1000 Chemical/Freeze-Quench apparatus with a Model 1019 Syringe Ram and a 715 Ram Controller (Update Instruments, Inc. Madison, WI) equipped with a water bath chiller circulator.
2. The Update Instruments system utilizes dismountable syringe barrels that can be assembled and loaded inside the glovebox. Syringes are available in 0.5, 1, and 2 mL volumes; these nominal volumes correspond to a 60 mm column in the syringe barrel. Connectors are made of PEEK tubing to minimize gas permeability. Because the inner diameter of the PEEK tubing is small (0.57 mm), only a minute section of the sample is exposed to air once the assembled syringe is loaded and transferred outside the glovebox; the small inner diameter also strongly limits fluid convection and diffusion of air inside the tubing.
3. The syringes are mounted vertically to the syringe ram in order to minimize the potential for trapping air bubbles inside the syringe barrels.

4. The ram controller defines the size (0.1–100 mm) and speed (0.8–8.0 cm/s) of the displacement of the syringe plungers. Up to four consecutive sequences can be programmed with variable displacement, speed, and delay time between sequences.
5. Aging loops are made of PEEK tubing of variable length.
6. The rapid mixer is a Wiskind grid T-shaped mixer with a relatively large dead volume (1.6 μL). It offers complete mixing even at low flow rates, leading to Reynolds numbers <2,000.
7. The nozzle aperture is of 250 μm.
8. Ethane gas (99 %).
9. Liquid nitrogen.
10. NMR tubes (7 in. long, 5 mm outer diameter, thin walled), glass funnels (10 cm tall, 20 mm diameter, 30 mL volume), and heat shrink wrap.
11. Packing Teflon block (9 in. long, 65 mm outer diameter, 6.5 in. long 6 mm inner diameter well).
12. Large stainless steel Dewar allowing complete immersion of the packing Teflon block in liquid nitrogen.

2.3 Low-Temperature Resonance Raman Spectroscopy

1. Custom McPherson 2061/207 spectrograph equipped with two sets of mirrors to offer a 0.67- or 1-m focal length. A wide choice of holographic gratings (3,600, 2,400, 1,800, 1,200, and 600 grooves per mm) allows for the acquisition of high-resolution spectra for laser excitations ranging from the near-UV to near-IR. The liquid nitrogen-cooled CCD detector (LN-1100PB, Princeton Instruments) is composed of 350 (vertical) by 1,100 (horizontal) pixels, with each pixel representing a detector area of $24 \times 24\ \mu m^2$. Vertical pixels are binned and read out as a single value, and the combination of entrance slit, focal length, and grating ruling number determines the ultimate resolution of the spectra. This value is measured experimentally using the bandwidth of the O–O stretching mode of O_2 gas at 1,555 cm^{-1}.
2. The 406- and 413-nm emission lines of a krypton ion laser (Innova 300C, Coherent) and the 442-nm emission of a helium–cadmium ion laser (Liconix 4240NB) are the most widely used excitations to study heme proteins.
3. Optical Dewar fitted with a cold-finger sample holder made of copper [24]. Direct measurements with a thermocouple inside a water-filled NMR tube confirm that the temperature of the sample remains at 110 K during the acquisition of RR spectra (Fig. 1).

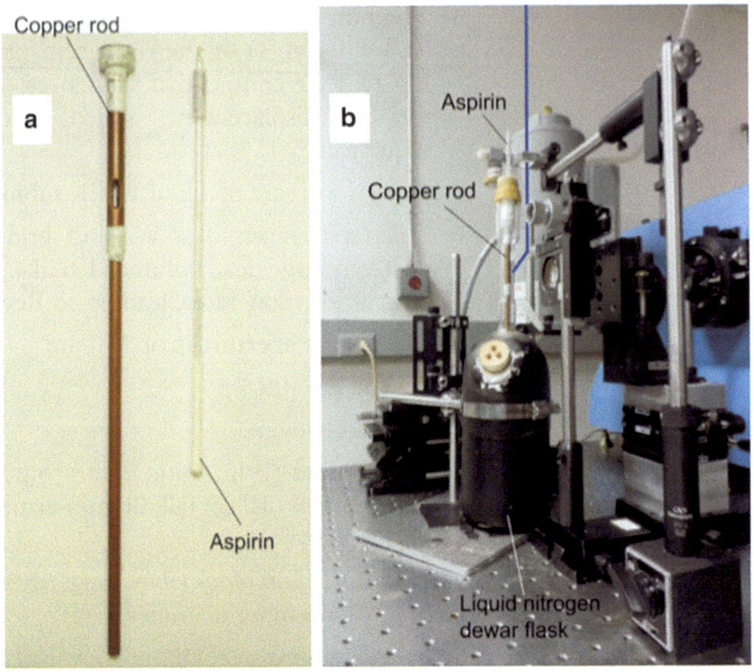

Fig. 1 Copper rod used as cold finger and aspirin powder in a flame-sealed NMR tube (7 in. long) (**a**). Setup layout of the backscattering geometry used for low-temperature RR measurements (**b**): the laser beam (*blue line*) is focused above the sample area with a 150-mm cylindrical focusing lens and reflected with a wide angle on the sample by a small mirror; the camera lens and sample assembly are mounted on manual XYZ stages; the supernotch or the edge filter is set in front of the entrance slit; the liquid nitrogen-filled CCD camera (*grey*) is mounted on the exit slit of the single-stage McPherson monochromator (*blue*)

3 Methods

3.1 Preparation of metMb Solution

1. Prepare a stock solution of metMb by dissolving metMb in phosphate buffer in the glovebox.
2. Remove insoluble residue and free hemin from the stock solution using a desalting spin column.
3. Calculate protein concentration on the basis of a 408-nm molar extinction coefficient, ε_{408}, of 188 mM^{-1} cm^{-1} using a UV-vis spectrophotometer (Cary 50, Varian Inc.) [23]. Dilute the myoglobin stock solution with phosphate buffer to a concentration of 0.6 mM.
4. Load an RFQ syringe with the metMb solution in the glovebox. The RFQ syringes and connecting tubing should be transferred to the glovebox the day before to ensure anaerobicity. After assembling the syringe, immerse the connecting tubing in the metMb solution and push/pull the plunger

multiple times to eliminate all bubbles inside the syringe barrel. The RR data presented herein were obtained using either 2 × 1 mL or 2 × 2 mL syringes.

3.2 NO Gas Purification and Saturated NO Solution Preparation

1. Purge ^{14}NO and ^{15}NO gases of higher oxides, N_2O_3 and NO_2, by incubation above a 1 M NaOH solution (*see* **Note 1**).
2. Fill a 10-mL glass vial with 7 mL phosphate buffer solution and tightly seal with a septum.
3. After drawing 3 mL from the vial headspace with a 5-mL gastight Hamilton syringe, bubble 4 mL of purified NO gas in the solution, through the septum, and mix for 5 min with the phosphate buffer solution. Repeat this gas exchange procedure twice.
4. Prepare a stock solution of dithionite using anaerobic buffer in the glovebox, and calculate its concentration using an ε_{315} of 6,900 M^{-1} cm^{-1} [25].
5. Fully reduce metMb by addition of ~5 mM sodium dithionite; the color of the solution should change from brown to red. Remove any excess reducing agent using a desalting spin column.
6. Put the deoxyMb solution in a UV-vis cuvette equipped with a tightly fitted septum. Confirm the complete conversion of metMb to deoxyMb by UV-vis spectroscopy, and calculate the concentration of deoxyMb using $\varepsilon_{435} = 121\ mM^{-1}cm^{-1}$ [23].
7. Use the deoxyMb solution to determine the concentration of the NO solution by adding substoichiometric amounts of the latter to the deoxyMb solution in an anaerobic UV-vis cuvette [26]. The NO concentration of a saturated NO solution is generally estimated to be ~1.8 mM [26].
8. Repeat this procedure with the ^{15}NO-saturated solution, and match the concentrations of the ^{14}NO and ^{15}NO solutions within 10 % by supplementing the less concentrated solution with additional gas in the headspace. Variations in NO concentration will result in different reaction rates, which in turn will lead to different levels of conversion for matching quenching times of reactions with ^{14}NO and ^{15}NO.
9. Load the RFQ syringe with the NO-saturated solution by attaching a needle to the connecting tubing. Push and pull the plunger with the syringe needle in the headspace of the NO solution vessel a few times before inserting the needle in the solution, and draw the desired volume of NO-saturated solution into the syringe.

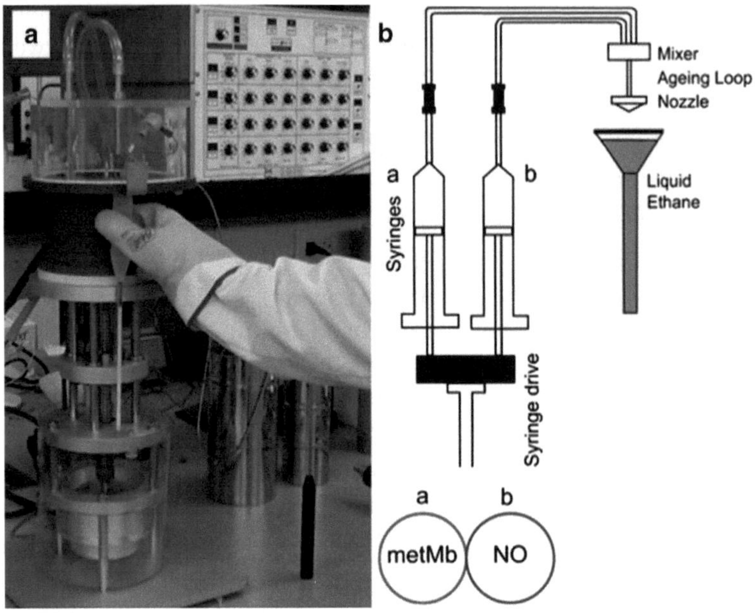

Fig. 2 Photograph (**a**) and schematic (**b**) of the Update Instruments Inc. System 1000 Chemical/Freeze-Quench apparatus

3.3 RFQ Sample Preparation of metMb Reaction with NO

A photograph and schematic of a System 1000 Chemical/Freeze-Quench apparatus are shown in Fig. 2.

1. Cool the packing Teflon block to −180 °C with liquid nitrogen in a stainless steel Dewar covered at all time with aluminum foil to minimize water vapor condensation. This step requires regular refilling with liquid nitrogen and takes approximately 30 min before the remaining liquid nitrogen can be poured out of the Dewar.
2. Join the NMR tubes and glass funnels vertically and coaxially with heat-shrink electric wrap to minimize problems during the collection of the sample jet and packing.
3. Bring the syringes from the glovebox, and mount them tightly to the syringe ram system (System 1000 Chemical/Freeze-Quench apparatus); attach the connecting tubes to the mixer and couple to the nozzle using the smallest aging loop.
4. Cool the sample assembly (i.e., syringes, connecting tubing, mixer, and nozzle) to 4 °C using a circulating water bath.
5. In order to confirm end product formation, a test shot with a minimal displacement (i.e., 1 mm) at high velocity is typically run and a UV-vis spectrum of the diluted aliquot is obtained; often, two to three shots are required if some bubbles have remained in the syringe/connector assembly. However, in the experiments presented here, the relatively low binding affinity

and high off rate of metMb(NO) preclude the detection of the complex as a resting state.

6. Wash the mixer, aging reactor, and nozzle with water and then ethanol, before thoroughly drying them with compressed air. Residual liquid in the nozzle prior to a shot will freeze rapidly when the sample funnel, filled with liquid ethane, is brought to collect the sample, and the resulting obstruction can lead to a failed shot, excessive back pressure, and potential shattering of a syringe barrel.
7. Set the parameters (displacement, velocity, and reactor size) prior to executing the RFQ program. Although NMR tubes require only 200 μL to fill a 1-cm height necessary for low-temperature RR analysis, 400 μL of mixed sample is needed to guarantee adequate packing and filling. In this study, time points of 6, 12, 20, 75, and 280 ms were chosen on the basis of reported stopped-flow kinetics ($k_{on} = 2.5 \pm 0.6 \times 10^4$ $M^{-1}\,s^{-1}$, $k_{off} = 2.71 \pm 0.01\ s^{-1}$ at 20 °C) [20].
8. Mixing two 1-mL syringes with a displacement velocity of 8 cm/s leads to a volumetric flow rate of 2.7 mL/s and a linear flow rate at the spraying nozzle of 1.3 m/s. With these experimental parameters, we estimate our shortest RFQ time at ~6 ms.
9. Because metMb is commercially available, sample collection at all time points may be accomplished with high displacement rates and increasing aging loops. High flow rates generate homogenous quenching of the reaction and uniform frozen solutions with reproducible packing. For experiments with valuable protein samples, however, residual loss in long aging loops becomes prohibitive, so time points beyond 200 ms may be achieved using repetitive shots with set delays.
10. Inside a chemical hood and behind protective shatterproof windows, liquefy ethane gas from a compressed tank (mp: −182 °C; bp: −89 °C) into a cold trap placed inside a stainless steel Dewar and immersed in liquid nitrogen (mp: −210 °C; bp: −195 °C) (*see* **Note 2**).
11. Fill the well of the precooled packing Teflon block with liquid ethane, insert an NMR tube assembly in the well, and quickly fill it with liquid ethane. Use a precooled packing rod to remove trapped air inside the NMR tube and top off the funnel with liquid ethane.
12. Execute the RFQ program which includes a 10-s delay, and remove the NMR tube assembly from the packing Teflon block, holding it within 2 mm of the RFQ spray nozzle (Fig. 2). Because the sample jet will splatter some liquid ethane outside the funnel, protective gloves and long sleeves are essential.

Fig. 3 A 6-ms RFQ sample before (**a**) and during (**b**) packing

13. Return the NMR tube assembly to the packing Teflon block immediately after sample collection, and let the frozen sample settle to the bottom of the funnel. Using a stainless steel rod with a Teflon tip closely matching the inner diameter of the NMR tube, start packing the frozen sample past the connection between the funnel and the NMR tube without attempting to push it to the bottom of the tube until a large portion of the sample is packed near the top of the tube. A second stainless steel rod with no Teflon tip and a smaller diameter may be required to push the accumulated samples to the bottom of the NMR tube.
14. Break the shrink wrap connection between the funnel and the NMR tube, releasing the excess liquid ethane in the funnel onto the side of the packing Teflon block; label the tube and store in liquid nitrogen until ready to analyze by RR spectroscopy.
15. Figure 3 shows a photograph of a 6-ms RFQ sample of [metMb + NO] before and after packing (notice the evidence of sample loss along the surface of the funnel after packing). Depending on the reaction time, the color of the frozen samples distinctively changes from brown to pink (Fig. 4).
16. As a control, collect an RFQ sample of the starting metMb sample at the shortest time point and without mixing. An RR spectrum of this control sample will be compared with that of resting metMb frozen slowly in liquid nitrogen. This comparison will confirm that the rapid freezing procedure and the exposure to cryosolvent have no impact on the RR spectrum of the starting material. A similar control should be carried out

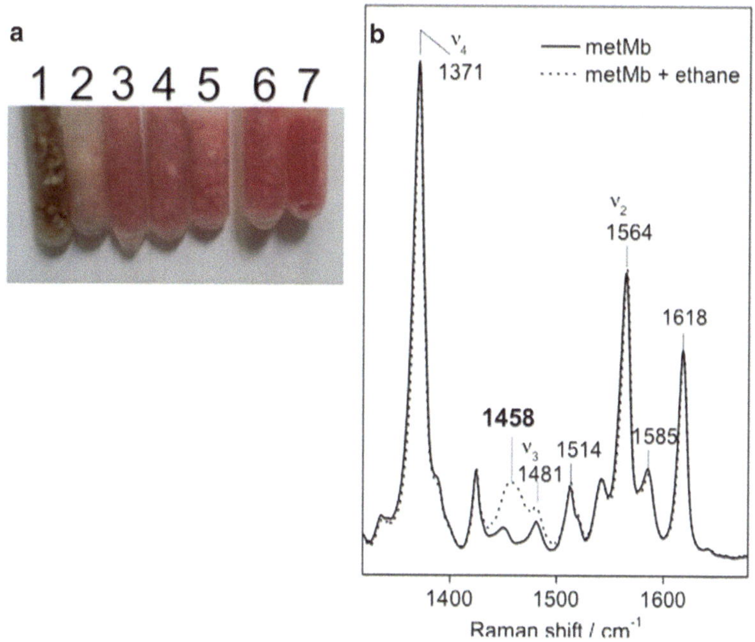

Fig. 4 Visual comparison of frozen samples of metMb (*1*), metMb(NO) adduct (*7*), and RFQ samples obtained at 6 ms (*2*), 12 ms (*3*), 20 ms (*4*), 75 ms (*5*), and 280 ms (6) (**a**). Low-temperature RR spectra of RFQ-metMb control before and after evaporation of liquid ethane (**b**) (λ_{exc} = 406 nm, 5 mW; monochromator settings: 100 μm slit, 0.67 m focal length, 2,400 groove/mm grating, 110 K sample temperature)

with the end product. We consider such controls to be essential even though we have never observed evidence of RFQ-induced changes in the RR spectra of resting metalloproteins.

17. After the last RFQ shot is acquired, residual solutions in the syringes should be capped and transferred back to the glovebox. Use UV-vis spectroscopy to confirm the integrity of the metMb solution and the deoxyMb assay to measure the NO concentrations in the NO-saturated solutions.

3.4 Low-Temperature RR Spectroscopy of RFQ Samples of [metMb + NO]

1. Using the 406-nm laser line of a krypton ion laser as the excitation source, set the power to <10 mW during the entire alignment procedure. Employ a prism monochromator (Applied Photophysics) or a single-band bandpass filter (Semrock Inc.) to eliminate plasma lines from the laser emission.
2. With a cylindrical lens and a small mirror, generate a vertical illumination (~200 μm wide, 1 cm tall) of the sample in the optical Dewar. Collect the backscattered light with a camera lens to image the illuminated area on the entrance slit of a single-stage monochromator (Fig. 1). Optimize this alignment first at room temperature using NMR tubes filled with fluorescent solutions (e.g., diluted rhodamine solutions).

3. Insert the aspirin powder standard, and set a supernotch filter (Kaiser Optics Inc.) or edge long-pass filter (RazorEdge filters, Semrock Inc.) in front of the entrance slit to attenuate the Rayleigh scattering (Fig. 1).
4. Optimize the choice of holographic ruled grating and focal length of the monochromator to achieve the desirable spectral window and resolution, and with the aspirin sample still in place, optimize the alignment using the CCD camera readout and store a spectrum for future spectral calibration. Frequency accuracy should reach ± 1 cm^{-1}, and spectra resolution should be below 8 cm^{-1} (*see* **Note 3**).
5. Once the calibration is achieved, add liquid nitrogen to the optical Dewar flask and continue to top off until the temperature is stabilized (~15 min). Set up a continuous flow of dry compressed air to avoid condensation on the outer surface of the Dewar during data acquisition.
6. Wipe the NMR tube containing the RFQ sample quickly and repeatedly to remove any condensation on the tube glass wall before inserting it into the copper cold finger. Cap the top of the optical Dewar with a drilled septum so as to permit access to the top of the NMR tube.
7. Assess the photosensitivity of all samples by comparing rapid acquisitions with minimal laser power, continuous sample spinning, and longer data acquisitions on static samples. If photosensitivity is observed, continuous rotation of the sample can be achieved with a variable-speed electric motor. In this study, the RFQ samples showed no evidence of photosensitivity with 5 mW laser power.
8. Acquire RR spectra for all RFQ samples under the same conditions. Determine the optimal data accumulation times to minimize the number of readouts of the CCD camera and to maximize the utilization of the camera's dynamic range; accumulate acquisitions to reach a high signal-to-noise ratio. Typically, for hemoproteins with Soret excitations, a total acquisition time of 10–20 min is sufficient to obtain high-quality spectra. In this study, accumulation times of 6 and 13 min were employed for the high-frequency and low-frequency RR spectra, respectively.
9. Because trapped intermediates may slowly decay, even at cryogenic temperatures, RR spectra should be collected as early as possible after the preparation of the RFQ samples. Ethane trapped as a solid in RFQ samples is readily detected by a non-resonant Raman band at 1,458 cm^{-1} which corresponds to the asymmetric CH_3 bending vibration (Fig. 4). The only other vibrational mode from ethane presenting significant intensity is the C–C stretch at 993 cm^{-1}. This paucity of

Raman signals from ethane is in sharp contrast with isopentane which presents many strong vibrations in the 300–1,700 cm^{-1} region.

10. After this initial collection of RR spectra, ethane can be evaporated from the frozen samples with a 2-h incubation of the samples at −80 °C. This rapid procedure to eliminate the trapped cryosolvent is another major incentive toward using ethane over isopentane, since the latter requires prolonged evacuation of the samples at −80 °C.

11. Low-temperature RR spectra of the starting metMb solution; RFQ samples trapped at 6, 12, 20, 75, and 280 ms; and a metMb(NO) adduct prepared with a 1-atm NO headspace are shown in Fig. 5.

12. For quantitative analyses of the RFQ-RR data, compare the RR spectra of starting metMb and the metMb(NO) adduct at 20 μM protein concentrations, using 100 mM sodium selenate as the intensity standard. In this study, the ν_1 symmetric stretching mode of selenate at 838 cm^{-1} shows that the intensity of the ν_{46} porphyrin mode in metMb and metMb(NO) at 930 cm^{-1} is unchanged. In the high-frequency region, the observed intensity at 1,584 cm^{-1}, despite the overlap of multiple modes, is unchanged between metMb and metMb(NO). Thus, these two frequencies were used to normalize the low-frequency and high-frequency RR spectra of the RFQ samples.

13. As reported previously [27, 28], the high-frequency RR spectrum of metMb includes a very strong band at 1,371 cm^{-1} which corresponds to the ν_4 oxidation marker band characteristic of Fe(III) hemes (Fig. 5a). The weak band at 1,481 cm^{-1} and the strong band at 1,564 cm^{-1} correspond to the ν_3 and ν_2 modes, respectively, and are characteristic of six coordinate high-spin (6cHS) species [11]. As expected, the RR spectra of the RFQ samples display a gradual conversion of these 6cHS marker bands to higher frequency bands characteristic of low-spin species. The most distinctive component attributable to the six-coordinate low-spin (6cLS) nitrosyl complex is the ν_{10} mode observed at 1,647 cm^{-1}. In contrast, the growth of the ν_3 of metMb(NO) at 1,511 cm^{-1} is more difficult to observe because of concomitant changes in the intensity of the porphyrin skeletal ν_{38} mode at 1,514 cm^{-1}. The ν_4 mode shift from 1,371 cm^{-1} in metMb to 1,375 cm^{-1} in metMb(NO) reflects the strong π-acid character of the nitrosyl ligand.

14. The low-frequency region of the RR spectra highlights the rise of two bands at 570 and 597 cm^{-1} (Fig. 5b), which downshift upon substitution with ^{15}NO and are assigned to the bending and stretching Fe^{III}–NO vibrations, respectively [29]. The rate of growth of these metal–ligand vibrations matches closely to

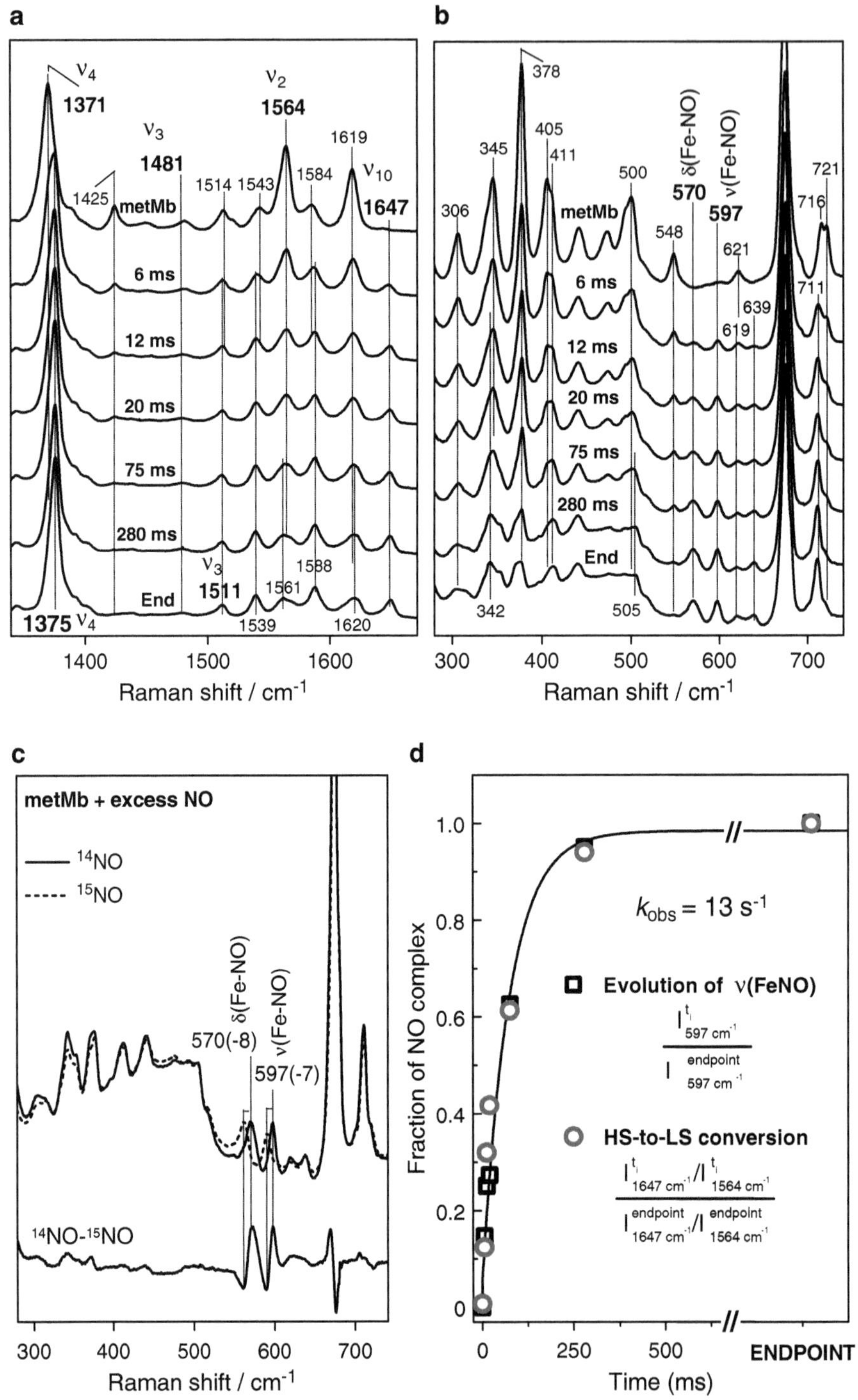

Fig. 5 High-frequency (**a**) and low-frequency (**b**) RR spectra of RFQ samples of [metMb + ^{14}NO] at 6, 20, 75, and 280 ms and of metMb and metMb(NO). Low-frequency RR spectra of metMb(^{14}NO) and metMb(^{15}NO) (**c**) (experimental conditions as in Fig. 4). Fraction of metMb(NO) in individual RFQ samples has been determined on the basis of the intensity of the Fe–NO stretching mode (*open black square*), high-frequency porphyrin skeletal modes (*open grey circle*), and a single-exponential fit (*black line*) (**d**)

the conversion rate deduced from the high-frequency modes. Indeed, plotting the peak intensity of the ν(Fe-NO) against reaction times yields an apparent rate of formation of the ferric-nitrosyl adduct of 13 s^{-1} (Fig. 5d), which is in good agreement with the 10 s^{-1} observed rate measured with 1 mM NO at 4 °C using stopped-flow absorption spectroscopy [20]. Using an I_{1647}/I_{1564} ratio for each RR spectrum as a reporter of the conversion from 6cHS metMb to 6cLS metMb(NO) leads to an equivalent observed rate (Fig. 5d).

4 Notes

1. NO must be handled with extreme caution. If inhaled, the colorless gas can be extremely harmful, causing respiratory tract irritation, skin irritation, and blood damage. On contact with air, NO undergoes spontaneous oxidation to the reddish brown nitrogen dioxide which is a highly poisonous gas. Exposure to even low concentrations of NO_2 can cause severe skin and eye burns and respiratory system irritation; severe overexposure may cause unconsciousness and death. To avoid risk, use small tanks of NO in a fume hood located in a well-ventilated room.
2. Ethane is a colorless, odorless, and flammable gas. Keep away ignition sources and work under a fume hood. Contact of liquid ethane with skin will result in severe cryogenic burns; protect hands and arms with cryogenic gloves and long sleeves.
3. Aspirin is stable toward laser irradiation and temperature. The RR spectrum of aspirin exhibits a large number of bands from 250 to 1,750 cm^{-1} at room temperature. It is important to note that these frequencies shift, albeit only slightly, with temperature; because they become sharper and better resolved at 110 K, we use these spectra as calibration files to convert pixels to wavenumbers.

Acknowledgments

This work was supported by the National Institute of Health (P.M.L., GM074785). H.M. acknowledges the financial support from the Japan Society for the Promotion of Science (Research Fellowship for Young Scientists).

References

1. Ballou DP, Palmer GA (1974) Practical rapid quenching instrument for the study of reaction mechanisms by electron paramagnetic resonance spectroscopy. Anal Chem 46:1248–1253
2. Bray RC (1961) Sudden freezing as a technique for the study of rapid reactions. Biochem J 81:189–193
3. Witt RK, Kemp JD (1937) The heat capacity of ethane from 15 K to the boiling point. The heat of fusion and the heat of vaporization. J Am Chem Soc 59:273–276
4. Guthrie GB, Huffman HM (1943) Thermal data. XVI. The heat capacity and entropy of isopentane. The absence of a reported anomaly. J Am Chem Soc 65:1139–1143
5. Schmidt B, Mahmud G, Soh S, Kim SH, Page T, O'Halloran TV, Grzybowski BA, Hoffman BM (2011) Design implementation, simulation, and visualization of a highly efficient RIM microfluidic mixer for rapid freeze-quench of biological samples. Appl Magn Reson 40:415–425
6. Cherpanov AV, de Vries S (2004) Microsecond freeze-hyperquenching: development of a new ultrafast micro-mixing and sampling technology and application to enzyme catalysis. Biochim Biophys Acta 1656:1–31
7. Lin Y, Gerfen GJ, Rousseau DL, Yeh SR (2003) Ultrafast microfluidic mixer and freeze-quenching device. Anal Chem 75:5381–5386
8. Tanaka M, Matsuura K, Yoshioka S, Takahashi S, Ishimori K, Hori H, Morishima I (2003) Activation of hydrogen peroxide in horseradish peroxidase occurs within approximately 200 micro s observed by a new freeze-quench device. Biophys J 84:1998–2004
9. Nakamoto K (1997) Infrared and Raman spectroscopy of inorganic and coordination compounds, vol A-B, 5th edn. Wiley, New York
10. Spiro TG, Czernuszewicz RS (1995) Resonance Raman spectroscopy of metalloproteins. Methods Enzymol 246:416–460
11. Spiro TG, Li XY (1988) Resonance Raman spectroscopy of metalloporphyrins. In: Spiro TG (ed) Biological applications of Raman spectroscopy. Vol. 3. Resonance Raman spectra of hemes and metalloproteins. Wiley, New York, pp 1–37
12. Oellerich S, Bill E, Hildebrandt P (2000) Freeze-quench resonance Raman and electron paramagnetic resonance spectroscopy for studying enzyme kinetics: application to azide binding to myoglobin. Appl Spectrosc 54:1480–1484
13. Lu S, Wiertz FGM, de Vries S, Moënne-Loccoz P (2005) Resonance Raman characterization of a high-spin six-coordinate iron(III) intermediate in metmyoglobin-azido complex formation trapped by microsecond freeze hyperquenching (MHQ). J Raman Spectrosc 36:359–362
14. Moënne-Loccoz P, Krebs C, Herlihy K, Edmondson DE, Theil EC, Huynh BH, Loehr TM (1999) The ferroxidase reaction of ferritin reveals a diferric μ-1,2 bridging peroxide intermediate in common with other O_2-activating non-heme diiron proteins. Biochemistry 38:5290–5295
15. Yukl ET, de Vries S, Moënne-Loccoz P (2009) The millisecond intermediate in the reaction of nitric oxide with oxymyoglobin is an iron(III)-nitrato complex, not a peroxynitrite. J Am Chem Soc 131:7234–7235
16. Gardner PR (2005) Nitric oxide dioxygenase function and mechanism of flavohemoglobin, hemoglobin, myoglobin and their associated reductases. J Inorg Biochem 99:247–266
17. Ouellet H, Ouellet Y, Richard C, Labarre M, Wittenberg B, Wittenberg J, Guertin M (2002) Truncated hemoglobin HbN protects *Mycobacterium bovis* from nitric oxide. Proc Natl Acad Sci U S A 99:5902–5907
18. Yukl ET, Ioanoviciu A, Sivaramakrishnan S, Nakano MM, Ortiz de Montellano PR, Moënne-Loccoz P (2011) Nitric oxide dioxygenation reaction in DevS and the initial response to nitric oxide in *Mycobacterium tuberculosis*. Biochemistry 50:1023–1028
19. Yukl ET, Jepkorir G, Alontaga AY, Pautsch L, Rodriguez JC, Rivera M, Moënne-Loccoz P (2010) Kinetic and spectroscopic studies of hemin acquisition in the hemophore $HasA_p$ from *Pseudomonas aeruginosa*. Biochemistry 49:6646–6654
20. Laverman LE, Wanat A, Oszajca J, Stochel G, Ford PC, van Eldik R (2001) Mechanistic studies on the reversible binding of nitric oxide to metmyoglobin. J Am Chem Soc 123:285–293
21. Sharma VS, Isaacson RA, John ME, Waterman MR, Chevion M (1983) Reaction of nitric oxide with heme proteins: studies on metmyoglobin, opossum methemoglobin, and microperoxidase. Biochemistry 22:3897–3902
22. Sharma VS, Traylor TG, Gardiner R, Mizukami H (1987) Reaction of nitric oxide with heme proteins and model compounds of hemoglobin. Biochemistry 26:3837–3843

23. Antonini E, Brunori M (1971) Hemoglobin and myoglobin in their reactions with ligands. North Holland, Amsterdam
24. Loehr TM, Sanders-Loehr J (1993) Techniques for obtaining resonance Raman spectra of metalloproteins. Methods Enzymol 226: 431–470
25. Creutz C, Sutin N (1974) Kinetics of the reactions of sodium dithionite with dioxygen and hydrogen peroxide. Inorg Chem 13: 2041–2043
26. Lim MD, Lorkovic IM, Ford PC (2005) The preparation of anaerobic nitric oxide solutions for the study of heme model systems in aqueous and nonaqueous media: some consequences of NOx impurities. Methods Enzymol 396:3–17
27. Morikis D, Champion PM, Springer BA, Egebey KD, Sligar SG (1990) Resonance Raman studies of iron spin and axial coordination in distal pocket mutants of ferric myoglobin. J Biol Chem 265:12143–12145
28. Hu S, Smith KM, Spiro TG (1996) Assignment of protoheme resonance Raman spectrum by heme labeling in myoglobin. J Am Chem Soc 118:12638–12646
29. Benko B, Yu NT (1983) Resonance Raman studies of nitric oxide binding to ferric and ferrous hemoproteins: detection of Fe (III)–NO stretching, Fe(III)–N–O bending, and Fe(II)–N–O bending vibrations. Proc Natl Acad Sci U S A 80:7042–7046

Chapter 9

A Practical Guide for Nuclear Resonance Vibrational Spectroscopy (NRVS) of Biochemical Samples and Model Compounds

Hongxin Wang, Esen Ercan Alp, Yoshitaka Yoda, and Stephen P. Cramer

Abstract

Nuclear resonance vibrational spectroscopy (NRVS) has been used by physicists for many years. However, it is still a relatively new technique for bioinorganic users. This technique yields a vibrational spectrum for a specific element, which can be easily interpreted. Furthermore, isotopic labeling allows for site-specific experiments. In this chapter, we discuss how to access specific beamlines, what kind of equipment is used in NRVS, and how the sample should be prepared and the data collected and analyzed.

Key words Nuclear resonance vibrational spectroscopy, NRVS, Metalloproteins, Mössbauer

1 Introduction

Nuclear resonance vibrational spectroscopy (NRVS) has been used by physicists for nearly two decades [1], but it is still a relatively new technique for bioinorganic users. Why is NRVS of interest to chemists and biologists? Among other things, it is an exciting tool because:

- NRVS yields a vibrational spectrum for a specific element
- The resulting vibrational spectrum is easily interpreted or calculated
- The isotopic sensitivity allows site-specific labeling experiments
- Low frequency acoustic and optical modes can be measured down to 8 cm^{-1}
- Raman and infrared silent modes can be observed
- Anisotropy of motion can be determined accurately

Before going further, we present an example to show what all the excitement is about (Fig. 1 left panel). In this case, we have a

Juan C. Fontecilla-Camps and Yvain Nicolet (eds.), *Metalloproteins: Methods and Protocols*,
Methods in Molecular Biology, vol. 1122, DOI 10.1007/978-1-62703-794-5_9,

small molecule where IR and Raman had difficulty identifying an Fe=O stretch. By recording the spectrum with natural abundance O and then ^{18}O, it was simple to identify the band at 831 cm^{-1} as the stretching mode of an Fe=O double bond. For biochemical users, the beauty of NRVS spectrum is not only that it will provide such vibrational frequencies, similar to an IR or Raman spectrum, but also that it only senses those normal modes in which the probe nucleus (usually ^{57}Fe) is moving. Although ^{57}Fe is by far the dominant isotope used for NRVS, there more than a dozen other elements are currently feasible (Fig. 1 right panel). In this chapter, we describe the procedures for scheduling an NRVS experiment, preparing the samples, conducting the measurement, processing the data, and interpreting the results (Table 1).

Once you have decided that NRVS will be helpful, you will need access to a beamline where such experiments can be done. Compared to X-ray diffraction or EXAFS, the choice of

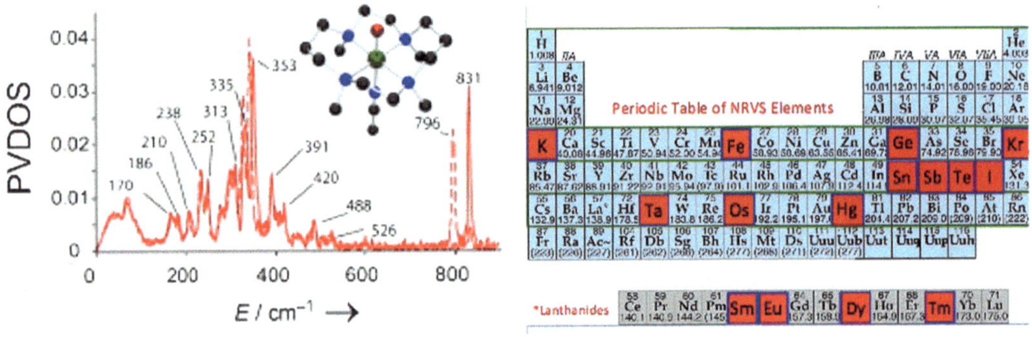

Fig. 1 *Left*: a generic NRVS spectrum. *Right*: the current NRVS periodic table

Table 1
World-wide summary of NRVS-capable beamlines

Facility/ beamline	*E* (GeV)	Emittance (nm rad)	Period (mm)	# of periods	Isotopes covered
ESRF/ID18	6	4	20/32		^{57}Fe, ^{151}Eu, ^{149}Sm, ^{119}Sn, ^{161}Dy, ^{121}Sb, ^{125}Te, ^{129}Xe
APS/16ID	7	3.3		144	^{57}Fe
APS/03ID	7	~3.1	27	88	^{57}Fe, ^{151}Eu, ^{83}Kr, ^{119}Sn, ^{161}Dy
SPring-8/09XU	8	~3.4	32	140	^{57}Fe, ^{151}Eu, ^{149}Sm, ^{119}Sn, ^{40}K
SPring-8/19LXU	8	~3.4	32	781	^{57}Fe
PETRA-III/P01	6	1	32	314	^{57}Fe

Fig. 2 Synchrotron radiation labs with NRVS Facilities. *Top left*: APS, *top right*: ESRF, *bottom left*: PETRA-III, *bottom right*: SPring-8

facilities is relatively narrow: APS (http://www.aps.anl.gov), ESRF (http://www.esrf.eu), SPring-8 (http://www.spring8.or.jp/en/), and PETRA-III (http://petra3.desy.de) (Fig. 2; *see* **Note 1**).

2 Materials

2.1 Sources and Monochromators

Regardless of which facility you choose, the experimental apparatus is pretty much the same. As illustrated in Fig. 3, the initial source of X-rays is an undulator in the storage ring, which produces intense X-rays with a spectral bandwidth of ~100 eV. Heat is removed and the bandwidth is reduced to ~1 eV by a high heat load monochromator (HHLM). Since the range of a typical NRVS experiment is typically ~0.1 eV, once optimized, there is usually no need to adjust the undulator or even the HHLM.

With the more manageable beam from the HHLM, a ~1 meV resolution beam is then produced by the high-resolution monochromator (HRM). These devices achieve a bandwidth about 1/1,000 of that of typical EXAFS monochromators. A high diffraction angle, asymmetric reflections, and combination of different single crystal reflections are used to achieve the high energy resolution and a wide acceptance angle in these monochromators.

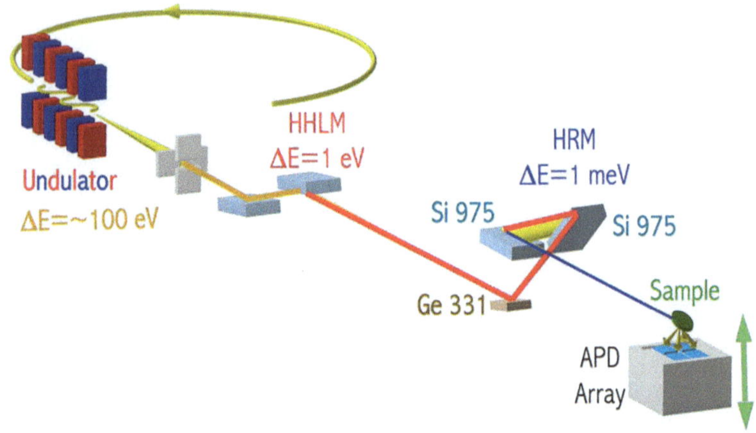

Fig. 3 A schematic overview of the NRVS beamline at SPring-8 BL09XU

However, there is a tradeoff between resolution and intensity that has to be considered in the NRVS experiments, because both factors are usually important. Higher resolution cuts the flux due to the narrower energy bandwidth and the lower reflectivity of the more asymmetric reflection.

A representative compromise between flux and resolution is the three bounce type HRM adopted at SPring-8 Beamline 09XU (Fig. 3). Ge 331 reflection with asymmtery factor of $b = 1/2$ is followed by two Si (975) reflections with the asymmetry factors of 1/10.8 and 18.0. Here the asymmetry factor is defined as $b = \sin(\theta_{in})/\sin(\theta_{out})$, where θ_{in} and θ_{out} are incident and outgoing angles from the crystal surface, respectively. (The Ge reflection is used to set the output beam almost horizontal.) The energy of the output beam is determined by the angle between the two Si (975) crystals, which is controlled by a piezoelectric linear stage.

2.2 Detectors and Electronics

One of the breakthroughs that permitted the development of the NRVS technique was the introduction of avalanche photodiode (APD) detectors with nanosecond response times. These devices are used to detect nuclear fluorescence and X-ray fluorescence (following internal conversion) from the sample. One can use timing electronics to distinguish these "signal" photons (which arise from a nuclear absorption event) from the much larger number of "background" photons that arise from X-ray scattering and conventional sample fluorescence. This is possible because NRVS "signal" photons arrive with a delay corresponding to the half-life of the nuclear excited state (about 100 ns for ^{57}Fe), while the "background" photons are essentially prompt and only appear during the ~70 ps length of the synchrotron radiation pulse (Fig. 4 left panel).

The electronics used to distinguish nuclear events from electronic scattering and X-ray fluorescence is also shown in Fig. 4, right panel. The important devices are the "bunch clock," which

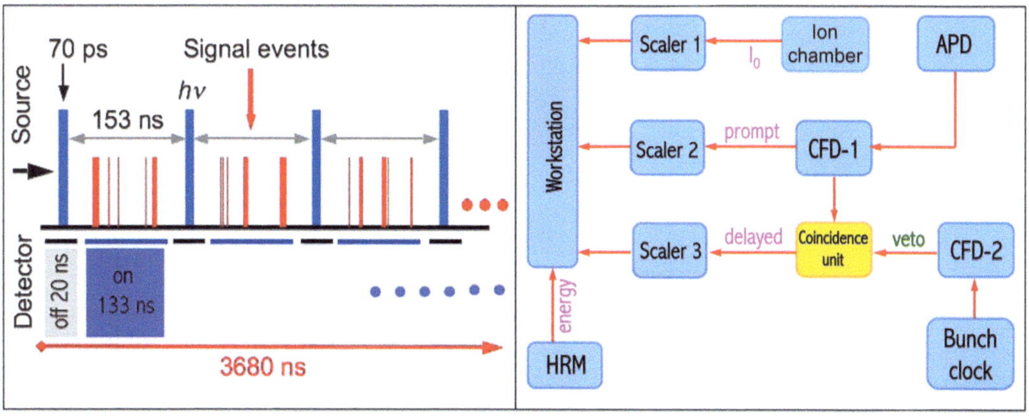

Fig. 4 *Left*: typical time structure of "signal" events. *Right*: electronics for time-gated NRVS detection. The delayed signal from CFD1 corresponds to nuclear events

tells the experimenter the initial $t = 0$ time point when the synchrotron pulse arrives, and a constant fraction discriminator (CFD) that is a very fast analyzer of photon arrivals. A veto interval X, where X is about 10–20 ns, is set around $t = 0$, to allow the APD to recover from the enormous background pulse. Then, events that occur after $t + X$ and before the next synchrotron pulse is accepted by the system and recorded by a counter.

Apart from the timing window, one can also set a CFD lower threshold to discriminate against electronic noise and an upper level to reject cosmic ray events. With these conditions set, a good beamline APD system can achieve dark count rates on the order of 1-5 $\times$ 10^{-3} s^{-1}, a limit set in part by residual particles traveling in bunches that should be empty. This background level is an important factor in setting a lower limit to sample concentrations that can be studied. The electronic windows will usually be set by a beamline scientist (*see* **Note 2**).

3 Methods

3.1 Isotope Enrichment

Natural abundance Fe contains 2.14 % ^{57}Fe. The procedures for enriching inorganic samples with ^{57}Fe (or other NRVS isotopes) will be as varied as the synthetic methods that are used (*see* **Note 3**). Enrichment of biological samples can also follow several routes, depending on the level of control over the site of interest. The most basic way to enrich a sample is to enrich the growth medium for the organism that is producing the sample. For the most common isotope, ^{57}Fe, one can dissolve elemental Fe in *aqua regia* and add to an Fe-free growth medium. Alternatively, one can purchase or prepare soluble ^{57}Fe-enriched salts, such as $^{57}FeCl_3$ or $^{57}FeSO_4$ (*see* **Note 4**).

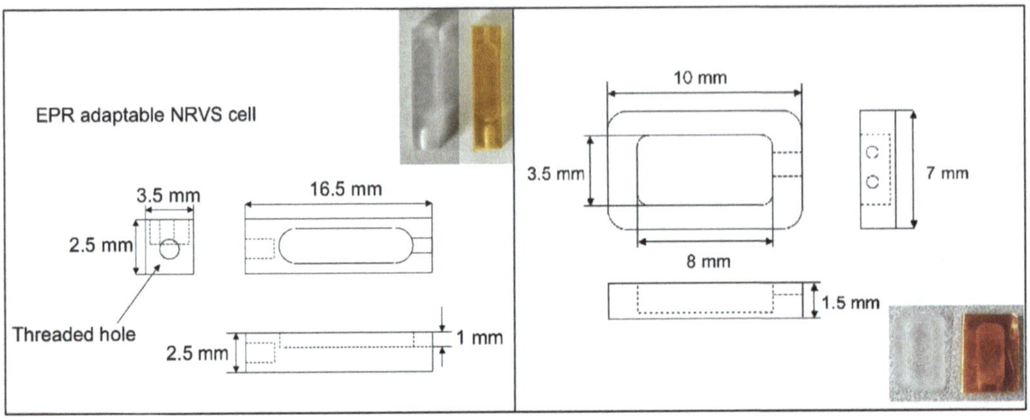

Fig. 5 Typical sample cells. *Left*: frozen solution cell. *Right*: model compound powder cell. The *inserted photos* are real NRVS sample holders with (yellow/brown) and without (grey) samples

3.2 Sample Cells

The sample path length for an NRVS experiment will depend on the energy of the nuclear resonance being used and the matrix surrounding the isotope of interest. For water, the $1/e$ path length at the ^{57}Fe resonance at 14.4 keV is almost 6 mm, so ideally one would like a ~1 cm sample to use most of the incoming photons. On the other hand, for the Fe Kα photons at 6.5 keV, the $1/e$ path length is barely more than 0.5 mm, so not much radiation from more than 1 mm below the surface will escape. Therefore, sample cells for ^{57}Fe metalloproteins are typically 10 mm in length, 3 mm in width, and 1 mm in depth, as shown in Fig. 5.

There is sometimes a benefit in using the same sample cell for NRVS and another spectroscopies, such as EXAFS, EPR, or Mössbauer. The cell for protein solutions, illustrated in Fig. 5 (left panel), is adaptable to a 5 mm in diameter EPR cavity, as well as other spectroscopies. An alternative sample cell in Fig. 5 (right panel) has a large surface area for easier loading and sealing of powder samples.

3.3 Sample Preparation, Shipping, and Mounting Procedures

For protein samples, the NRVS cell shown in Fig. 5 (left panel) will first be sealed with a window made of Kapton tape. A solution sample is then injected through the entry hole using a Hamilton syringe, and the hole is then sealed with a bit of grease. Such a loaded sample will be frozen with liquid nitrogen (LN_2) and shipped to the beamline with dry LN_2 shipper.

Powder samples can be loaded into the cell in Fig. 5 (right panel) with a spatula. Especially when working in a glove-box, an anti-static device is helpful to avoid static built-up that can make powders fly. After the cell is full, it is sealed with a piece of Kapton tape. Depending on their sensitivity, such prepared samples can be either double-sealed and shipped at room temperature or frozen and shipped with a dry shipper.

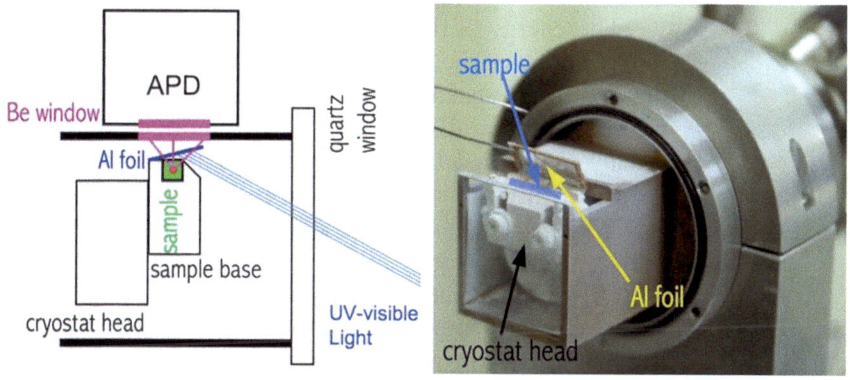

Fig. 6 *Left*: schematic illustration of a photolysis NRVS experiment. The X-ray beam is perpendicular to the page surface. *Right*: a photo showing the photolysis setup (around the sample) at SPring-8 BL09XU

Most biochemical samples are run at low temperatures, mounted on a cold-finger in the beamline cryostat. The samples are either attached using screws or using a cryogenic adhesive. In the latter procedure, the samples are briefly warmed up above the effective melting point of the cryogenic adhesive, which is ~175 K for low temperature grease and 147 K for 1-proanol [2]. Once cooled below those temperatures, the samples are held firmly in place, usually with better thermal contact than provided by mechanical screws or clamps.

3.4 NRVS Data Collection

3.4.1 Temperature Control

For most biological NRVS experiments, it will be best to run the sample at low temperatures. This will protect the samples against radiation damage, and it will also provide a cleaner NRVS measurement with sharper features. All of the beamlines provide a cryostat for these purposes (*see* **Note 5**).

3.4.2 Photolysis

In situ photolysis combined with NRVS can be a powerful tool. Early on, beautiful work was done on myoglobin-CO photolyzed at the beamline [3]. For such in situ photolysis experiments, samples can be illuminated from outside of the experimental chamber. In one design (Fig. 6), an aluminized Mylar film is placed above the sample to reflect the light directly onto the sample surface and also to protect the APD detector.

3.4.3 Monochromator Calibration and Stability

The *precision* and *accuracy* of the NRVS energy scale cannot be taken for granted. By precision we mean the scan-to-scan reproducibility of the spectral features, such as the elastic peak position, which is used as the energy zero for the vibrational spectrum. By accuracy we mean the true energies of NRVS peaks as opposed to what the computer is reporting. While the precision of most beamlines is often exquisite, accurate vibrational energies usually require a post data-collection correction.

The energy of the X-rays from the monochromators depends on the angles of the crystals and their atomic *d*-spacings, hence the precision of a spectrum depends on the reproducibility of these factors. The angular position of the monochromator crystals is usually monitored by "encoders" that provide $\pm 2.5 \times 10^{-6°}$ or 45 nrad control and the beam position is also controlled to better than a microradian, so in a modern beamline the diffraction angle is not a source of error.

However, the remaining factor, the atomic *d*-spacing, is directly sensitive to the temperature of the crystals. For Si, the coefficient of thermal expansion at room temperature is 2.56×10^{-6} K^{-1}. The consequences are that a temperature drift of 0.03 K will cause an energy shift of 9 cm^{-1}, which corresponds to the resolution of most beamlines. Something as innocuous as entering the monochromator hutch could change the crystal temperature by 0.1°, and this might require hours to re-equilibrate. So, as far as a protocol goes,

- Check the zero of the energy scale by including a fine scan over the elastic peak in each scan and monitor its reproducibility
- Do not enter the monochromator hutch unless absolutely necessary
- Plan ahead so monochromator hutch entry coincides with sample change or other downtime

Even with a reproducible elastic peak position, the energy scale for the NRVS spectrum cannot be taken for granted. As an example, in Fig. 7 (top two panels, top curves), we show the spectra for the same $(NEt_4)(FeCl_4)$ sample run at different beamlines. The Fe–Cl stretch peak varies by ~4 meV, less than one-millionth the incident energy of 14.4 keV. However, ~4 meV is also ~32 cm^{-1}, nearly a 10 % error on the vibrational energy scale. By knowing from IR spectroscopy that the correct peak is at 380 cm^{-1}, a linear correction factor of about 0.920–1.005 brings all of the spectra into very good alignment (Fig. 7 top two panels, bottom curves). So, as far as a protocol goes,

- Do not assume the calibration is constant or correct
- Bring a sample with a known set of vibrational frequencies and run this intermittently
- If possible, employ simultaneous calibration with an inline standard

The simultaneous calibration approach is illustrated in Fig. 7, bottom left panel. It employs the same tactics as the 3-ion-chamber geometry used for simultaneous EXAFS calibrations. In this approach, a secondary standard sample and an APD detector are placed downstream of the experiment chamber. An ideal secondary standard will have a strong and sharp peak, even at room

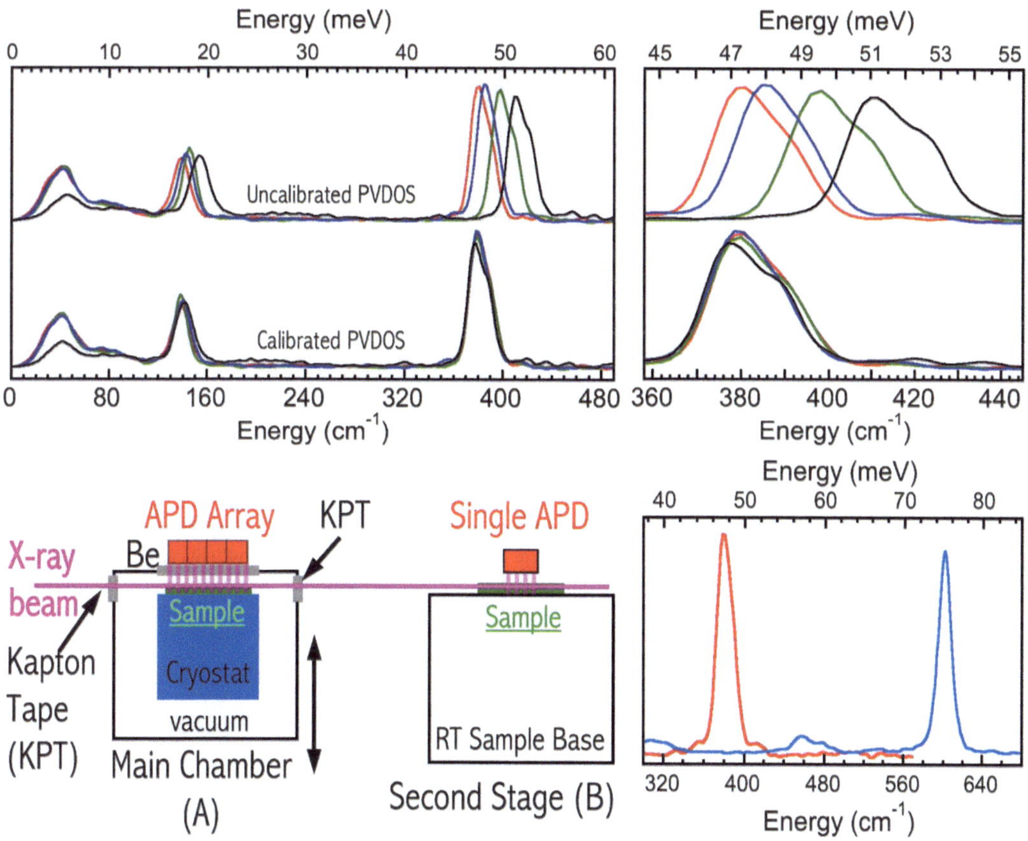

Fig. 7 *Top left*: NRVS spectra for $(NEt_4)(FeCl_4)$ before (*upper*) and after (*lower*) recalibration. Data are, respectively, from APS (*red line*), ESRF (*dark blue line*), SPring-8 (*green line*), and PETRA-III (*black line*). *Top right*: close-up of Fe–Cl stretching region before (*upper*) and after (*lower*) recalibration. *Bottom left*: schematic of dual sample operation for online calibration. *Bottom right*: overlap of spectra for $(NEt_4)(FeCl_4)$ (*red line*) and $(NH_4)_2MgFe(CN)_6$ (*light blue line*) calibration standards

temperature. A nice standard for ^{57}Fe work is $(NH_4)_2MgFe(CN)_6$, which exhibits peaks out to 602 cm^{-1} (Fig. 7, bottom right panel) and does not require a cryostat.

3.4.4 Scan Parameters and Detecting Weak Features

Unlike FT-IR or dispersive Raman spectroscopy, where the entire spectrum is recorded at the same time, NRVS data is collected one point at a time. Therefore, as in an EXAFS experiment, one has to decide a scan range, step size, and weighting scheme appropriate for the sample of interest. Also like EXAFS, experience has shown that it is best to average a dozen individual scans that take on the order of 1 h each, rather than attempting a single 12-h scan.

A typical scan will start in the anti-Stokes region, pass through the elastic line, and then continue to the highest energy for which vibrational modes can be seen. The step size is dictated by the beamline resolution. As with other spectroscopy you typically

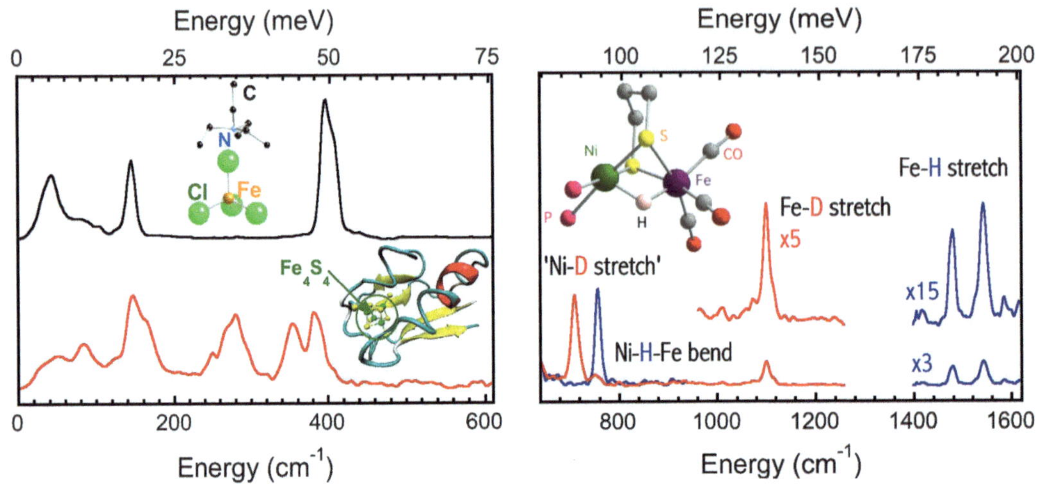

Fig. 8 *Left*: typical equal weight NRVS scans [*black line* for $(NEt_4)(FeCl_4)$ and *red line* for *P. furiosis* 4Fe-4S protein]. *Right*: NRVS spectrum for an Ni(H/D)Fe model complex that used a scan heavily weight in the Fe–H region

want at least 3–4 points per linewidth interval, hence a step size of ~0.25–0.35 eV. Although it might seem wasteful, you need a good anti-Stokes region for confirmation of the sample temperature, and you need the elastic line to check the scan-to-scan energy reproducibility. For many model compounds and simple proteins, equal counting times are appropriate, and if the overall scan range is ~100 meV, then a reasonable scan uses about 5 s per point (Fig. 8, left panel).

In some cases, the NRVS experiment is designed to detect very weak features in the overall spectrum. Examples are the search for Fe–CO bands in a protein where only 1 out of 12 Fe binds CO [4] or intrinsically weak bands such as Fe–H stretching modes (Fig. 8, right panel). In this case a weighted or even discontinuous scan approach is appropriate. Even in these cases, it is important to record and have adequate statistics in the elastic and low energy region.

3.4.5 Limits of Feasibility

The first question that comes up in most conversations about biochemical NRVS is—what concentration do you need? Of course, a spectroscopist will answer—as high as possible. Here we try to be more quantitative about what is currently feasible. Thus, in Fig. 9 (left panel), we show data from SPring-8 for typical count rates on the elastic peak as a function of ^{57}Fe concentration in protein samples—the slope corresponds to about 30 counts per second per mM ^{57}Fe concentration (*see* **Note 6**).

Although the elastic peak might yield 300 cps for a 10 mM ^{57}Fe sample, the count rates for particular vibrational modes in an NRVS spectrum are much lower (Fig. 9, right panel), in part because the inelastic intensity is spread over hundreds of meV. Depending on the symmetry of the Fe sites, the same 10 mM sample might

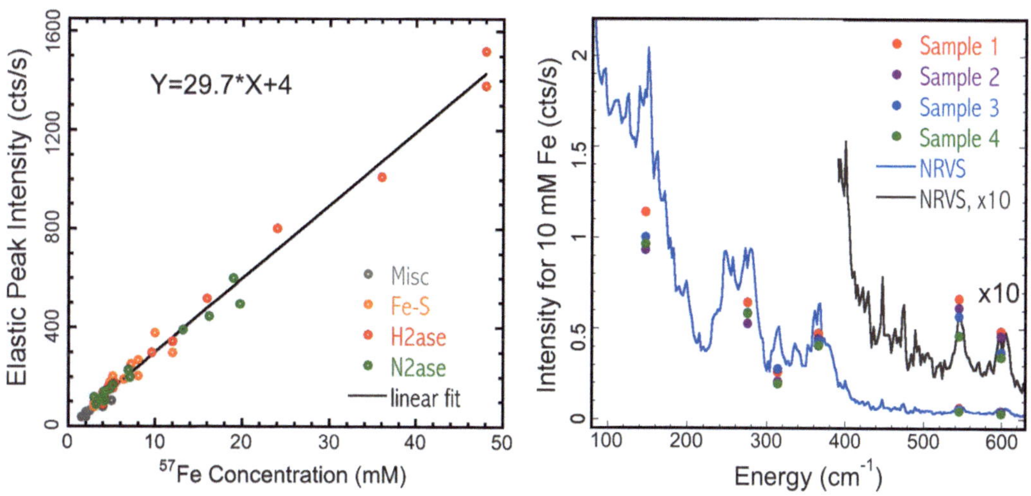

Fig. 9 *Left*: the relationship between the elastic peak intensity (cts/s) and the ^{57}Fe concentration measured at SPring-8 BL09XU. *Right*: the NiFe H_2ase NRVS spectrum, being scaled to an assumed 10 mM Fe concentration, and showing the Fe–S (*blue*) and Fe–CO (×10, *black*) vibrational features. The symbols are the intensity distribution for the corresponding peaks in four measurements (on four different NiR samples)

only yield one count per second (or less) on a vibrational peak. Since with photon statistics the signal-to-noise goes as the $\sqrt{S}$, then assuming that an S/N of 10 is desired, one only needs to count 100 photons for adequate statistics. At 1 cps that corresponds to 100 s per data point, or about 12 h (including overhead) for a 400 point spectrum.

If this calculation could be extrapolated to 1 mM Fe, a 1 mM sample would take 5 days—a long time but not impossible. Unfortunately, there is additional noise from the electronics that contributes a background rate on the order of 0.01 cps.

3.5 NRVS Data Processing

There are several software packages available for NRVS data processing, including PHOENIX [5] and DOS [6].

In general, spectral analysis was performed following the published procedure using the PHOENIX software package, [5, 7] where the observed raw NRVS spectra were calibrated to the nuclear resonant peak position (E_0), normalized to I_0, summed all the raw spectra, and converted to the ^{57}Fe partial vibrational density of states (PVDOS or DOS for abbreviation). The last step (raw NRVS → PVDOS) actually includes the following steps (Fig. 10, left panel) [5]: (1) deconvolute the summed raw spectrum to obtain a theoretical spectrum without linewidth (a → b); (2) removal of the resonant peak and convert the spectrum into a transition probability (b → c); (3) derive the single photon transition probabilities (c → d); and (4) convert to PVDOS (d → e).

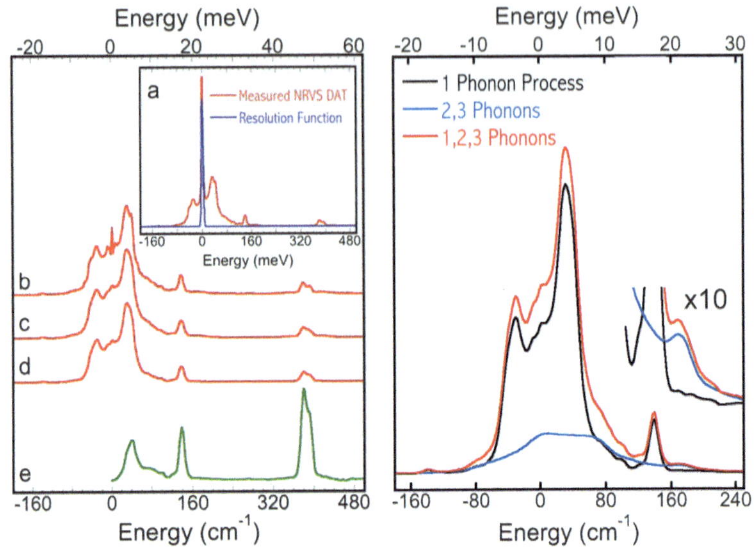

Fig. 10 *Left*: flow chart for NRVS analysis ($a \rightarrow e$), using $(NEt_4)(FeCl_4)$ as an example. *Right*: the one-phonon vs. multi-phonon spectra to show the artifact if multi-phonon effect is not removed

In the right panel of Fig. 10, it is obvious that multi-phonon effects exist in the NRVS spectra. For example, the artifact peak at 21.5 meV might be accidentally considered as a vibrational mode if multi-phonon effect is not removed during an NRVS analysis.

3.6 Interpretation

Interpreting the NRVS, once you have it, can be as simple as pointing to a peak, or as complex as a spin DFT calculation. But that is beyond the scope of this chapter.

In summary, synchrotron radiation rings, undulators, and NRVS beamlines continue to improve. As storage rings continue to improve and new beamlines become available, there are likely another 1–2 orders of magnitude improvement in store. The biochemical applications of this technique are only just beginning.

4 Notes

1. Depending on the lab, proposals are accepted one, two, or three times a year. As with other SR experiments, it is best to contact a beamline scientist first to see if your proposal is a good match for the beamline.
2. Still, it is important for the biochemical user to know that such pulse processing exists, and that it is critical to the success of the experiment. If there is an unacceptably large background count rate, the window settings are one item to look at.

3. Although highly discouraged by beamline scientists, in some cases unenriched samples have to be run, like naturally isolated molecules, or meteorites. Of course, the experiment will take 50 times longer.
4. Examples of isotope vendors are as follows: Isoflex (http://www.isoflex.com), Cambridge Isotope Labs (http://www.isotope.com), Advanced Materials Technologies (http://www.isotope-amt.com), and National Isotope Development Center (http://www.isotopes.gov/).
5. The general rule is that the lower the temperature, the better the data. So, when using an LHe flow cryostat, higher flow rates are encouraged. However, given the current price and shortages of liquid He, you can expect a shift towards use of closed-cycle cryo-coolers. In addition, as discussed above, sample loading medium and method are also critical to real sample temperatures [2].
6. This correlation would not hold for model compounds, because in such cases the signal from the elastic peak saturates and is no longer linear with concentration.

Acknowledgements

This work was funded by NIH grants GM-65440, EB-001962, and the DOE Office of Biological and Environmental Research. NRVS spectra were measured at APS Beamline 03ID, ESRF Beamline ID18, PETRA Beamline P01, and SPring-8 Beamlines 09XU, and BL19LXU. We also thank all our colleagues and collaborators for their assistances.

References

1. Sturhahn W et al (1995) Phonon density of states measured by inelastic nuclear resonant scattering. Phys Rev Lett 74:3832–3835
2. Wang H et al (2012) Real sample temperature: a critical issue in the experiments of nuclear resonant vibrational spectroscopy on biological samples. J Synchrotron Radiat 19:257–263
3. Sage JT et al (2001) Long-range reactive dynamics in myoglobin. Phys Rev Lett 86: 4966–4969
4. Kamali S et al (2013) Observation of the Fe-CN and Fe-CO vibrations in the active site of [NiFe] hydrogenase by nuclear resonance vibrational spectroscopy. Angew Chem-Int Ed 52:724–728
5. Sturhahn W (2000) CONUSS and PHOENIX: evaluation of nuclear resonant scattering data. Hyperfine Interact 125:149–172
6. Kohn VG, Chumakov AI (2000) DOS: evaluation of phonon density of states from nuclear resonant inelastic absorption. Hyperfine Interact 125:205–221
7. Smith MC et al (2005) Normal mode analysis of $[FeCl_4]^-$ and $[Fe_2S_2Cl_4]^{2-}$ via vibrational Mössbauer, resonance Raman, and FT-IR spectroscopy. Inorg Chem 44:5562–5570

Chapter 10

Study of Metalloproteins Using Continuous Wave Electron Paramagnetic Resonance (EPR)

Serge Gambarelli and Vincent Maurel

Abstract

Electron paramagnetic resonance (EPR) is an invaluable tool when studying systems with paramagnetic centers. It is a sensitive spectroscopic method, which can be used with dilute samples in aqueous buffer solutions. Here, we describe the basic procedure for recording an X-band EPR spectrum of a metalloprotein sample at low temperature. We also discuss basic optimization techniques to provide spectra with a high signal to noise ratio and minimum distortion.

Key words Electron paramagnetic resonance (EPR), Paramagnetic species, Saturation, Parameter optimization, X-band, Metalloprotein

1 Introduction

Electron paramagnetic resonance (EPR) is a method of choice when studying paramagnetic systems. The sample is placed in a magnetic field, where due to the Zeeman effect, electron spin energy levels are no longer degenerate. It is possible to induce a transition between spin energy levels using microwaves if the energy between two levels is equal to the energy of the incident photon (Fig. 1). The nomenclature used to designate the different energy levels is modelled on that used widely in the literature relating to magnetic resonance.

EPR can only detect systems with a non-null electronic spin. Ordinary molecules, with only fully filled electronic shells, exhibit no signal. Thus, EPR spectroscopy is particularly suitable for the study of some metalloprotein metallic centers in solution: the solvent, the buffer, and the protein backbone are EPR-silent, while the paramagnetic center is easily detected. Nevertheless, not all metallic ions and redox states are suitable for study by EPR. For instance,

Juan C. Fontecilla-Camps and Yvain Nicolet (eds.), *Metalloproteins: Methods and Protocols*, Methods in Molecular Biology, vol. 1122, DOI 10.1007/978-1-62703-794-5_10,

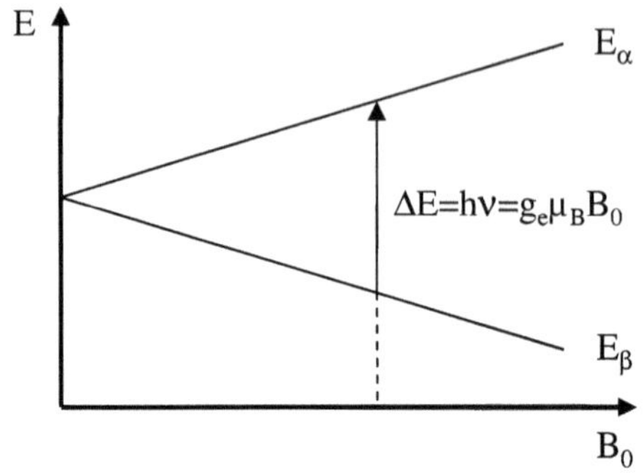

Fig. 1 Energy levels are split due to the electronic Zeeman effect in an $S = 1/2$ paramagnetic system

Co^{2+}, Fe^{3+}, Mn^{2+}, and most of Fe^{2+} are paramagnetic and EPR-active, whereas Zn^{2+}, Cu^{+} are diamagnetic and EPR-silent.

Detailed analysis of EPR spectra can provide a wealth of information on the metallic center in these systems, but the methods involved are beyond the scope of this chapter [1–6]. Rather, we focus on the basic parameters used to analyze EPR spectra.

- The first of these parameters is the g value, which indicates the position of a line in the spectrum. The g value is analogous to the chemical shift in NMR (μ_B is the Bohr magneton, ν the microwave frequency, h the Planck constant, and B_0 the magnetic field) and is described by Eq. 1.

$$h\nu = E_\alpha - E_\beta = g_e \mu_B B_0 \quad (1)$$

  The g value is essential to characterize the active center of a metalloprotein. A number of articles have been published describing how to analyze and compute g values for various families of systems [7–10].

- The second essential parameter is known as hyperfine coupling (denoted a_n in the following scheme). Hyperfine coupling is due to the interaction between the electronic spin and the nuclear spins in the molecule studied. It is analogous to J coupling in NMR. Hyperfine coupling gives clues as to which atoms surround the metallic ions (Fig. 2).

  It is important to note that most spectra are acquired from systems in a disordered state at a cryogenic temperature where no molecular tumbling is possible (i.e., frozen solution). As a consequence, EPR spectra correspond to a superposition of all the spectra for the molecules in all possible directions (powder spectrum). Powder spectra generally have a complex shape since the g value and the values for hyperfine interactions vary depending on the orientation of the molecule relative to the B_0 magnetic field applied. Computer simulations are generally

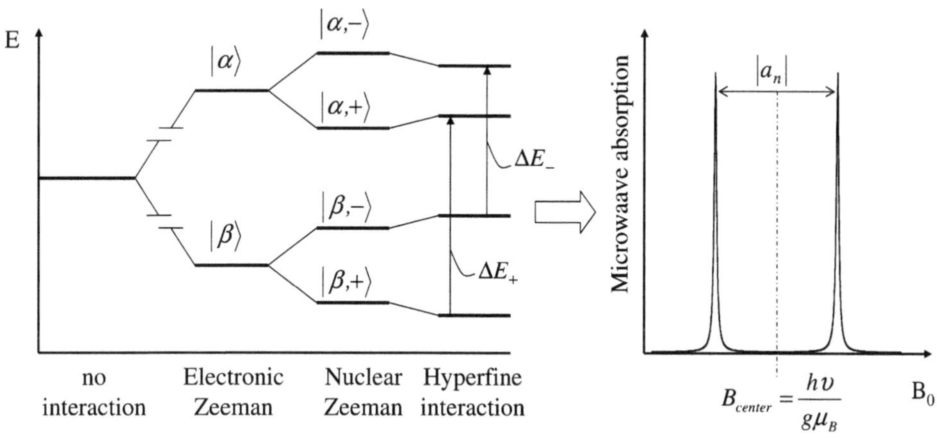

Fig. 2 Energy levels splitting and the corresponding EPR spectrum for an $S = 1/2$ paramagnetic system interacting with an $I = 1/2$ nucleus (for instance a proton). In the absence of the $I = 1/2$ nucleus, a single line would be observed in the EPR spectrum at $B_0 = h\nu/g\mu_B$. The hyperfine interaction splits this line into two lines separated by a distance of a_n

necessary to unravel all the parameters involved in EPR spectra for metalloproteins [11, 12].

In theory, all paramagnetic systems can be studied by EPR. In practice, the systems can be ranked in ascending degrees of difficulty:

- One-half systems are the easiest to study. They generally exhibit a modest anisotropy with *g* values around 2. Among the best known systems of this type are the Cu^{2+} metalloproteins and some 2Fe–2S and 4Fe–4S clusters [2, 13, 14].
- Half-integer spin systems with $S > 1/2$ tend to exhibit a greater anisotropy and stronger relaxation; *g* values can be very different from 2. The most common examples of this type of system are Fe^{3+} metalloproteins.
- Integer systems (with $S = 1, 2, 3 \ldots$). Due to strong interactions specific to paramagnetic systems with two or more single electrons (called zero-field splitting), these systems cannot generally be studied effectively with an X-band (9.5 GHz) commercial spectrometer. These systems are sometimes called "EPR-silent" paramagnetic systems. However, the term is not adequate as they can be clearly detected using a high field–high frequency EPR spectrometer [15].

2 Materials

2.1 EPR Principles

A spectrometer consists of a microwave source operating at a constant frequency (around 9–9.5 GHz for commercial X-band spectrometers), an electromagnet producing a variable magnetic field

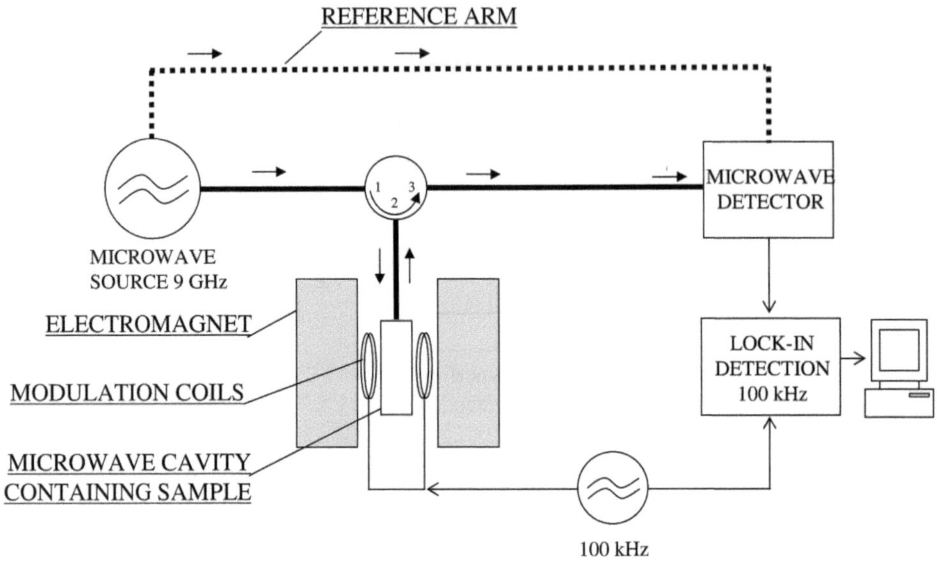

Fig. 3 Schematic view of an EPR spectrometer

(between 0 and 1.5 T for commercial spectrometers) and a cavity into which the sample in placed. This cavity is usually cooled by a cryogenic system to a temperature between 4 and 300 K. Liquid helium temperatures are often required to obtain a measurable EPR signal, thus cryogenics is an essential part of the EPR spectrometer. Sample holders are tubes of 4 mm external diameter and 3 mm internal diameter, generally made from the purest fused quartz available. Maximum sensitivity is obtained when samples are placed at the center of the cavity (Fig. 3).

During EPR acquisition, the magnetic field produced by the electromagnet is slowly changed and the microwave absorption is measured regularly. An additional oscillating magnetic field (field modulation) with a frequency of approximately 100 kHz and an amplitude of approximately 10 G is added to the main magnetic field. This modulation is connected to lock-in detection, which enhances the spectrometer's sensitivity. In the most favorable cases (EPR spectra with only a few narrow lines), this allows samples containing only ~10^{10} spins, (i.e., ~10^{-13} mol of paramagnetic centers) to be studied easily. Another consequence of lock-in detection is that the recorded EPR spectrum appears as the derivative of the spectrum that would be obtained by directly measuring microwave absorption.

The intensity of the EPR signal is closely linked to the temperature of the sample and the power of microwaves entering the microwave cavity (noted P_{mw}). These two parameters must be explored and optimized together to maximize signal intensity.

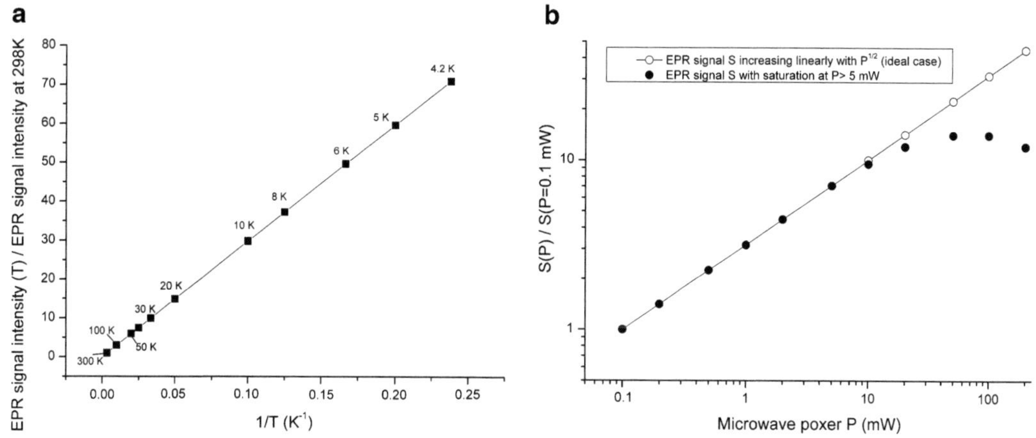

Fig. 4 (**a**) Effect of temperature on EPR signal (*left panel*). Expected increase of EPR signal intensity, as temperature drops due to Curie's Law, in the absence of saturation (ideal case). (**b**) Effect of microwave power on EPR signal intensity (*right panel*). EPR signal intensity increases with increasing microwave power at a given temperature (linear increase with $\sqrt{P_{mw}}$). *Open circles*: ideal case without saturation effect; *filled circles*: simulation of a pronounced saturation effect occurring at $P > 5$ mW. The optimal microwave power setting for recording the EPR spectrum without saturation would be 5 mW at the temperature considered here

Due to Curie's law, which applies for all paramagnetic samples, the EPR signal tends to be inversely proportional to the temperature ($1/T$ in Kelvin). Thus, lowering the temperature should produce a higher EPR signal (*see* Fig. 4a). However, Curie's law may not apply for polynuclear systems where ferromagnetic or antiferromagnetic coupling occurs.

EPR signal intensity also tends to be proportional to $\sqrt{P_{mw}}$ (*see* Fig. 4b). Thus, as a first approximation, the temperature should be set as low as possible and P_{mw} as high as possible to maximize signal.

Unfortunately, in most cases, at a temperature of 4 K and P_{mw} of 200 mW (maximum value for commercial spectrometers), EPR signal intensity is not maximal, due to saturation of the EPR transition. This saturation effect occurs in an EPR experiment if the rate of absorption of microwave photons (promoting electron spins from the β state to the α state) is too great relative to the rate of relaxation (spontaneous transitions from the α states to the β states). When this happens, the system is no longer close to the thermal equilibrium, and a distorted EPR spectrum is produced. Lowering the temperature makes relaxation less efficient, and raising the microwave power increases the rate of microwave photons, thus both tend to favor the undesired saturation effect. Because of this, for each individual sample, a compromise must be found between lowering the temperature and increasing microwave power as far as possible to maximize the signal intensity, without inducing the saturation effect.

3 Methods

3.1 Sample Preparation and Introduction

1. Fill a clean EPR tube (typically suprazil quartz, 4 mm external diameter, 3 mm internal diameter) with 100 μL of the selected metalloprotein solution in a suitable buffer (*see* **Notes 1–3**).
2. Samples should be prepared in a glovebox, once placed in the sample tube, flush the tube with helium, argon, or nitrogen gas to remove oxygen; close it with a septum. Insert a needle attached to a balloon filled with He gas (*see* **Notes 4** and **5**).
3. Freeze the solution by plunging the EPR tube slowly into liquid nitrogen in a small Dewar (*see* **Notes 6** and 7).
4. Introduce the sample into the cavity without defrosting it. Please follow the guidelines from the supplier of the cryogenic equipment.
5. Couple the cavity. A general procedure is described here (*see* **Note 8**). In the tune mode and at low power (30 dB attenuation), select the microwave frequency by setting it equal to the cavity frequency. Switch to operating mode. Set the bias to zero and increase the power slowly. The detector diode signal should be at zero even at high power. If this is not the case, adjust the cavity coupling using the iris buttons to decrease the diode signal. When this step has been successfully completed, decrease the power to 40 dB and adjust the bias and phase to optimize the detector diode signal. Increasing the power should not change the detector diode signal. If it does, adjust using the iris buttons.

3.2 Optimal Spectrometer Parameters

It is difficult to give precise rules for optimizing the parameters for an unknown metalloprotein. If the system has been previously studied, the best option is to begin with the parameters described in the literature. Nevertheless, you should still optimize the parameters for your own spectrometer since it could be different from the one used in other laboratories. A general procedure is given here. For an initial spectrum, use the following guidelines:

1. The optimal temperature for studying your sample depends strongly on the nature of the metalloprotein. In general, monometallic systems with an $S = 1/2$ spin (Cu^{2+}) should be studied at higher temperatures than polymetallic coupled systems with $S > 1/2$. Decreasing the temperature tends to increase the signal intensity and decrease the rate of relaxation, hence increasing the saturation. A temperature of 5–10 K is a good compromise for a first pass (*see* **Note 9**).
2. The optimal power also depends on the metalloprotein to be studied, and the temperature selected. The higher the power, the more intense the signal. A starting power of 10 dB is generally a good initial condition (*see* **Note 10**).

3. The amplitude modulation must be lower than the minimum linewidth for the system. With commercial systems, amplitude modulation is generally less than 20 G. A larger value increases the signal but can lead to baseline distortion or lineshape distortion. Since the linewidths in EPR spectra of metalloproteins tend to be broad (>10 G); an initial value of 10 G is a good starting point.
4. The gain is the parameter that adapts the signal amplification at the level of the receiver electronics. The highest possible value should be chosen, without saturating the receiver equipment. If the receiver saturates (receiver indicator at maximum), simply decrease the gain value (*see* **Note 11**).
5. Magnetic field range and other parameters: Unless the sample is well known, it is very important to scan the widest magnetic field range possible (in X-band spectrometers, this is usually from 0 up to 1.5 T). The acquisition time must be sufficiently slow for the electromagnet to change the magnetic field at the required speed (5 min for a 1 T scan is appropriate). The number of points must be sufficient: the distance between successive points must be less than the amplitude modulation value (for instance, with an amplitude modulation of 10 G, a point must be recorded at least every 10 G) (*see* **Notes 12** and **13**).

3.3 Parameter Optimization

1. Signal intensity is very dependent on the sample's position in the cavity. Move the sample by steps of approximately 2 mm in the cavity and record the spectrum, without changing any other parameters. The optimal sample position is that producing a maximum intensity signal (*see* **Note 14**).
2. Find the best temperature/microwave power balance. Remember that with all other parameters identical, increasing the microwave power by a factor N will increase the signal intensity by a factor $\sqrt{(N)}$ (*see* Fig. 4b). If the signal does not follow this rule, this indicates microwave saturation (*see* **Note 15**).
3. As stated above, amplitude modulation should be less than the minimum linewidth present in the system. For commercial systems, this is less than 20 G. A larger value will increase the signal but can lead to baseline distortion or lineshape distortion. Examine the spectrum and determine the feature with the smallest linewidth. If the amplitude modulation is greater than or within the same order of magnitude as this linewidth, decrease the amplitude modulation and acquire another spectrum.

If no signal is produced, it is useful to check some points before suspecting a major instrumental problem or concluding that the sample does not produce an EPR spectrum. The following are common reasons for a lack of signal:

- Inappropriate tube position in the cavity.
- Temperature problem. A very common problem causing major disruption is the presence of a steep temperature gradient between the thermocouple and the sample. If, for some reason, the temperature indicated differs from the actual temperature, and if the signal for the sample is very temperature sensitive, this can drastically reduce the signal. Check the signal intensity with a known sample at the same temperature.
- Poor sample structure. In some case freezing the sample produces crystals of ice instead of a glassy frozen solution. If this occurs, it may be useful to warm the sample to room temperature, add some (more) glycerol, and refreeze it.

3.4 Baseline Correction and Cavity Signal Subtraction

Sometimes, paramagnetic impurities are present in the EPR cavity. This can be due to accidental contamination (e.g., from a broken sample holder) or due to the materials from which the cavity is made. The best way to deal with these parasitic signals is to remove them before acquiring the spectrum for the sample of interest. In general, EPR suppliers provide a standard procedure for cavity cleaning. If this does not solve the problem or is not possible, you should proceed as follows:

1. Acquire an EPR spectrum of an EPR sample holder with a buffer solution at exactly the same position in the cavity, and with the same experimental parameters as for the sample (*see* **Note 16**).
2. Subtract the two spectra to remove the spurious signal. This can be done using standard EPR software or another scientific program.

3.5 Absolute and Relative Intensity

For several reasons (saturation phenomenon and transition probability in particular), absolute intensity (spin counting) involves a rather complex procedure and generally gives a result with a relative error of 10–20 %. If a very precise result is required, another spectroscopic method should be chosen or a specialized procedure developed. Due to the intrinsic difficulties of absolute signal quantification with paramagnetic systems where $S > 1/2$, we will treat only the $S = 1/2$ case.

1. Choose a reference sample with a known concentration in paramagnetic centers and for which the spectrum is known. Ideally, this reference sample should have almost the same spectrum as the system to be studied. Its concentration should also be comparable. For instance, a monometallic copper protein can be compared to a copper EDTA reference sample.

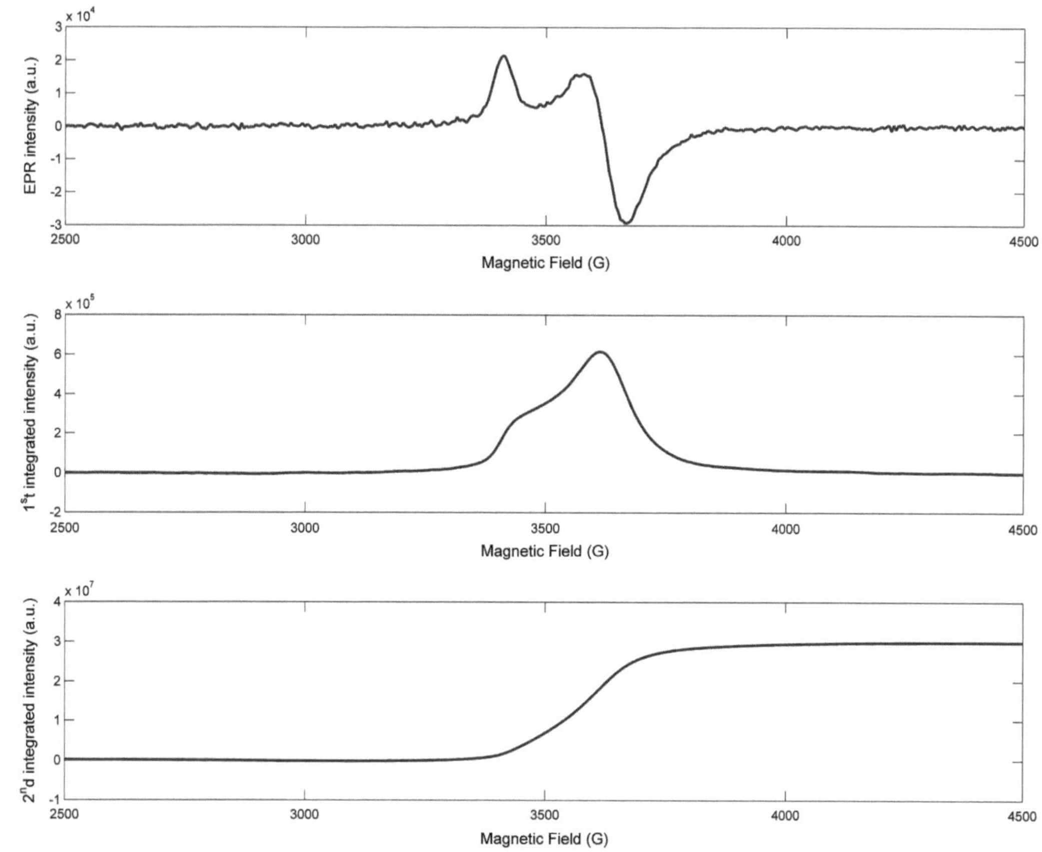

Fig. 5 Double integration of an EPR spectrum for quantitative analysis. The experimental spectrum ($T = 10$ K, 10 dB microwave power) for a metalloprotein containing an iron–sulfur cluster with $S = 1/2$ is shown (*upper panel*). Due to lock-in detection, this spectrum must be integrated to determine the spectrum corresponding to microwave absorption in the EPR experiment (*middle panel*). A second round of integration provides a value (here 3×10^7) which is proportional to the number of $S = 1/2$ spins in the sample. This value can be compared to the value obtained for a reference sample studied in the same experimental settings. Note that the spectrum is from a 4Fe–4S cluster protein [14]

For complex systems, no obvious choice may be available, and copper EDTA is a compromise.

2. Choose a set of conditions (temperature, power, modulation, gain) for which neither the reference sample nor the metalloprotein sample are saturated (gain and power).
3. Maintain a constant magnetic field range, number of points, acquisition time, and more generally all experimental parameters when acquiring spectra for the reference sample and the metalloprotein sample.
4. Perform double integration using EPR software. Due to field modulation, the EPR signal is acquired as the derivative of the signal. To obtain the signal intensity (like in an NMR experiment), one must perform a first integration (to obtain the signal) and then a second one (*see* **Note 17** and Fig. 5).

3.6 Measuring g Values

As mentioned above, the g value or g tensor, if anisotropy is present, indicates the position of a line in the spectrum. To compute it, both the field of resonance of the line and the frequency of the microwave power used to acquire the spectrum must be known. The microwave frequency used for each spectrum should therefore be recorded using a frequency meter (*see* **Note 18**).

The computation of a g value may be erroneous due to the fact that the resonance field recorded on an EPR spectrum is not the real field and exhibits some errors (*see* **Note 19**). If the g value in a spectrum must be precisely measured, it is best to compare it to a reference system with a known g value. Several systems have been used, but for biological systems, a free radical or Mn^{2+} ions are generally used. The overall procedure involves recording the metalloprotein spectrum and a reference spectrum over the same magnetic field range, acquisition time, number of points, and microwave frequency. The two samples should be placed at the same position in the spectrometer's cavity. The unknown g value can be calculated using the following formula (where g_{ref} is the g value for the reference system, and B_{ref} is the magnetic field value found in the spectrum, while B is the magnetic field value for the system studied)

$$gB = g_{\text{ref}} B_{\text{ref}}.$$

4 Notes

1. Since EPR can only detect paramagnetic species, buffers commonly used in chemical biology are generally suitable. Ions from the first two rows of the periodic table are diamagnetic, along with phosphate, amine, and organic acid derivatives.
2. The EPR tube should be clean and made of suprasyl quartz. This is important as some metallic ions, even present in very minute quantities, can produce a strong signal. For especially demanding experiments (very dilute, expensive, or difficult-to-obtain sample), it is best to use a new suprasyl X-band tube. Alternatively, tubes should be thoroughly cleaned with ethanol, washed very carefully with distilled water to remove the ethanol, and then washed again with a 35 % HCl solution. The tube should be extensively rinsed with milliQ water.
3. In the X-band cavity of most commercial spectrometer, the sensitive zone is around 1 cm long. With an internal tube diameter of 3 mm, this equates to a usable volume of around 100 μL. Increasing the volume of the sample will only lead to a marginal increase in signal intensity.
4. Simply closing an EPR tube with a septum can lead to sample destruction: at low temperatures, a small leak in the septum will

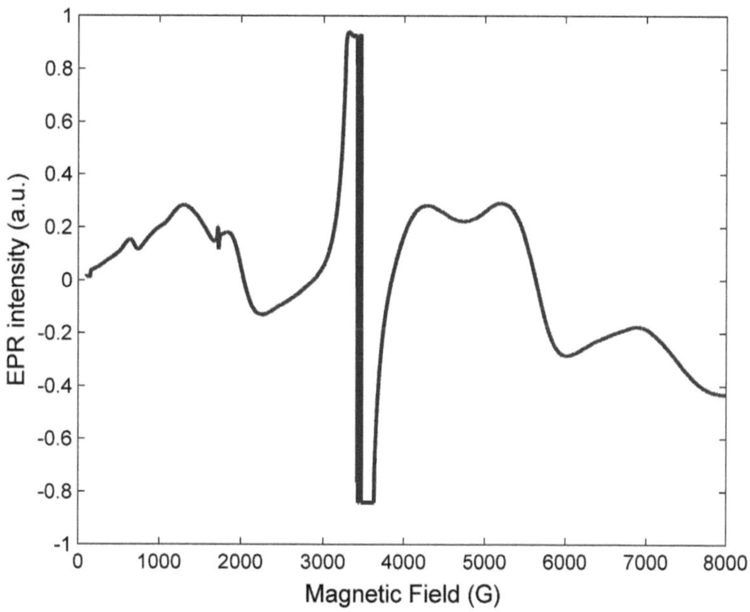

Fig. 6 EPR spectrum recorded at $T = 10$ K for a frozen sample solution containing an $S = 1/2$ spin system and solid oxygen. The strongest signal, between 3,200 and 3,800 G, is due to the $S = 1/2$ system under study. All other broad features at lower and higher magnetic fields are due to solid oxygen present in the sample

lead to accumulation of liquid or solid nitrogen/oxygen in the bottom of the tube. A sudden increase in temperature will induce a sudden surge in pressure and destruction of the sample.

5. The EPR spectrum for liquid/solid oxygen consists of very broad, featureless lines (*see* Fig. 6) mainly observed at high field, high power, and low temperature in X-band. If these lines are suspected in the spectrum, remove the sample, purge it, and repeat the measurement.
6. Be careful when freezing an EPR tube or handling it. Breakage can lead to projection of small glass pieces. Be sure to wear suitable safety goggles to protect your eyes.
7. In general, good sample structure can be obtained with the metalloprotein in its buffer solution. The ideal structure is devoid of ice crystals and is generally optically transparent. If possible, adding 20–30 % volume glycerol substantially increases the glass quality and is strongly recommended.
8. Please consult the documentation of your spectrometer for additional information and contact an experienced spectroscopist if in doubt. You can permanently damage the spectrometer during this step if inappropriate actions are performed.

9. Depending on your cryogenic equipment, the temperature on your temperature recorder is not measured exactly where you place the sample. A systematic temperature error may thus be present, with the sample temperature generally higher than the recorded temperature. The higher the He gas flow in the cryostat, the smaller the discrepancy.
10. The power available on the sample depends on the cavity coupling, the cavity quality factor, *Q*, and the cavity type. This can make comparison difficult when using a new spectrometer/ cavity, or when comparing your results to those in the literature.
11. Be careful to distinguish between saturation due to an excessive gain value and saturation due to excessive microwave power. Gain saturation is a purely electronic phenomenon in the spectrometer, distorting only intense lines. Microwave saturation is a physical phenomenon distorting the whole spectrum.
12. Acquiring a spectrum too fast may lead to problems, while a very slow acquisition time makes the spectrum sensitive to various instabilities, in particular temperature. If the temperature changes during a scan, the overall shape of the spectrum will be distorted. If necessary, it is better to acquire several spectra and add them.
13. Even with well-known samples, it is a good idea to perform the broadest scan possible to rule out spectrometer contamination, partial sample degradation, etc. It is also very important to record enough of the "baseline" at low and high fields around the spectrum. This baseline is required for many signal analysis procedures. The low field and high field baseline can each represent up to 15 % of the total field range.
14. Do not forget to adjust the coupling for each position. Except for very small samples, expect the maximum to be rather broad. Remember that changing the sample position in a cavity can lead to temperature errors if the He flow is slow.
15. Even with microwave saturation, signal intensity may increase with an increase in microwave power. However, the signal is distorted, and quantification becomes impossible. It is important to choose non-saturating conditions, even if it leads to a slightly smaller signal to noise ratio.
16. Never use an empty cavity to do these experiments. The coupling, cavity frequency is strongly changed when a tube is placed in a cavity. As described before, the most important parameters to keep constant are the power and the temperature. If for some reason, the gain is different, it is possible to correct it since it is simply a multiplicative factor.

17. In this process, special attention should be paid to baseline correction. It is a good idea to perform baseline correction and cavity signal subtraction *before* the double integration procedure.
18. In general, frequency does not change much when samples are of the same nature and size, and when the spectrometer is operated at the same temperature.
19. Increasing the acquisition time improves the magnetic field because the electromagnet and power supply have the time to settle.

References

1. Calle C et al (2006) Pulse EPR methods for studying chemical and biological samples containing transition metals. Helv Chim Acta 89:2495–2521
2. Van Doorslaer S, Cereghetti GM, Glockshuber R, Schweiger A (2001) Unraveling the Cu2+ binding sites in the C-terminal domain of the murine prion protein: a pulse EPR and ENDOR study. J Phys Chem B 105:1631–1639
3. Hagen W (2009) Biomolecular EPR spectroscopy. CRC, Boca Raton, FL
4. Schweiger A, Jeschke G (2001) Principles of pulse electron paramagnetic resonance. Oxford University Press, UK
5. Ubbink M, Worrall JAR, Canters GW, Groenen EJJ, Huber M (2002) Paramagnetic resonance of biological metal centers. Annu Rev Biophys Biomol Struct 31:393–422
6. Lyubenova S et al (2010) Multifrequency pulsed electron paramagnetic resonance on metalloproteins. Acc Chem Res 43: 181–189
7. Gambarelli S, Mouesca JM (2004) Correlation between the magnetic g tensors and the local cysteine geometries for a series of reduced [2Fe-2S*] protein clusters. A quantum chemical density functional theory and structural analysis. Inorg Chem 43:1441–1451
8. Neese F (2001) Prediction of electron paramagnetic resonance g values using coupled perturbed Hartree-Fock and Kohn-Sham theory. J Chem Phys 115:11080–11096
9. Neese F, Solomon EI (1998) Calculation of zero-field splittings, g-values, and the relativistic nephelauxetic effect in transition metal complexes. Application to high-spin ferric complexes. Inorg Chem 37:6568–6582
10. Orio M, Mouesca JM (2008) Variation of average g values and effective exchange coupling constants among [2Fe-2S] clusters: a density functional theory study of the impact of localization (trapping forces) versus delocalization (double-exchange) as competing factors. Inorg Chem 47:5394–5416
11. Stoll S, Britt RD (2009) General and efficient simulation of pulse EPR spectra. Phys Chem Chem Phys 11:6614–6625
12. Stoll S, Schweiger A (2006) EasySpin, a comprehensive software package for spectral simulation and analysis in EPR. J Magn Reson 178:42–55
13. Lubitz W, Reijerse E, van Gastel M (2007) [NiFe] and [FeFe] hydrogenases studied by advanced magnetic resonance techniques. Chem Rev 107:4331–4365
14. Perche-Letuvee P et al (2012) 4-Demethylwyosine synthase from Pyrococcus abyssi is a radical-S-adenosyl-L-methionine enzyme with an additional [4Fe-4S](+2) cluster that interacts with the pyruvate co-substrate. J Biol Chem 287:41174–41185
15. Goldberg DP et al (1997) EPR spectra from "EPR-silent" species: High-field EPR spectroscopy of manganese(III) porphyrins. J Am Chem Soc 119:8722–8723

Chapter 11

Mössbauer Spectroscopy

Martin Clémancey, Geneviève Blondin, Jean-Marc Latour, and Ricardo Garcia-Serres

Abstract

Given its ability to detect all iron centers, to identify their electronic structures, and to quantify the ratios of the different iron forms present in a sample, many researchers turn to Mössbauer spectroscopy when wanting to address structural and mechanistic questions involving iron proteins. Yet, this technique applied to biochemistry is provided by only a few dedicated teams in the world. Technical difficulties ranging from sample preparation to data analysis and interpretation make necessary the collaboration between biochemists and Mössbauer spectroscopists. This chapter will be confined to iron Mössbauer. It will focus on giving biologists and biochemists the keys to understand what essential information Mössbauer spectroscopy can yield, and how to engage in successful collaborations with spectroscopists. After introducing the basic principles of a Mössbauer experiment, we will describe first how to prepare a suitable Mössbauer sample, then how this technique is applied to the identification of different iron species inside proteins.

Key words Mössbauer spectroscopy, Isomer shift, Quadrupole splitting, Hyperfine interactions, Iron proteins, Heme, Non-heme, Diiron, Iron–sulfur clusters

1 Introduction

For an in-depth theoretical introduction, the reader can refer to one of the many reviews that are available on iron Mössbauer spectroscopy [1, 2]. A block diagram of a typical Mössbauer spectrometer is shown in Fig. 1.

The underlying physical process behind Mössbauer spectroscopy is the absorption by a nucleus of a gamma radiation emitted by another decaying nucleus of the same isotope. This absorption called "recoilless gamma absorption" happens only under very precise conditions. One of them is that both the emitter and the absorber nuclei are embedded in a solid matrix.

Protein solutions need therefore to be frozen and spectra recorded at low temperatures. Another condition is that the energy as well as the lifetime of the excited nuclear state lies within a certain range. This implies that only certain isotopes present an observable

Juan C. Fontecilla-Camps and Yvain Nicolet (eds.), *Metalloproteins: Methods and Protocols*,
Methods in Molecular Biology, vol. 1122, DOI 10.1007/978-1-62703-794-5_11,

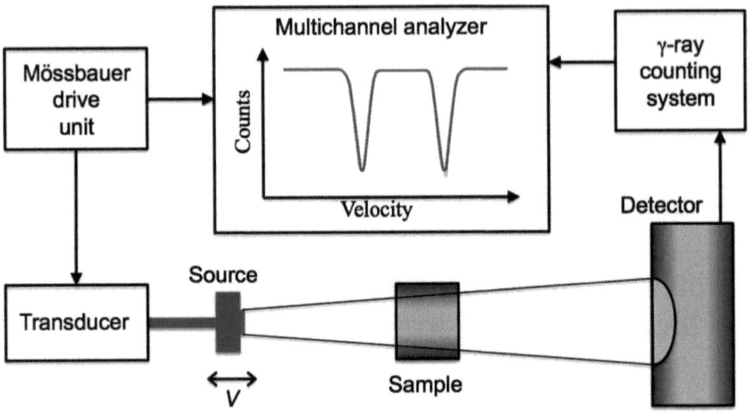

Fig. 1 Schematic block diagram of a Mössbauer spectrometer

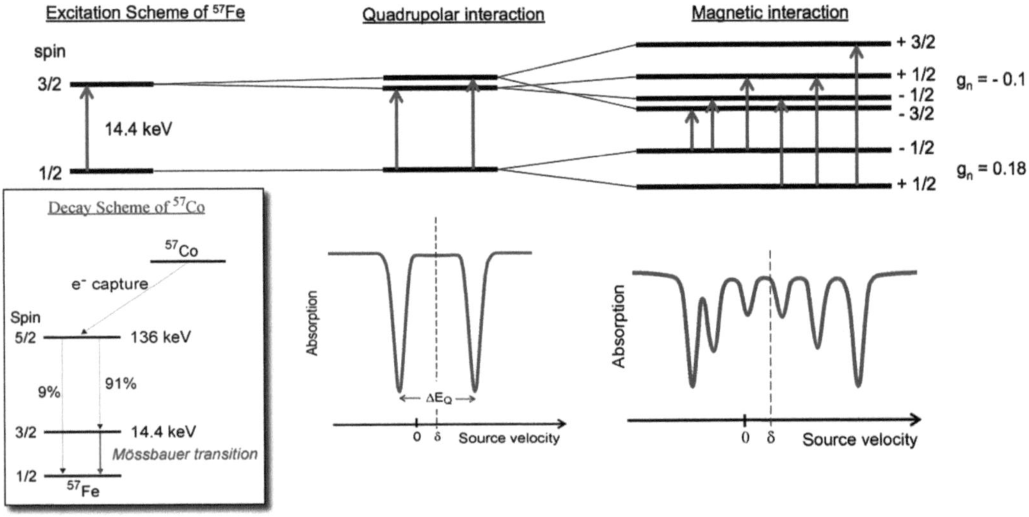

Fig. 2 The Mössbauer transitions in the presence or absence of a magnetic field at the ^{57}Fe nucleus

Mössbauer effect. Luckily for metal biologists, iron has one of the best suited isotopes for Mössbauer spectroscopy, ^{57}Fe. For this isotope, the Mössbauer transition has an energy of 14.4 keV (Fig. 2, left). This radiation is emitted by decaying ^{57}Co nuclei, which have a half-life of 271.8 days.

One of the main issues when measuring Mössbauer spectra of proteins is to obtain a good signal to noise ratio with little material. The signal intensity depends, among other factors, upon a resonance factor called "Lamb-Mössbauer factor" or "recoil-free fraction," which is a decreasing function of temperature. Samples must therefore be measured at the lowest possible temperature. For practical purposes, the temperatures of 4.2 or 78 K are often used, although a good cryostatic device will allow the operator to set any temperature he wishes. The choice of 4.2 K has the additional advantage that at this temperature, all iron nuclei within a given frozen solution have the same cross-section, implying that the cumulated absorption (area between a given signal and the baseline) is strictly proportional to the number of ^{57}Fe nuclei that account for that signal [3]. In other words, Mössbauer is a quantitative technique in terms of relative abundance of individual iron species.

If nuclear states would not be affected by electron distribution, Mössbauer spectroscopy would not have many applications. It is the interactions between the nucleus on one hand and the charge distribution and electron spin moment on the other hand that make this technique interesting to biochemists. For any process to be sensitive to the interaction between the nuclear spin and the electronic spin, this process has to be slower than the nuclear Larmor precession. Luckily, Mössbauer relaxation time (~10^{-7} s) is always at least one order of magnitude larger than the ^{57}Fe Larmor precession time, meaning that the interaction between ^{57}Fe nuclear spin and electronic spin can be detected by Mössbauer spectroscopy. On the other hand, electronic spin relaxation can be faster than ^{57}Fe Larmor precession, especially at higher temperatures. When this is the case, each iron nucleus will appear as if subjected to a single magnetic field created by the thermal average of all electronic states. This is commonly called "fast relaxation limit." In the absence of an external magnetic field, spins up and spins down have the same probability; therefore the thermal average of the electronic states will yield zero magnetic field at the nucleus. Thus, if a system is fast-relaxing in the absence of an applied magnetic field, its Mössbauer spectrum will not reflect any magnetic interaction.

However, since iron centers in proteins are isolated from each other by the protein backbone, electronic spin relaxation is essentially a spin-lattice relaxation process, which can be slowed down by decreasing the sample temperature. Therefore, at 4.2 K proteins behave usually in the "slow-relaxation" limit, where each electronic spin state is observed individually. The Mössbauer spectrum of a slow-relaxing system will be the thermally weighed superposition of spectra of the individual spin states.

2 Materials

2.1 Experimental Setup

A typical Mössbauer experiment consists in sending a gamma radiation through a sample containing an ^{57}Fe absorber, and onto a gamma detector, usually a proportional counter. The source is a cobalt isotope, ^{57}Co, dispersed in a solid matrix (for example rhodium). ^{57}Co decays to ^{57}Fe, thereby emitting several gamma frequencies.

In order to scan the energy axis, the 14.4 keV energy of the γ-rays is modulated by Doppler effect, by imprinting a back and forth motion to the source. The velocity transducer is synchronized with the gamma counting system through a multi-channel analyzer that assigns to a given velocity the corresponding number of gamma counts. The Mössbauer spectrum is then displayed in gamma counts (*see* **Note 1**) as a function of source velocity (mm/s). Figure 1 illustrates the experimental setup.

As stated above, the sample is generally placed inside a cryostat to regulate its temperature. This cryostat may be equipped with a superconducting cryomagnet that will allow the recording of Mössbauer spectra in applied magnetic fields, usually up to 8 T.

2.2 ^{57}Fe Enrichment

^{57}Fe is present in natural iron with 2.2 % abundance. With few exceptions, biological samples must be enriched in order to be sufficiently concentrated in ^{57}Fe.

When purifying a holo-protein, care has to be taken that the cells integrate as much ^{57}Fe and as little ^{56}Fe as possible. Cells are grown in minimal media such as M9. Enriched iron solution is added just before induction, so that the overexpressed protein integrates the freshly added iron salt (*see* **Note 2**). If the protein is purified without the metal, reconstitution is to be performed with a ^{57}Fe-enriched iron salt. Fe^{II} sulfate, Mohr's salt, Fe^{II} or Fe^{III} chloride, are routinely used (*see* **Note 3**).

2.3 Sample Size and Thickness

Proteins being a costly resource, we will try to place the source as close as possible to the sample, in order to minimize the sample cross section that is needed to catch all γ-rays within a given solid angle. The sample is usually inside a cryostat, while the source and the detector are outside. The cryostat tail, which the gamma beam has to cross, is usually 10–30 cm in diameter. The sample cross section has to be chosen taking in account the distance and the section of the γ-beam that we wish to collect. It is necessarily a small section, since catching γ-rays that deviate from the direction of the source motion adds uncertainty in the energy scale. As a result, the sample will have a cross-section of ~1 cm in diameter (*see* **Note 4**). Care has to be taken to cut off all γ-rays not passing through the sample. Sample cups are made to avoid any useless volume, and therefore exactly adopt the shape of the section of γ-beam that is detected (*see* Fig. 1 and **Note 5**). However, there is no absolute constraint on the sample shape, and cups can be adapted to fit different kinds of samples or experiments (*see* **Note 6**).

The ideal sample thickness has been thoroughly studied for various types of absorbers, the aim being to have a maximal signal to noise ratio in a minimal time. Too thin, the sample will not absorb enough γ-rays to yield a suitable signal. If it is too thick, too many counts will be lost through nonresonant absorption or diffusion.

In the case of proteins, the buffer solution is accountable for most of the diffusion, and therefore the ideal sample thickness depends little on the sample concentration (*see* **Note 7**).

2.4 Sample Concentration

It is the proportionality between iron concentration and percent absorption that allows us to quantitate the different iron species based on their signal integration. If the percent absorption is too high (>15 %), it is no longer proportional to ^{57}Fe concentration, and the lines become broadened. This, however, is seldom the case with biological samples, where the difficulty is rather to obtain sufficiently concentrated samples. We know from Poisson's statistics that the noise is proportional to the square root of the number of counts. Therefore, in the linear range, signal-to-noise (S/N) ratio is proportional to the square of the concentration at constant number of counts. Conversely, the time that is necessary to accumulate a spectrum with an acceptable S/N ratio increases rapidly with decreasing ^{57}Fe concentration: If a sample that is 1 mM in ^{57}Fe yields a good Mössbauer spectrum after one day of accumulation, a 0.2 mM sample will require 25 days of accumulation to achieve the same S/N ratio (*see* **Note 8**).

2.5 Velocity Range

Since the energy of the γ-rays varies linearly with source velocity, the energy range is scanned sequentially, one channel at a time. This means that the time spent scanning baseline channels is wasted in terms of data acquisition. Therefore, widening the scanned velocity window results in loss of resolution, regardless of the number of channels used. On the other hand, it is necessary to have a flat baseline on the left and on the right of the spectrum in order not to miss any absorption and to have a 100 % transmission reference. The scanned velocity range must therefore be very carefully chosen, so as to be the narrowest comprising all absorptions. If the spread of the expected spectrum is unknown, a preliminary spectrum measured in a broad velocity range often proves useful. In common iron Mössbauer setups, the widest velocity range is ± 12 mm/s, since even the broadest spectra will fall within this range.

3 Methods

3.1 Spectral Features

Absorption lines are detected as decreases in transmission. Their shape is naturally a Lorentzian with a full width at half maximum (FWHM) of ~0.22 mm/s. However, very often inhomogeneity effects lead to greater line widths and modified peak profiles. Additionally, substantial line broadening may occur as a result of weak hyperfine interactions in paramagnetic integer-spin systems.

In the absence of magnetic interactions, a Mössbauer spectrum generally consists of two equal-intensity lines centered on a velocity called "isomer shift", noted δ, and separated by a gap called "quadrupole splitting", noted ΔE_Q (Fig. 2, middle). These two parameters are characteristic of the oxidation state of the iron and of its immediate environment. The isomer shift has to be specified relative to a reference, to which is arbitrarily assigned the value $\delta = 0$. Generally this reference is taken to be the centroid of the spectrum of metallic Fe at 298 K. Very thin and pure iron foils are manufactured specifically as Mössbauer references.

In the presence of magnetic interactions, Mössbauer spectra present a six-line pattern (Fig. 2, right), which depends on many parameters, including electric field gradient, spin multiplicity, zero-field splitting, and hyperfine coupling. These parameters can be obtained by finding the theoretical spectrum that best reproduces the experimental data, via a spin Hamiltonian simulation.

3.2 Choice of Temperature and Applied Magnetic Field

The temperature of 4.2 K is usually convenient because it produces a high recoil-free fraction, and it is low enough to be in the slow relaxation mode and to isolate the ground state of most coupled systems. At this temperature, each paramagnetic state will yield a six-line spectrum via hyperfine splitting. Magnetically split spectra give information about a wider set of parameters compared to doublet spectra. However, it sometimes proves useful to raise the temperature in order to be in fast-relaxation mode and have a better determination of parameters δ and ΔE_Q as the spectrum collapses into a quadrupole doublet. To this end, the temperature of 78 K can be sufficient, but for some systems it may be necessary to go to 200 K or higher.

Regardless of the presence or absence of hyperfine splitting, the application of an external magnetic field of a few teslas is sufficient to split both the electronic and the nuclear spin states through Zeeman interaction, unveiling the full set of Mössbauer parameters. In the case of diamagnetic ($S = 0$) centers, the splitting of the Mössbauer lines corresponds exactly to the applied field. In such case, one applied magnetic field is enough to ascertain the spin ground state and to determine the electric field gradient parameters. In the case of paramagnetic centers, a wide range of applied temperatures and fields (*see* **Note 9**) may be needed to investigate the electronic configuration of the iron centers and to determine the whole set of parameters.

In the case of multi-iron centers, knowing the ground spin state tells whether the different iron sites are coupled together or not, and if the coupling is ferromagnetic or antiferromagnetic. The magnitude of the coupling interaction can sometimes be estimated by varying the applied magnetic field and the temperature.

3.3 Spectroscopic Signatures of Common Types of Iron Proteins

Many books and detailed reviews exist on Mössbauer spectroscopy of the different types of iron proteins [1, 2]. Biological samples usually contain several types of iron species, each of which has a specific Mössbauer signature. The recorded Mössbauer spectrum is a weighed superposition of all these signatures. Its analysis can therefore be difficult and usually requires the expertise of a spectroscopist. The scope of this section is to present a digest of the easily recognizable features of the most common types of iron centers, in order for nonspecialists to be able to identify the iron species that may be present in their samples. Unless otherwise stated, the Mössbauer spectra are assumed to be measured at 4.2 K in a weak magnetic field (<0.1 T) applied parallel to the γ-rays.

3.3.1 Monoiron Proteins

Proteins containing mononuclear iron centers comprise heme proteins and non-heme monoiron centers. Porphyrin ligands provide a particular environment for the iron, exerting a strong ligand field in the plane of the porphyrin. Non-heme monoiron centers are commonly found in oxygenases associated to α-ketoglutarate (α-KG) or pterin cofactors, aromatic dioxygenases, and peptidases or bound to transferrin. Their environment is mostly composed of histidines, aspartates/glutamates, and tyrosines and provides a weak ligand field. As a result, non-heme centers will usually be high spin, while low-spin configurations will be more frequent in heme proteins. Non-heme monoiron centers coordinated by cysteines will be discussed in the "FeS cluster proteins" section. Even if technically not clusters, their *S* coordination confers them some specific spectroscopic features of FeS clusters.

All oxidation states from Fe^{II} to Fe^{IV} are commonly found in the biochemical cycles of monoiron proteins.

Ferrous Centers

Ferrous centers can be low spin or high spin. The low-spin ferrous state is diamagnetic and occurs often in hemes. Its Mössbauer spectrum is a quadrupole doublet with $\delta \approx 0.23$–0.45 mm/s and $\Delta E_Q \approx 0.4$–1.3 mm/s. The high-spin ($S = 2$) state generally appears at 4.2 K as a possibly asymmetric doublet with a large line width and nuclear parameters in the range $\delta \approx 0.6$–1.3 mm/s and $\Delta E_Q \approx 1.4$–4.2 mm/s. Small variations in the quadrupole splitting may reflect the binding of cofactors or substrates to the metal. Temperature dependence of the quadrupole splitting is interpreted as indicating the occurrence of a low-lying excited state.

Ferric Centers

Ferric hemes in proteins display at 4.2 K magnetically split Mössbauer spectra. The high-spin ($S = 5/2$) state yields a recognizable six-line pattern spread over a 12 mm/s energy range (trace A in Fig. 3 right), whereas the low-spin ($S = 1/2$) state displays overlapping absorptions over an 8 mm/s energy range (trace B in Fig. 3, right). When a ferric center is antiferromagnetically coupled to a radical, it appears as a doublet with $\delta \approx 0.21$–0.30 mm/s and $\Delta E_Q \approx 1.5$–2.3 mm/s (low-spin ferric) or $\delta \approx 0.3$–0.5 mm/s and $\Delta E_Q \approx 0.3$–2.0 mm/s (high-spin ferric).

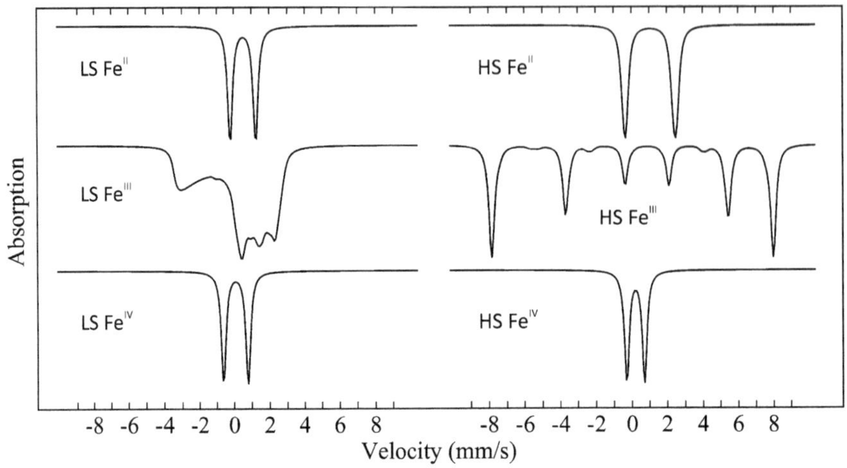

Fig. 3 Theoretical simulations of typical Mössbauer spectra of mononuclear iron centers in heme (*left*) and nonheme (*right*) iron proteins

Oxoferryl Centers

The Fe^{IV} oxidation state is commonly found in the catalytic cycle of heme-containing hydroxylase enzymes. The active hydroxylating intermediate is an oxoferryl heme coupled to a porphyrin radical, where the metal center is found to possess an $S = 1$ ground state in all systems investigated to date. The corresponding Mössbauer doublet is readily identifiable from its low isomer shift (0.03–0.17 mm/s). Several $Fe^{IV}{=}O$ intermediates have also been characterized in oxygenases and O_2-dependent halogenases thanks to rapid-mix freeze-quench techniques. All of these species are high spin $S = 2$ and appear at 4.2 K as a doublet with nuclear parameters in the range $\delta \approx 0.1$–0.3 mm/s and $\Delta E_Q \approx 0.7$–2 mm/s.

3.3.2 Diiron Proteins

We consider in this section carboxylate-bridged diiron proteins that do not contain sulfide ligands on the irons. In this group we find among others hydroxylase enzymes, ferritins, hemerythrin, diiron purple acid phosphatase, the R2 subunit of ribonucleotide reductase. The Mössbauer features of diiron centers depend strongly on the oxidation states of both iron centers (Fig. 4).

Diferrous

This is the form that binds an O_2 molecule in hydroxylases and hemerythrin. It contains two high-spin ferrous ions that yield two equal-intensity doublets in the Mössbauer spectrum. Since the environments of both irons are very similar, the two doublets most often overlap, forming a single doublet.

Mixed-Valent $Fe^{II}Fe^{III}$

It can be obtained by one-electron reduction of the diferric form or directly upon purification in some cases. It consists of two antiferromagnetically coupled ions, yielding an $S = 1/2$ ground state. Its Mössbauer spectrum is therefore a six-line pattern, spread over an 8 mm/s energy range.

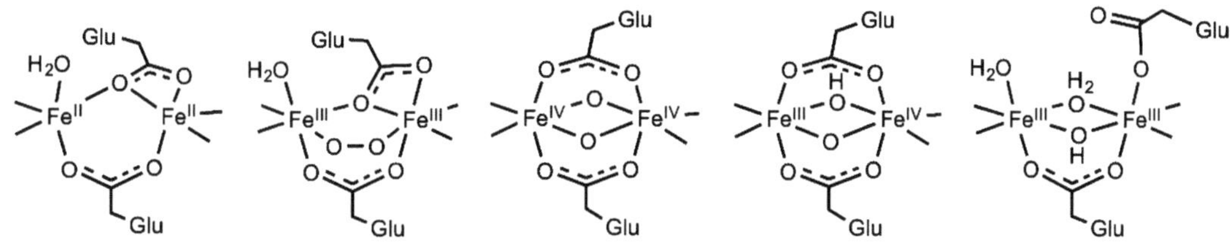

Fig. 4 Chemical structures of catalytically relevant forms of the diiron center of methane monooxygenase hydroxylase (MMOH) [4]

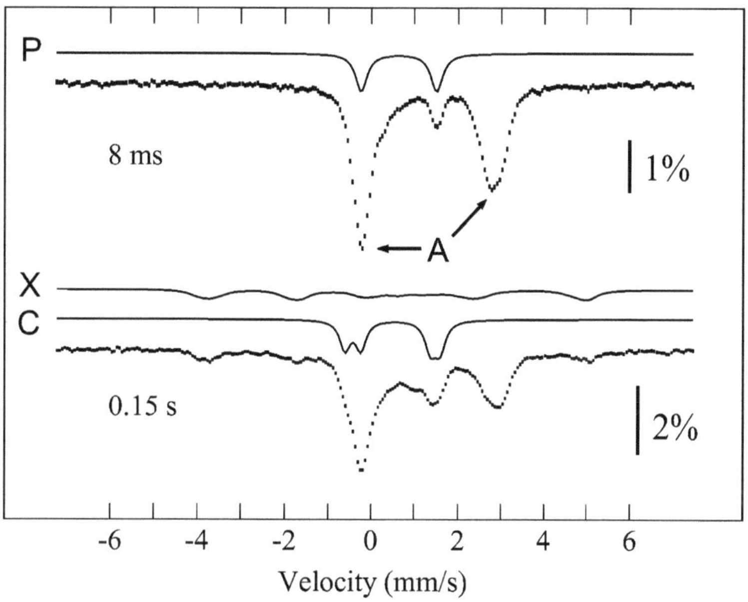

Fig. 5 Mössbauer spectra of diiron proteins. Samples from the reaction of diferrous mouse R2 (doublet A) with O_2, freeze-quenched at different reaction times. Intermediate P is apparent at 8 ms, while at 0.15 s intermediate X as well as the final product (trace C) are detected. Adapted from ref. 7

Diferric

The diferric form contains two high-spin $S = 5/2$ ferric ions, antiferromagnetically coupled into a diamagnetic ($S = 0$) ground state. The resulting Mössbauer spectrum is therefore composed of two equal-intensity doublets that may or may not be resolved, depending on the bridging ligand: a hydroxo bridge yields doublets that appear often as one, with $\delta \approx 0.50$ mm/s and $\Delta E_Q < 1.2$ mm/s, whereas a μ-oxo bridge yields two doublets with $\delta \approx 0.50$ mm/s and fairly distinct quadrupole splittings, where $\Delta E_{Q1} > 1.2$ mm/s and $\Delta E_{Q2} > 2.0$ mm/s (Fig. 5, trace C). Interconversion between these two forms is possible and pH-dependent [5].

The μ-peroxo diferric forms display equivalent iron sites that appear as a single doublet (Fig. 5, trace P) with $\delta > 0.60$ mm/s and $\Delta E_Q > 1.06$ mm/s [4] in all but one reported case [6].

Mixed-Valent $Fe^{III}Fe^{IV}$

This form corresponds to the intermediate commonly called R in the catalytic cycle of methane monooxygenase hydroxylase (MMOH) or X in that of ribonucleotide reductase R2 [7]. It is a paramagnetic $S = 1/2$ species, and its Mössbauer spectrum (Fig. 5, trace X) is similar to that of mixed-valent $Fe^{II}Fe^{III}$ centers, except for a much lower isomer shift (Table 1).

Diferryl $Fe^{IV}Fe^{IV}$

The diferryl form of diiron hydroxylases (commonly called Q) is the reactive species that is able to abstract a hydrogen atom from the substrate and create a reactive radical. It is a diamagnetic ($S = 0$) center displaying a single quadrupole doublet (from two equivalent Fe^{IV} sites) with a low isomer shift.

3.3.3 Iron–Sulfur Clusters

Iron–sulfur proteins contain one or several iron centers ligated by cysteinate ligands and clustered with sulfide ions. The most common geometries of FeS centers are mono-iron in tetrahedral cysteinate coordination (rubredoxin type), [2Fe–2S] clusters, and [4Fe–4S] clusters (Fig. 6).

These centers bear a charge, resulting from the positive charge of the iron ion, and the negative charges of the cysteinate and sulfide ligands. Their Mössbauer spectra depend both on the type of cluster and on its net charge. All these types of centers possess iron sites with tetrahedral sulfur ligation and high-spin configurations. Their quadrupolar parameters (δ and ΔE_Q) depend mostly on the oxidation state of each iron site: Fe^{2+}, Fe^{3+}, or delocalized $Fe^{2.5+}$.

Rubredoxin Type

$Fe^{II}(Cys)_4$ centers in rubredoxins (Rd) and desulforedoxins display a quadrupole doublet with $\delta \approx 0.70$ mm/s and $\Delta E_Q \approx 3.1$–3.6 mm/s (Fig. 7 left, doublet) characteristic of ferrous ions with sulfur ligation. Fe^{III} rubredoxins display a magnetic spectrum (Fig. 7 left, six-line pattern) with $\delta \approx 0.25$ mm/s and $\Delta E_Q \approx 0.5$–0.8 mm/s characteristic of ferric ions with sulfur ligation.

[2Fe–2S]

These centers are found mostly in two oxidation states: $[2Fe–2S]^{+/2+}$, although the diferrous (neutral) state has also been characterized [8, 9]. Clusters in the 0 and 2+ oxidation states consist of two antiferromagnetically coupled ferrous or ferric ions, respectively, resulting in a diamagnetic ($S = 0$) ground state. The Mössbauer spectrum of one such cluster is therefore composed of two equal intensity quadrupole doublets that often appear as a single one. On the other hand, $[2Fe–2S]^{+}$ clusters comprise an Fe^{2+}

Table 1
Reported Mössbauer parameter ranges of common types of iron centers in proteins at 4.2 K

Protein	LS Fe^{II} hemes	HS Fe^{II} hemes	LS Fe^{III} hemes	HS Fe^{III} hemes	$S = 1$ Fe^{IV} hemes	$\{FeNO\}^6$ hemes
δ (mm/s)	0.23–0.45	0.6–1.0	0.21–0.30	0.40–0.45	0.03–0.17	~0.15
ΔE_Q (mm/s)	0.4–1.3	1.4–4.2	1.5–2.3	0.4–2.0	1.0–2.5	~1.15

Protein	Fe^{II} nonheme	Fe^{III} nonheme	Fe^{IV} nonheme	$\{FeNO\}^7$ nonheme
δ (mm/s)	1.1–1.3	0.3–0.5	0.1–0.3	~0.2
ΔE_Q (mm/s)	2.5–3.5	0.3–1.5	0.7–1.2	~1.0

Protein	Fe^{II} in diiron	Fe^{III} in diiron	Fe^{III} in (μ-O^{2-}) diferric	Fe^{III} in (μ-peroxo) diferric	Fe^{IV} in diiron	$\{FeNO\}^7$ in diiron
δ (mm/s)	1.14–1.32	0.45–0.55	0.45–0.55	0.62–0.68	0.14–0.26	~0.7
ΔE_Q (mm/s)	2.4–3.1	0.5–1.9	1.5–2.3	1.0–1.9	0.5–0.6	0.5–1.2

Protein	$Fe^{II}(Cys)_4$	$Fe^{III}(Cys)_4$	Fe^{II} in [2Fe–2S]	Fe^{III} in [2Fe–2S]	Fe^{II} in [4Fe–4S]	$Fe^{2.5+}$ in [4Fe–4S], [3Fe–4S]	Fe^{III} in [4Fe–4S], [3Fe–4S]
δ (mm/s)	0.69–0.79	0.24–0.27	0.62–0.81	0.24–0.35	0.59–0.71	0.40–0.53	0.24–0.35
ΔE_Q (mm/s)	3.15–3.55	0.50–0.75	2.3–3.3	0.5–1.2	0.9–2.6	0.8–1.4	0.5–1.2

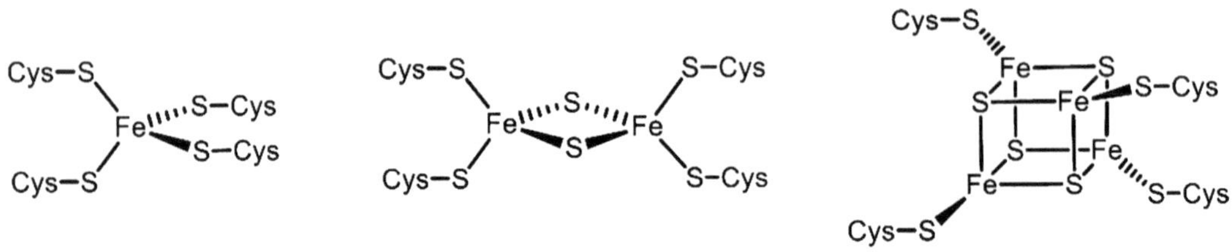

Fig. 6 Chemical structure of Rubredoxin centers, [2Fe–2S] clusters and [4Fe–4S] clusters

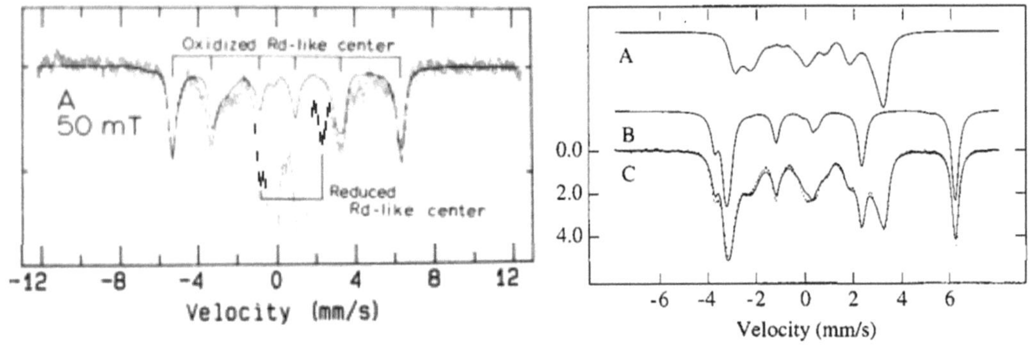

Fig. 7 *Left*: Spectrum of a sample of Rubrerythrin. Contributions from a reduced (Fe^{II}) Rd-like center (doublet) and from an oxidized (Fe^{III}) Rd-like center (sextet and *solid line*) are indicated. From ref. 15. *Right*: Spectrum of Fd from *Clostridium pasteurianum* showing an admixture of $S = 1/2$ $[2Fe–2S]^+$ clusters (trace A) and $S = 9/2$ $[2Fe–2S]^+$ clusters (trace B). The convolution of both contributions is overlaid with the experimental data (trace C). From ref. 16

antiferromagnetically coupled to an Fe^{3+}, and their Mössbauer spectrum (Fig. 7 right, trace a) results from the superposition of the two corresponding $S = 1/2$ six-line spectra. In some rare cases where the cluster is highly symmetric, double exchange dominates over the antiferromagnetic coupling [10], resulting in a delocalized iron pair with two equivalent $Fe^{2.5+}$ ions and an $S = 9/2$ ground state. In such cases the spectrum has a broader energy spread than those of $S = 1/2$ $[2Fe–2S]^+$ clusters and presents only six lines (Fig. 7, right, trace B).

Rieske proteins contain a [2Fe–2S] cluster with a particular coordination, where one iron is bound to the protein through two cysteines and the other one through two histidines. In the diferrous form, the iron with a mixed sulfur/nitrogen coordination sphere has a higher isomer shift than Cys-coordinated iron, making the site assignment straightforward [11–14]. Moreover, the His ligands are more electronegative and less covalent than the cysteines. The $[2Fe–2S]^+$ center has therefore a localized valence and contains a ferrous His-ligated site antiferromagnetically coupled to a ferric Cys-ligated site, resulting in an $S = 1/2$ ground state.

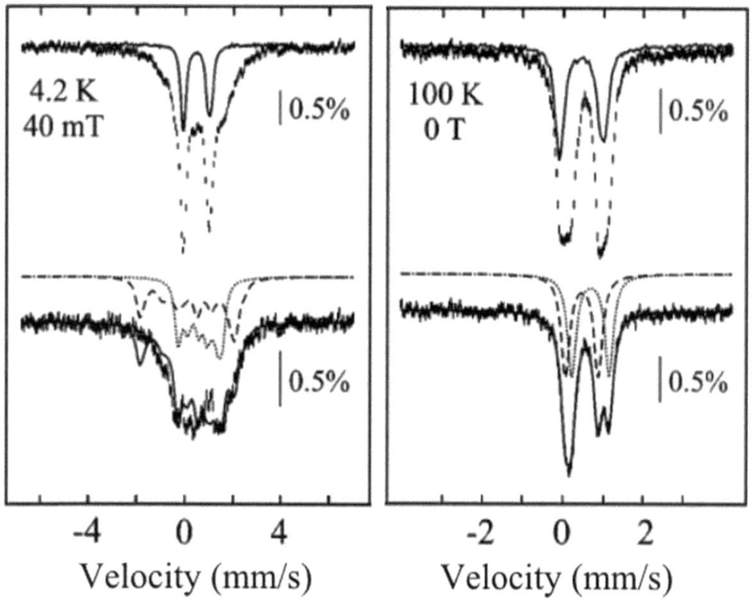

Fig. 8 *Left*: Spectra of Heliobacterial Reaction Center showing a doublet corresponding to a $[4Fe–4S]^{2+}$ cluster (*top*, *solid line*). The absorption remaining after subtraction of the doublet (*bottom*) corresponds to an $S = 3/2$ $[4Fe–4S]^{+}$ cluster. Simulations for the diferrous (*dotted line*) and mixed-valent (*dashed line*) sites are represented above the spectrum. *Right*: Same spectra recorded at high temperature. The contribution from the $[4Fe–4S]^{+}$ cluster collapses into two quadrupole doublets under the effect of fast relaxation. From ref. 21

[4Fe–4S]

These centers are found in four oxidation states: $[4Fe–4S]^{0/+/2+/3+}$, the two intermediate states being by far the most common.

Clusters in the neutral state consist of four ferrous ions and have a total spin $S = 4$. The resulting spectrum is a broad quadrupole doublet with $\delta \approx 0.7$ mm/s and $\Delta E_Q \approx 2$ mm/s that is the convolution of four equal-intensity doublets.

Clusters in the 1+ state consist of two ferrous ions coupled to a delocalized ($2Fe^{2.5+}$) pair. Depending on the coupling, the resulting ground state can be $S = 1/2$ or $S = 3/2$ [17]. The Mössbauer spectrum is therefore a superposition of two six-line spectra, one for the ferrous pair and one for the delocalized pair. The absorption range extends from −2 mm/s to +3 mm/s for $S = 1/2$ clusters and is slightly narrower for $S = 3/2$ clusters (Fig. 8). The similarity between the spectra of the two types of $[4Fe–4S]^{+}$ clusters makes it difficult to differentiate them, and calls for spectra recorded in high applied magnetic fields.

Clusters in the 2+ state consist of two delocalized ($2Fe^{2.5+}$) pairs, antiferromagnetically coupled into an $S = 0$ ground state. The resulting spectrum is the convolution of four quadrupole doublets with $\delta \approx 0.44$ mm/s and $\Delta E_Q \approx 1$ mm/s that appear as a single, sometimes asymmetric doublet.

Clusters in the 3+ state have been characterized, among others, in High-Potential Iron-sulfur Proteins (HiPIPs) [18, 19]. They consist of two ferric ions coupled to a delocalized ($2Fe^{2.5+}$) pair, yielding an $S = 1/2$ ground state [20]. The resulting Mössbauer spectrum is similar to that of $S = 1/2$ $[4Fe–4S]^+$ clusters but is globally shifted to lower velocities. Simulations of spectra measured in several applied fields are, however needed to pinpoint spectral parameters and ascertain the nature of the cluster.

[3Fe–4S]

[4Fe–4S] clusters, which are the active centers of many enzymes, can be degraded into [3Fe–4S] clusters upon loss of one iron. In some cases, this degradation is reversible and the [4Fe–4S] cluster can be reconstituted by the addition of Fe^{2+} in reducing conditions. [3Fe–4S] clusters in proteins are found in one of two oxidation states: $[3Fe–4S]^{0/+}$.

$[3Fe–4S]^+$ clusters may be cuboidal or linear. The cubic form has the shape of a [4Fe–4S] cluster where simply one iron has been removed. A rearrangement is necessary to form the linear cluster, where the middle iron is bound to four S^{2-} ligands and each of the other two irons is bound to two Cys and two S^{2-} (Fig. 9). In aconitase, conversion from cuboidal to linear takes place at high pH (>9.5) [22]. These two forms are easily distinguishable from their Mössbauer spectra. The cuboidal form has an $S = 1/2$ ground state, resulting from strong exchange coupling between all three iron sites [23], whereas in the linear form, each one of the two terminal iron sites is coupled to the central iron site, resulting in an $S = 5/2$ ground state [22]. Their Mössbauer spectra are therefore the convolution of three equal-intensity six-line spectra extending over a 6 mm/s velocity range for the $S = 1/2$ clusters (Fig. 10, left), and over a 10 mm/s velocity range for the linear $S = 5/2$ clusters (Fig. 10, middle).

$[3Fe–4S]^0$ clusters exhibit an $S = 2$ ground state that is the result of antiferromagnetic coupling between an $S = 5/2$ ferric site and a delocalized ($S = 9/2$) $Fe^{2.5+}$ pair. In zero field, the Mössbauer spectrum consists of two quadrupole doublets in a 2:1 integration ratio. The most intense doublet has $\delta \approx 0.44$ mm/s and $\Delta E_Q \approx 1$ mm/s, typical for $Fe^{2.5+}$ with S ligation, while the minor doublet ($\delta \approx 0.30$ mm/s) corresponds to a high-spin ferric site.

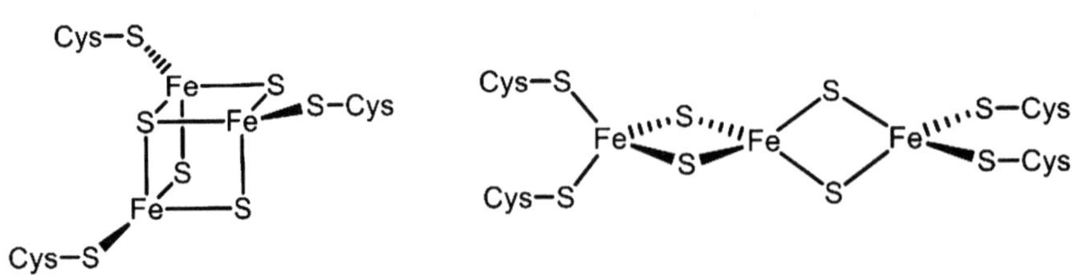

Fig. 9 Two possible chemical structures for $[3Fe–4S]^+$ clusters

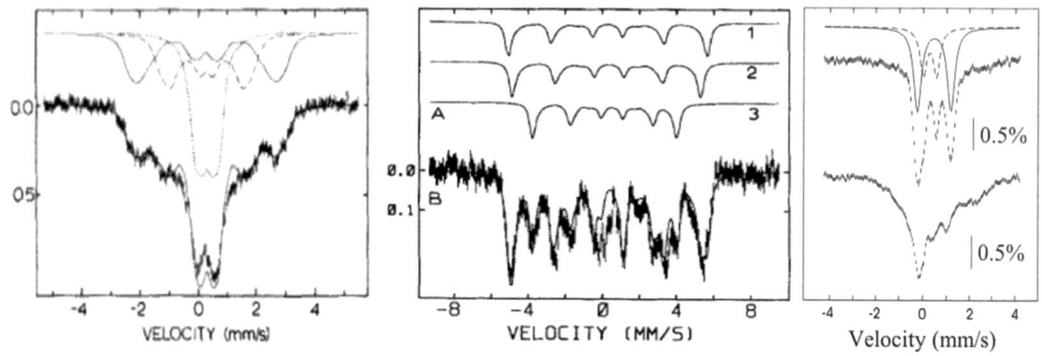

Fig. 10 *Left*: 1.5 K Mössbauer spectrum of Ferredoxin II from *D. gigas* revealing a cuboidal, $S = 1/2$, $[3Fe–4S]^+$ cluster. The *lines* above the experimental data are the simulations for each of the three ferric sites. From ref. 24. *Middle*: 1.3 K spectrum of the linear, $S = 5/2$, $[3Fe–4S]^+$ cluster of purple aconitase from beef heart. Traces 1–3 represent simulations for each one of the three ferric sites. From ref. 22. *Right*: 4.2 K spectra of the human enzyme MOCS1A measured in zero (*top*) or in a 500 G (*bottom*) magnetic field applied parallel to the γ-rays. From ref. 25

As all three irons share the same $S = 2$ ground state, both doublets are very sensitive to applied magnetic fields, broadening beyond recognition in fields of only a few hundred Gauss (Fig. 10, right).

4 Notes

1. Since the number of counts is proportional to the acquisition time, the ordinate on the vertical axis is often expressed relative to the number of counts for a channel on the baseline. This way the ordinate becomes a relative magnitude, labeled "% transmission," independent of acquisition time.
2. The iron solution is prepared by dissolving 25 mg of commercially available metallic ^{57}Fe (powder or cut foil) in 0.3 mL of 6 M hydrochloric acid. Complete dissolution can take several hours. Add 0.3 mL of the iron solution to 16 L of cell culture. Owing to the very large dilution factor, the pH does not need to be corrected.
3. A very common problem after a reconstitution is the presence in the samples of aggregates of ferric oxides or hydroxides. At very high concentrations they can be visually detected by a brownish color of an otherwise colorless sample. The formation of aggregates cannot always be avoided by working in an anaerobic glove box. Additional purifications on size-exclusion columns are recommended in order to obtain a good Mössbauer sample [26].
4. The purification of a protein can be very time and resource consuming. In extreme cases, only a fraction of the amount

needed for Mössbauer spectroscopy can be prepared. In such cases it is still possible to measure a spectrum by reducing the inner diameter of the sample cup. Recording times will be accordingly longer.

5. A typical sample cup will therefore be a Delrin® or nylon cylinder of ~1 cm diameter, with a conical receptacle carved inside. Circular grooves may be carved inside the cup to prevent the frozen solution from falling off. Proteins are mostly studied in aqueous buffers, which cause the sample to expand upon freezing, thus clamping it to the grooves. Another side effect of the sample expansion is a convex meniscus, which is not a problem, as opposed to the hole in the middle created by a concave meniscus (organic solvents).

6. When measuring organelles or whole cells, the samples are in the form of a solid paste that has to be inserted into the Mössbauer cups. Such pastes have a tendency to disaggregate when frozen, which can lead to loss of the sample. A workaround can be to disperse the paste in a small amount of water for it to form an ice block that holds, but one can also use sample cups with caps to prevent the sample from spilling.

   A special setup is used to make samples of short-lived intermediates trapped by rapidly freeze-quenching a reaction. In such cases, samples are collected in the form of finely dispersed droplets that are frozen by coming in contact with a cold organic fluid such as isopentane. The droplets are subsequently packed into a special Mössbauer cup equipped with a screw cap to hold them in place [27].

   As a general rule, the spectroscopist should have access to a workshop that allows him to prepare suitable sample cups for a variety of specific cases. Custom cups can also be made to allow for the combined use of other spectroscopies (Raman, EXAFS, etc.) on the same samples.

7. The ideal sample thickness was found to be between 5 and 8 mm high in the cup, corresponding to 0.4–0.6 mL in a ~ 10 mm inside diameter cup. Atoms with high molecular weight ($Z > 10$), including phosphorus, sulfur, chlorine, and bromine, absorb the 14.4 keV radiation and therefore also decrease the ideal sample thickness. However, the effect is barely noticeable if these atoms are present in the millimolar concentration range.

8. When assessing the required iron concentration, other factors come in play, such as the complexity of the spectrum, presence of impurities, type of iron center, type of signal expected (doublet vs. magnetically split spectrum), need of a strong applied field, activity of the source, which need to be discussed with the spectroscopist.

9. When recording a spectrum in an applied magnetic field, the angle between the magnetic field vector and the direction of the γ-rays has an influence on the relative absorption intensities within the six-line pattern. This effect can be used to identify the Mössbauer transitions, by measuring spectra in two different orientations. The angles of 0° (called parallel orientation) and 90° (perpendicular orientation) are typically used.

References

1. Gütlich P, Bill E, Trautwein A (2011) Mössbauer spectroscopy and transition metal chemistry: fundamentals and application. Springer, Berlin
2. Schunemann V, Winkler H (2000) Structure and dynamics of biomolecules studied by Mossbauer spectroscopy. Rep Prog Phys 63:263–353
3. Debrunner PG, Dwivedi A, Pederson T (1979) Recoilless fraction of iron proteins in frozen solution. J Phys Colloques 40:531–533
4. Merkx M, Kopp DA, Sazinsky MH, Blazyk JL, Muller J, Lippard SJ (2001) Dioxygen activation and methane hydroxylation by soluble methane monooxygenase: a tale of two irons and three proteins. Angew Chem Int Ed 40:2782–2807
5. Fox BG, Hendrich MP, Surerus KK, Andersson KK, Froland WA, Lipscomb JD, Munck E (1993) Mossbauer, Epr, and Endor studies of the hydroxylase and reductase components of methane monooxygenase from methylosinus-trichosporium Ob3b. J Am Chem Soc 115:3688–3701
6. Murray LJ, Garcia-Serres R, Naik S, Huynh BH, Lippard SJ (2006) Dioxygen activation at non-heme diiron centers: characterization of intermediates in a mutant form of toluene/o-xylene monooxygenase hydroxylase. J Am Chem Soc 128:7458–7459
7. Yun D, Garcia-Serres R, Chicalese BM, An YH, Huynh BH, Bollinger JM (2007) (mu-1,2-peroxo)diiron(III/III) complex as a precursor to the Diiron(III/IV) intermediate X in the assembly of the iron-radical cofactor of ribonucleotide reductase from mouse. Biochemistry 46:1925–1932
8. Leggate EJ, Bill E, Essigke T, Ullmann GM, Hirst J (2004) Formation and characterization of an all-ferrous Rieske cluster and stabilization of the [2Fe-2S](0) core by protonation. Proc Natl Acad Sci U S A 101:10913–10918
9. Yoo SJ, Meyer J, Munck E (1999) Mossbauer evidence for a diferrous [2Fe-2S] cluster in a ferredoxin from Aquifex aeolicus. J Am Chem Soc 121:10450–10451
10. Achim C, Golinelli MP, Bominaar EL, Meyer J, Munck E (1996) Mossbauer study of Cys56Ser mutant 2Fe ferredoxin from Clostridium pasteurianum: evidence for double exchange in an [Fe2S2](+) cluster. J Am Chem Soc 118:8168–8169
11. Fee JA, Findling KL, Yoshida T, Hille R, Tarr GE, Hearshen DO, Dunham WR, Day EP, Kent TA, Munck E (1984) Purification and characterization of the Rieske iron-sulfur protein from Thermus-thermophilus—evidence for a [2fe-2s] cluster having non-cysteine ligands. J Biol Chem 259:124–133
12. Pikus JD, Studts JM, Achim C, Kauffmann KE, Munck E, Steffan RJ, McClay K, Fox BG (1996) Recombinant toluene-4-monooxygenase: catalytic and Mossbauer studies of the purified diiron and Rieske components of a four-protein complex. Biochemistry 35:9106–9119
13. Schunemann V, Trautwein AX, Illerhaus J, Haehnel W (1999) Mossbauer and electron paramagnetic resonance studies of the cytochrome bf complex. Biochemistry 38:8981–8991
14. Tinberg CE, Tonzetich ZJ, Wang HX, Do LH, Yoda Y, Cramer SP, Lippard SJ (2010) Characterization of iron dinitrosyl species formed in the reaction of nitric oxide with a biological Rieske center. J Am Chem Soc 132:18168–18176
15. Legall J, Prickril BC, Moura I, Xavier AV, Moura JJG, Huynh BH (1988) Isolation and characterization of rubrerythrin, a non-heme iron protein from desulfovibrio-vulgaris that contains rubredoxin centers and a hemerythrin-like binuclear iron cluster. Biochemistry 27:1636–1642
16. Achim C, Bominaar EL, Meyer J, Peterson J, Munck E (1999) Observation and interpretation of temperature-dependent valence delocalization in the [2Fe-2S](+) cluster of a ferredoxin from Clostridium pasteurianum. J Am Chem Soc 121:3704–3714

17. Lindahl PA, Day EP, Kent TA, Ormejohnson WH, Munck E (1985) Mossbauer, electron-paramagnetic-Res, and magnetization studies of the Azotobacter-Vinelandii Fe protein—evidence for a [4fe-4s]1+ cluster with spin S = 3/2. J Biol Chem 260:1160–1173
18. Bertini I, Campos AP, Luchinat C, Teixeira M (1993) A Mossbauer investigation of oxidized Fe4s4 Hipip-Ii from Ectothiorohodospira-Halophila. J Inorg Biochem 52:227–234
19. Middleton P, Dickson DPE, Johnson CE, Rush JD (1980) Interpretation of the Mossbauer-spectra of the high-potential iron protein from chromatium. Eur J Biochem 104:289–296
20. Bominaar EL, Borshch SA, Girerd JJ (1994) Double-exchange and vibronic coupling in mixed-valence systems—electronic-structure of [Fe4s4](3+) clusters in high-potential iron protein and related models. J Am Chem Soc 116:5362–5372
21. Heinnickel M, Agalarov R, Svensen N, Krebs C, Golbeck JH (2006) Identification of FX in the heliobacterial reaction center as a [4Fe-4S] cluster with an S = (3)/(2) ground spin state. Biochemistry 45:6756–6764
22. Kennedy MC, Kent TA, Emptage M, Merkle H, Beinert H, Munck E (1984) Evidence for the formation of a linear (3Fe-4S] cluster in partially unfolded Aconitase. J Biol Chem 259:4463–4471
23. Krebs C, Henshaw TF, Cheek J, Huynh BH, Broderick JB (2000) Conversion of 3Fe-4S to 4Fe-4S clusters in native pyruvate formate-lyase activating enzyme: Mossbauer characterization and implications for mechanism. J Am Chem Soc 122:12497–12506
24. Huynh BH, Moura JJ, Moura I, Kent TA, LeGall J, Xavier AV, Munck E (1980) Evidence for a three-iron center in a ferredoxin from Desulfovibrio gigas. Mossbauer and EPR studies. J Biol Chem 255:3242–3244
25. Hanzelmann P, Hernandez HL, Menzel C, Garcia-Serres R, Huynh BH, Johnson MK, Mendel RR, Schindelin H (2004) Characterization of MOCS1A, an oxygen-sensitive iron-sulfur protein involved in human molybdenum cofactor biosynthesis. J Biol Chem 279:34721–34732
26. Lanz ND, Grove TL, Gogonea CB, Lee KH, Krebs C, Booker SJ (2012) RImN and AtsB as models for the overproduction and characterization radical SAM proteins. Methods Enzymol 516:125–152
27. Bollinger JM, Tong WH, Ravi N, Huynh BH, Edmondson DE, Stubbe J (1995) Use of rapid kinetics methods to study the assembly of the diferric-tyrosyl radical cofactor of Escherichia-Coli ribonucleotide reductase. Method Enzymol 258:278–303

Chapter 12

X-Ray Absorption Spectroscopy of Metalloproteins

Jesse Ward, Emily Ollmann, Evan Maxey, and Lydia A. Finney

Abstract

Metalloproteins are enormously important in biology. While a variety of techniques exist for studying metals in biology, X-ray absorption spectroscopy is particularly useful in that it can determine the local electronic and physical structure around the metal center, and is one of the few avenues for studying "spectroscopically silent" metal ions like Zn(II) and Cu(I) that have completely filled valence bands. While X-ray absorption near-edge structure (XANES) and extended X-ray absorption fine structure (EXAFS) are useful for studying metalloprotein structure, they suffer the limitation that the detected signal is an average of all the various metal centers in the sample, which limits its usefulness for studying metal centers in situ or in cell lysates. It would be desirable to be able to separate the various proteins in a mixture prior to performing X-ray absorption studies, so that the derived signal is from one species only. Here we describe a method for performing X-ray absorption spectroscopy on protein bands following electrophoretic separation and western blotting.

Key words X-ray absorption spectroscopy, Metalloprotein, XANES, EXAFS, Imaging, Electrophoresis, Copper, Zinc

1 Introduction

Metalloprotein complexes are enormously important in biology and medicine. Proteins can bind transition metals such as iron, copper, and zinc and perform such diverse activities as catalysis, electron transfer, structure stabilization, and oxygen transport. While a variety of techniques exist for studying metals in biology, X-ray absorption spectroscopy has proven to be particularly useful for studying the local electronic and physical structure around the metal site. This is particularly true for the so-called "spectroscopically silent" metal ions such as Zn(II) and Cu(I), which have a closed-shell d^{10} electron configuration, and are thus unable to be studied by many other bioinorganic techniques like magnetic circular dichroism, electron paramagnetic resonance, and UV/Vis absorption spectroscopy [1].

Juan C. Fontecilla-Camps and Yvain Nicolet (eds.), *Metalloproteins: Methods and Protocols*, Methods in Molecular Biology, vol. 1122, DOI 10.1007/978-1-62703-794-5_12,

X-ray absorption spectra are somewhat arbitrarily divided into two regions: the X-ray absorption near-edge structure (XANES) and extended X-ray absorption fine structure (EXAFS). The XANES spectrum is defined as everything within approximately 50 eV of the metal center's absorption edge—the sudden jump in the X-ray absorption cross section that occurs when the energy of the incident X-rays exceeds the binding energy of one of the electrons in the sample [1, 2]. Features in the XANES spectrum depend on a variety of factors [3]. The edge position yields information about the metal center's oxidation state, with a higher oxidation state requiring more energy to remove a bound electron [4]. When present, the pre-edge peak allows determination of site symmetry. For absorption across a metal center's K-edge, an intense pre-edge feature implies an active site with a high degree of p–d mixing in the valence shell (as in tetrahedral sites), whereas a weak pre-edge feature is correlated with a site with inversion symmetry (square planar or octahedral) [5]. The intensity of the most prominent absorption feature, the so-called "white line," also depends on site symmetry and the size of the ligand "cage" [6]. Finally, various features in the XANES spectrum can be ascribed to transitions to empty electronic bound states [3, 7].

The EXAFS signal is considered to be everything in the absorption spectrum higher in energy than the XANES. In contrast to the XANES features, the EXAFS signal is due exclusively to the physical structure of the metal center site [3]. Electrons can be liberated from the metal center if an incident photon has high enough energy. In principle, for a completely isolated atom the X-ray absorption cross section would decrease monotonically after the edge as the incident energy is increased. However, the presence of neighboring atoms causes a modulation of the absorption cross section. This modulation is caused by scattering of the photoelectron off the neighboring atoms, such that the final state is not merely a photoelectron wave, but a superposition of outgoing and backscattered photoelectron waves. The modulation of the absorption is determined by whether this wave undergoes constructive or destructive interference at the absorbing atom, as determined by the photoelectron energy and the site geometry. Constructive interference causes an increase in absorption cross section, while destructive interference causes a decrease [3]. The extracted EXAFS spectrum resembles a damped sine wave, with a frequency that depends on the metal–ligand distance, amplitude that depends on coordination number, and phase that depends on atom identity.

While X-ray absorption techniques provide useful information about metal centers, they are limited by the fact that the overall signal for a particular metal will include signal from every metal center type present in the sample. Thus, for a mixture of two or more metalloproteins, the X-ray absorption signal will represent a weighted average of the signals of the two pure components [1]. This limits the usefulness of X-ray absorption techniques for studying metal centers in bulk cell samples or in cell lysates, since except in a few rare cases where one type of metal center overwhelmingly predominates (e.g., heme iron in red blood cells), cells contain a mixture of a variety of different metalloproteins. While in some cases site-selective EXAFS can be used to independently study the structures of two different metal sites in a sample with a mixture of metal centers [8], in general this technique requires a high-resolution wavelength-dispersive detector and a much more concentrated sample than one can normally obtain in a biological sample. It would be desirable to combine X-ray absorption studies with separation techniques so that individual metalloproteins can be studied separately (*see* **Note 1**).

Recently, a procedure has been described that allows the separation of metalloproteins in a complex mixture by electrophoretic methods and subsequent imaging of the separated protein bands through synchrotron X-ray fluorescence [9–11]. Performing X-ray absorption studies on the separated bands is a straightforward extension of the technique. Here, we describe a procedure for setting up, collecting, and analyzing X-ray absorption data sets over separated metalloproteins.

2 Materials

Prepare all solutions using ultrapure water (with a resistance of 18 MΩ at room temperature), and analytical grade reagents. Prepare and store all reagents at room temperature unless otherwise indicated. Diligently follow all waste disposal regulations when disposing waste materials. All gel tanks, cassettes, spatulas, scrapers, etc. should be rinsed with ultrapure water and allowed to dry before use.

2.1 Native Polyacrylamide Gel Electrophoresis

1. Separating gels: 10-well 4–15 %Tris-Cl gels (Bio-Rad) were used. Store at 4 °C.
2. Running buffer: Tris-Glycine buffer (25 mM Tris, 192 mM glycine, pH 8.3) was prepared from a 10× stock obtained from Bio-Rad. Store stock at 4 °C.
3. Loading buffer: 2× Native Tris-Cl buffer (500 mM Tris, 20 % glycerol, pH 6.8) (*see* **Note 2**).
4. Prestained molecular weight marker (Precision Plus Protein Kaliedoscope, Bio-Rad).
5. Equine alcohol dehydrogenase (Sigma), bovine carbonic anhydrase (Sigma), bovine superoxide dismutase (Sigma), rabbit zinc metallothionein (Enzo Life Sciences).
6. Glycerol.

2.2 Electroblotting

1. PVDF membranes (Immobilon, Millipore—*see* **Note 3**).
2. Blotting buffer: same as running buffer above, do not add methanol.

3. Plastic tweezers.
4. One large (9″ × 13″), shallow plastic bin, and two small (5″ × 7″) shallow plastic bins.
5. Blotting Roller (Invitrogen Corporation, Carlsbad, CA).
6. Pre-cut blotting papers.
7. Methanol for activating the PVDF.
8. Bio-Rad Mini-PROTEAN wet transfer unit, including plastic sandwich, frozen ice pack, and two foam pieces.
9. Magnetic stir bar and stirrer.

2.3 X-Ray Imaging and Absorption Spectroscopy

1. Arrange in advance for use of public resources at an appropriate synchrotron facility. To use dedicated instrumentation facilitated by the X-ray Science Division of the Advanced Photon Source (APS), beamtime is awarded on a competitive basis through a general user proposal system (*see* **Note 4**).
2. Small plastic container for sample transport. In instances where one is concerned about oxidation of the metal centers, samples may also be transferred in a vacuum desiccator.

3 Methods

During the native PAGE and transfer steps, gloves should be worn, but the gel and the membranes should not be touched by hand, even with gloved hands, whenever possible. If the membrane is touched with hands or gloves, the XRF image will show fingerprints of zinc and other metals. Likewise, the membrane should not come into contact with any metal implements. Everything should be minimally manipulated with plastic, preferably Teflon, tweezers.

3.1 Native PAGE

1. Dissolve the samples, or dilute, to about 10 mg/mL stock in loading buffer. Keep cold.
2. Assemble gel for electrophoresis according to manufacturer instructions. Fill with running buffer. Load each lane with about 30 μL total protein or molecular weight marker, taking care not to overfill the lane (*see* **Note 5**).
3. Electrophorese the samples at 100 mV for about 1 h or until the lowest molecular weight marker of the standard reaches the bottom of the gel. Meanwhile, refrigerate 1 L of running buffer for the blotting step.

3.2 Transfer

1. Prepare the large shallow bin with about 200–300 mL of the refrigerated transfer buffer. In one of the small bins, activate the PVDF with methanol by dipping it for about 30 s, then equilibrate it with a small amount of transfer buffer. The PVDF will

float—you can just flip it over once or twice to make sure both sides contact liquid. Place a small amount of the transfer buffer into the third bin.

2. Crack open the gel cassette with a small plastic scraper or spatula. Using this tool, coax the gel to gently fall into the small bin of buffer.
3. In the large bin of buffer, begin to assemble the transfer cassette, which is assembled as in a standard western Blot. Place the dark side of the sandwich down into the buffer. Then layer the foam pad, blotting paper, gel, PVDF, blotting paper. Using the roller, and wearing gloves, with the sandwich completely submerged in buffer, gently roll over the sandwich to squeeze any air bubbles that might be between the PVDF and gel out (*see* **Note 6**). Keeping the sandwich submerged, add the last layer of foam and press the top down into the buffer to close the cassette. Latch it. Transfer it into the transfer unit, and add the buffer from the bin.
4. Add the ice pack, and stir bar, and then pour cold transfer buffer into the unit until it just reaches the bottom of the lip of the ice pack. With the stir bar spinning, run the transfer at 0.35 A for 1 h.
5. Following the transfer, disassemble the sandwich using plastic tweezers. Carefully, with the PVDF immersed in buffer if necessary, ensure that all pieces of gel are removed from the membrane (*see* **Note 7**). Allow the membrane to rest on blotting paper, and to completely dry.

3.3 X-Ray Absorption Spectroscopy

1. Using tweezers, place the membrane in a sealed, clean plastic dish, such as a petri dish or tissue culture plate. For samples one wishes to protect from air, place the sample dish in a vacuum desiccator and apply vacuum overnight. Close the stopcock and ship sealed.
2. Transport the membrane to the APS high-throughput imaging facility, currently at beamline 8-BM-B, or an equivalent facility (arrangement for beamtime in advance is necessary; contact the responsible beamline scientist for more information).
3. Tape the membrane to an acrylic mount provided at APS, making sure that the tape does not contact any parts of interest on the membrane, and that the membrane does not come into contact with anything dusty (to facilitate this, place a large, clean Kimwipe or sheet of blotting paper to cover your working surface). Keep covered, for example with the top of a petri dish, if work is stopped.
4. The experimental setup is as follows: Monochromatic X-rays are collimated with a 0.5 mm pinhole. An air-filled ion chamber (I0) is placed immediately after the pinhole array to measure

the incident intensity. The helium-filled sample box is placed downstream of the I0 ion chamber, at a 45° angle to the incident beam. A four-element Vortex GE detector is pointed towards the point where the beam strikes the sample, at a 90° angle to the incident beam to minimize the elastic scattering signal. The second and third air-filled ion chambers (I1 and I2, respectively) are placed downstream of the sample box, with a thin foil (approximately 10 μm thickness) of the element of interest placed between them for calibration purposes.

5. Mount the acrylic holder to the linear motion sample stages of the metalloproteomics instrument. Close the sample chamber door, and flush the sample chamber with helium for 30 min at 5 L/min. Once flushed, turn down the flow of helium to 300 mL/min.
6. Set the monochromator to a value well above the absorption edge of the element of interest (e.g., 10 keV when one is looking for zinc, which has an absorption edge at 9,659 eV).
7. Perform a 2D raster scan of a western blot, detecting the full fluorescence spectra from the sample at each point. A scan with an 80 mm × 30 mm area, 0.5 mm step sizes, and 5 s dwell times per pixel takes approximately 13.5 h to complete. Scan parameters can be scaled up or down as needed.
8. The X-ray map of the western blot can be visualized by plotting the intensities of the characteristic fluorescence lines as a function of position.
9. Identify proteins to subject to X-ray absorption studies by examination of the X-ray maps.
10. Prior to performing X-ray absorption scans, ascertain that the X-ray beam is properly calibrated. Set the monochromator to the presumptive X-ray absorption edge energy (tabulated in the literature [12]) and perform a scan between 100 eV below the edge to 100 eV above the edge, with 0.3 eV steps, and 1 s dwell time per energy point. Energy resolution or dwell time can be decreased to perform a quicker, coarser scan if it is suspected that energy calibration may be far off. Plot the calibration foil absorption spectrum as the natural logarithm of the ratio of the signals from I1 and I2 (ln(I1/I2)). By convention, the edge energy is taken to be the first inflection point of the absorption spectrum [3] (*see* **Note 8**).
11. Because the metal concentration in biological samples is relatively low compared to metal foils or pure, powdered compounds, the absorption contrast of the sample is very weak, and a plot of the natural logarithm of the ratio of the ion chamber signals before and after the sample (ln(I0/I1)) is unlikely to be informative. Instead, usually what is recorded is

the intensity of a fluorescence line, I_F, ($K\alpha$ for 3d transition metals) normalized to the incident beam intensity, I0, so that the detected signal becomes $I_F/I0$. This is fluorescence-detected X-ray absorption spectroscopy. Fluorescence intensity is directly proportional to the metal center's absorption cross section for sufficiently thin samples [13].

12. To perform a XANES scan of a metalloprotein, move the western blot so that the incident beam is striking the protein band of interest. Set up a scan between 100 eV below the edge and 150 eV above the edge and scan the incident energy in 0.3 eV steps, with 1 s dwell time per energy point. It is best to perform multiple sweeps of the incident energy and average the scans until a sufficient signal-to-noise ratio is achieved. Doing this allows one to check for the effects of beam damage on the sample (e.g., photoreduction of manganese, iron, or copper complexes) by comparing the signal derived from successive energy sweeps. Alternatively, if beam damage is not a primary concern, the dwell time per energy point may be adjusted to collect all the necessary data in a single sweep.
13. Performing an EXAFS scan requires more planning:
    (a) First, it is necessary to determine what the upper limit of the energy scan should be, as this will affect the maximum spatial resolution of the EXAFS Fourier transform. Two "shells" of atoms of similar identity around a metal center will be resolvable only if the difference between their distances, δR, is large enough to cause a "beat" in the EXAFS signal [1], which occurs if $\delta R \geq \pi/2k_{max}$. The parameter k_{max} is the maximum photoelectron wavevector k—essentially photoelectron momentum—and k is related to incident energy E by the formula $k = \sqrt[2]{2m_e(E - E0)/\hbar^2}$, where m_e is the electron mass, E0 is the edge energy, and ħ is the reduced Planck constant. Thus, if one requires the ability to resolve two shells of atoms with 0.1 Å separation, it is necessary to collect to approximately $k_{max} = 16$ Å^{-1}, or $E_{max} - E0 = 967$ eV (for example, for zinc $E_{max} \approx 10.6$ keV).
    (b) Once E_{max} has been determined, it is necessary to decide on the appropriate dwell time per energy point. The EXAFS amplitude is approximately 10 % of the absorption signal at $k = 2$ Å^{-1} and drops off as $1/k^3$, so at $k = 16$ Å^{-1}, the EXAFS signal represents 0.02 % of the total absorption signal. If N is the number of fluorescence counts at a given value of k, the signal-to-noise ratio is proportional to $N^{1/2}$. For the EXAFS amplitude to at least equal the signal-to-noise ratio at $k = 16$ Å^{-1}, one requires $N^{1/2} = 1/0.0002$, or $N = 25$ million counts. If the fluorescence intensity at that point is 10^6 counts per second, a 25 s dwell time (or 25 sweeps with 1 second dwell) is required to obtain sufficient statistics.

(c) EXAFS spectra can be collected most efficiently by dividing the scan into three regions. The first region is the pre-edge region, defined as everything below about 30 eV of the edge (e.g., everything up to 9,630 eV for zinc, where E0 = 9,659 eV). This region can be done with relatively low energy resolution (10 eV per step) and short dwell times (1 s per energy point), and is included mostly for background subtraction purposes. The second region is the edge itself, which is everything within about 30 eV of the edge (9,630–9,690 eV for zinc). This region should be done with sufficient resolution to detect all the main peaks in the edge region (about 0.3 eV resolution), but since the signal is relatively strong in this region, one can use a relatively short dwell time (1 s per energy point). Finally, the EXAFS region is everything above 30 eV above the edge. To facilitate the Fourier transform step to determine the pseudo-radial distribution function, it is helpful to express the energy in this region in terms of k, the photoelectron wavevector, using the equation $k = \sqrt[2]{2m_e(E - E0)/\hbar^2}$ as before. Typically 0.05 Å^{-1} steps are used. This effectively scales up the energy resolution from below 1 eV per step at the start of the EXAFS region ($k = 2$ Å^{-1}) to 6 eV and above towards the high end (around $k = 16$ Å^{-1}). Because the magnitude of the EXAFS oscillations decreases as $1/k^3$, the dwell time per energy point can be scaled up as k^3, so that the majority of the total scan time is spent in the high-energy side of the EXAFS region, where the signal to noise ratio is the smallest.

3.4 Data Analysis

1. A variety of software packages exist to facilitate analysis of X-ray absorption data. A partial list can be viewed here: http://xafs.org/Software. See the documentation of specific software packages for detailed instructions. The authors have used and can recommend EXAFSPAK [14], SixPACK [15], and FEFF [16] for analysis of X-ray absorption data (*see* **Note 9**). What follows is a general description of data analysis procedures. A more detailed description of the procedures and potential pitfalls of XAS data analysis can be found in the literature [1–3, 17, 18].
2. Analysis of both XANES and EXAFS data usually begins by subtracting a model function for the shape of the pre-edge region (everything under 30 eV of the edge). For transmission data (absorption expressed as ln(I0/I1)), the pre-edge is usually modeled as a straight line, while for fluorescence-detected X-ray absorption (absorption expressed as I_F/I0), the pre-edge is modeled as a Gaussian tail due to the presence of scatter

signal in the fluorescence window. Background subtraction ensures a flat baseline.

3. The baseline of the post-edge/EXAFS region (what the absorption coefficient would be for an isolated atom) is modeled using a spline function. Usually this consists of a curve divided into three or more regions, with a cubic or quartic polynomial fit in each region, and the polynomials connected smoothly between regions. Spline parameters should be adjusted such that the spline essentially matches the average X-ray absorption at higher energies and does not deviate wildly in the pre-edge region. The normalized XANES spectrum can be plotted as the background-subtracted spectrum normalized to the spline fit function.
4. XANES spectra are usually analyzed by comparing features in the unknown spectra to features in the spectra derived from model compounds of known structure, or by comparison to values in the literature. In addition, programs exist for modeling XANES spectra by calculating the electronic structure and predicted absorption spectrum through Fermi's Golden Rule [19, 20] or through low-energy scattering [21], with various degrees of success.
5. The EXAFS can be extracted by subtracting the spline fit function from the background-subtracted signal in the post-edge/EXAFS region and normalizing to the spline fit function. The energy in this region is typically expressed as k, the photoelectron wavevector, using the equation $k = \sqrt[2]{2m_e(E - E0)/\hbar^2}$ as before. Because the magnitude of the EXAFS oscillations drops off as $1/k^3$, the EXAFS signal is usually weighted by k^3 in order to emphasize the oscillations in the high-energy region.
6. The EXAFS data has the general form of a damped sine wave and is modeled by the EXAFS equation, $\chi(k) = \sum_j \frac{N_j f_j(k) e^{-2k^2\sigma_j^2}}{kR_j^2} \sin\left(2kR_j + \varphi_j(k)\right)$. $\chi(k)$ is the EXAFS signal, and the sum is taken over j "shells" (groupings of similar atoms). N_j is the number of atoms in shell j, $f_j(k)$ is the backscattering amplitude, R_j is the absorber–scatterer distance (bond length), σ_j is the Debye-Waller factor (thermal disorder), and $\varphi_j(k)$ is an atom-dependent phase function. The function is fit by making an initial guess of the atoms surrounding the absorbing atoms (which determines $\varphi_j(k)$), and N_j, R_j, and σ_j are allowed to float until the best fit is reached (*see* **Note 10**).
7. A pseudo-radial distribution function (RDF) can be computed by Fourier transforming the EXAFS spectrum. This function is considered a "pseudo" RDF because the peaks will be shifted to

lower bond length by about 0.5 Å compared to the "true" values due to the phase shift factor $\varphi_j(k)$. The presence of a significant peak lower than about 1.0 Å in the pseudo-RDF usually indicates a poor fit of the baseline function in the post-edge/EXAFS region, and the spline function should be recalculated with either a higher spline order or with more spline regions. For most metalloproteins, the presence of significant peaks higher than about 4.0 Å in the pseudo-RDF indicates significant amounts of noise in the data, usually at the high-energy end of the EXAFS spectrum, and if that is the case the pseudo-RDF should be recalculated using a shorter k range (truncate the data set to use a lower k_{max}).

8. If the protein structure is known ahead of time and one wishes simply to make structural refinements, the initial guess can be constructed from a Protein Data Bank file. In the case where the protein of interest is an unknown, usually initial guesses can be constructed by considering the most common types of structures a particular metal center occupies in metal-binding proteins. For example, zinc is most commonly found in either catalytic sites that are four-coordinate tetrahedral, bound to histidine, aspartate, glutamate, and/or water, co-catalytic sites that are four-, five-, or six-coordinate and bound through histidine, aspartate, or glutamate, or structural sites that are four-coordinate tetrahedral with mixed histidine and cysteine ligation [22] (*see* **Note 11**).
9. Goodness-of-fit is often judged by the parameter F, the sum of squares of the differences between the data and the fit. For a fit incorporating multiple shells, a reduced chi-squared statistic, F, should be computed to compensate for the improvement of fit quality that comes with performing a fit with more parameters. F can be defined as F/ν, where $\nu = N_{idp} - N_{par}$, N_{par} is the number of variable parameters in the fit, and N_{idp} is the number of independent data points [23]. For an EXAFS spectrum, N_{idp} is estimated as $2\Delta k \Delta R/\pi$, where Δk is the k range used for computing the pseudo-radial distribution function, and ΔR is the range of "useful" (non-noise) peaks in the RDF [1]. The "useful" range of peaks ΔR varies from sample to sample, but is typically around 2 Å [1].
10. Fits should be checked for physical reasonableness. N_j, R_j, and σ_j should all be positive numbers, and the bond lengths R_j should make sense given the coordination number N_j [1]. For a biological sample, values for N for the first shell will typically be between 4 and 6.

3.5 Example Data Set

As an example of how these concepts can be used in practice, we prepared a western blot of selected metalloproteins (carbonic anhydrase, Cu/Zn superoxide dismutase, metallothionein, and alcohol

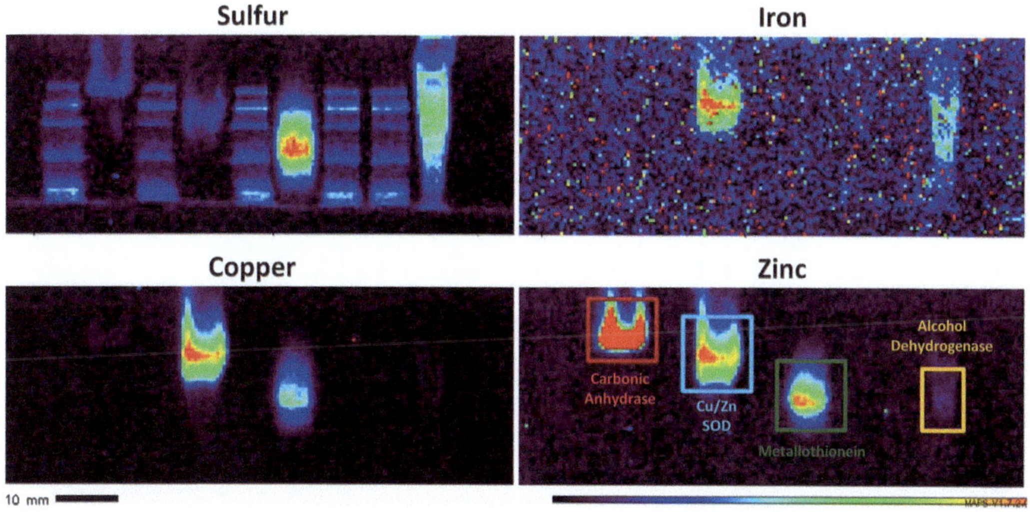

Fig. 1 2D X-ray fluorescence mapping of a western blot of selected proteins separated by native PAGE. The lanes are: 1. Marker ladder, 2. Carbonic anhydrase, 3. Marker ladder, 4. Cu/Zn SOD, 5. Marker ladder, 6. Metallothionein, 7. Marker ladder, 8. Marker ladder, 9. Alcohol dehydrogenase. Zinc is present in the carbonic anhydrase, Cu/Zn SOD, metallothionein, and alcohol dehydrogenase bands, as expected. There are significant levels of copper in the Cu/Zn SOD and metallothionein bands, and iron is present in Cu/Zn SOD and alcohol dehydrogenase bands

dehydrogenase), separated by native PAGE, and collected a 2D X-ray fluorescence image at Beamline 8-BM-B at the Advanced Photon Source. The incident energy was 10 keV, and the scan parameters were 80 mm × 30 mm area, 0.5 mm step sizes, and 5 s dwell times per pixel, taking approximately 13.5 h to complete. The result can be seen in Fig. 1. Sulfur bands are present and serve as markers for protein with significant cysteine and/or methionine content. Significant amounts of zinc can be seen in all four proteins, as expected, and significant amounts of copper were present in the Cu/Zn SOD and metallothionein bands. Detectable levels of iron were found in the Cu/Zn SOD and alcohol dehydrogenase bands.

The zinc XANES spectra were collected over the carbonic anhydrase and metallothionein bands. Incident energy was scanned between 9,600 and 9,800 eV in 0.3 eV steps and 30 s dwell time per energy point. Two energy sweeps were performed for each sample, for a total of 60 s collection time per energy point. Energy sweeps were averaged together for each band. The XANES spectra are compared in Fig. 2 following pre-processing and normalization as described above. The carbonic anhydrase spectrum shows a prominent white line at 9,667.5 eV, and a broader, secondary peak at 9,713.7 eV, and qualitatively looks similar to what has been reported in the literature previously [24]. The metallothionein scan shows more structure. Like the carbonic anhydrase scan,

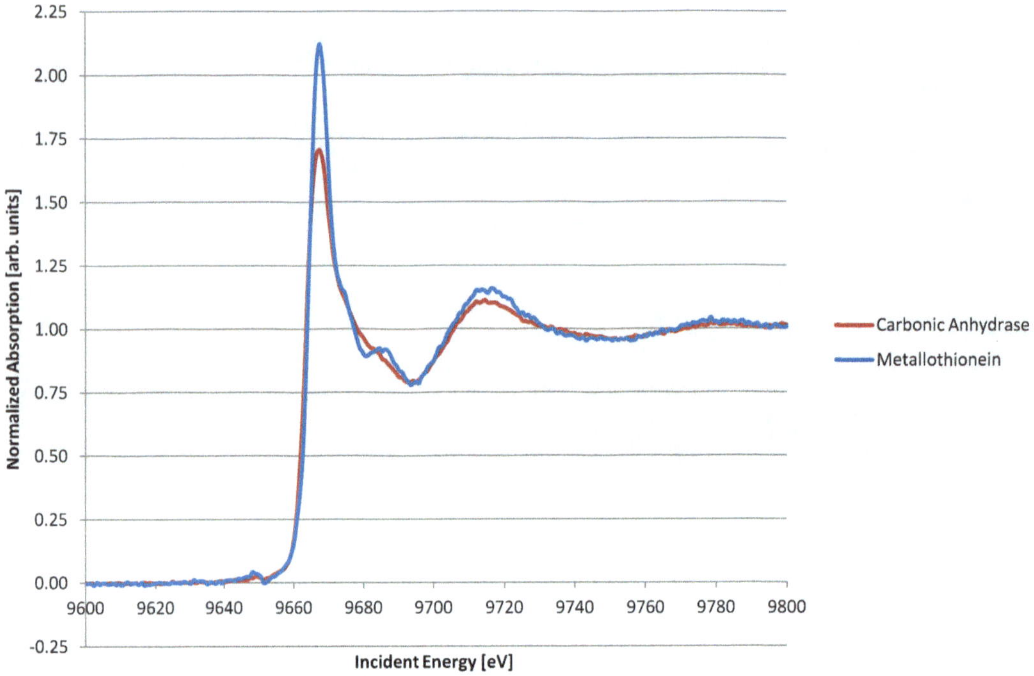

Fig. 2 Comparison of fluorescence-detected XANES scans of the Zn edge in the carbonic anhydrase and metallothionein bands. The lineshapes differ from each other and are consistent with what has been seen previously in the literature, suggesting that XANES can be used to successfully "fingerprint" these two PAGE-separated proteins

there is a white line at 9,667.5 eV and a broader peak at 9,713.7 eV; however, there is also a shoulder off the white line at 9,675.7 eV and a small peak at 9,686.1 eV. The general shape is also similar to what has been reported in the literature previously [25]. The results suggest that XANES can successfully be used for "fingerprinting" these two proteins in the western blot.

As a demonstration of EXAFS analysis on separated proteins, the copper edge on the Cu/Zn SOD band was scanned between 8,859 and 9,580 eV (out to $k = 12.6$ Å^{-1}), with energy steps of 0.35 eV and 30 s dwell time per energy point, for a single sweep. The EXAFS spectrum was extracted following preprocessing and normalization, and the pseudo-radial distribution function was calculated by Fourier transforming over a k range of 1–9 Å^{-1}. An initial guess for the best fit was constructed by inspection of a PDB file for Cu/Zn SOD [26]. The Cu active site of Cu/Zn SOD contains one water oxygen at 1.96 Å and three histidine nitrogens at an average distance of 2.02 Å. Because $k_{max} = 9$ Å^{-1} for this data set, the maximum spatial resolution for this data set is roughly $\delta R = \pi/(2 \times 9$ Å$^{-1}) = 0.17$ Å, insufficient to resolve the oxygen and the nitrogens, and since EXAFS typically has trouble distinguishing

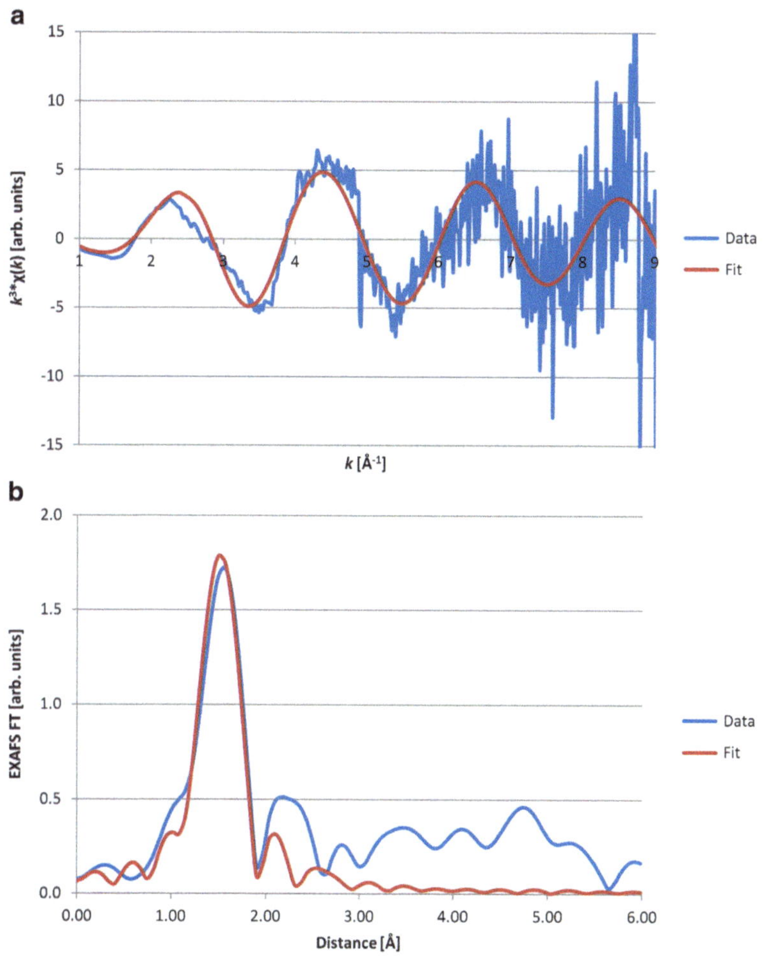

Fig. 3 (**a**) k^3-weighted copper EXAFS collected over the Cu/Zn SOD band, with the best fit of a four-nitrogen model at 1.95 Å. (**b**) Fourier transform of the Cu EXAFS in (**a**), showing the pseudo radial distribution function, along with the best fit of a four-nitrogen model

between oxygen and nitrogen ligands [1], all four ligands were grouped into a single "shell." The single shell fit was set up using a four-coordinate nitrogen shell at an average distance of 2.00 Å. Allowing R and σ to float resulted in convergence at $N = 4$, $R = 1.95$ Å, and $\sigma^2 = 4.3 \times 10^{-3}$ Å^2. Although the data could in principle support the inclusion of a second shell (for $\Delta R = 2$ Å and $\Delta k = 8$ Å^{-1}, $N_{\text{idp}} \approx 10$), increasing the number of nitrogens in the first shell or adding a second shell and allowing R and σ to float for each shell significantly increased the reduced goodness-of-fit parameter F, so these fits were rejected. The extracted EXAFS, Fourier transform, and fits can be seen in Fig. 3.

4 Notes

1. In considering this method, one should be aware that metals may be lost from proteins to some extent during electrophoresis—they are, after all, coordination complexes under an applied voltage. Given this, one must be aware of the caveat that not all metalloproteins in a given sample may be observed by this method. False negatives are possible.
2. Concentrated hydrochloric acid dissolves contaminant metals present in plastics and glass into solution. Use trace-element grade hydrochloric acid, and, to avoid contamination of stock solution, pour out small quantities for each use, rather than inserting a pipette into the stock bottle.
3. The Immobilon PVDF from Millipore is a membrane that we have found to consistently have low x-ray fluorescence background. Other PVDF we examined sometimes had quite high background—sources should be evaluated prior to use. Nitrocellulose we examined not only was less rigid, making it more difficult to keep the sample-to-detector distance constant over the entire sample, but also had higher X-ray fluorescence background.
4. Information on the APS, and the general user proposal system may be found at http://www.aps.anl.gov/Users/Scientific_Access/General_User/. Instrumentation is available for use free of charge for non-proprietary work on a competitive basis through the general user proposal system, and awarded time includes the support of staff scientists. The instrumentation currently at beamline 8-BM-B of the APS was developed by the authors for use by the user community.
5. The amount of protein loaded should be optimized for each sample type. In this case, since the proteins are commercially available, for demonstration purposes we chose a rather large quantity.
6. If a blotting roller is not available, a 10 mL plastic disposable pipette can be cut into about a 4 in. section and used instead.
7. It is important that the PVDF remain flat. If gel adheres to the PVDF following transfer, do not scrape the surface of the PVDF with the tweezers. Instead, try to gently remove the gel by rinsing with buffer.
8. In addition to running scans of a metal foil for calibration, it is also advised to scan several well-characterized model compounds for comparison. This is particularly true if one wishes to determine the oxidation state of their unknown—in this case, one should scan a variety of model compounds with various oxidation states in order to determine the exact effect

oxidation state has on the edge energy. Note that a powder sample will have to be both uniform and have the correct thickness to prevent hole effects and self-absorption effects, both of which will artificially dampen the EXAFS oscillations [18]. For a given sample cell thickness, the density of the model compound should be adjusted such that the sample thickness corresponds to 2–3 absorption lengths—the distance into the sample an X-ray beam can travel before it is attenuated by a factor of $1/e$. For a typical transition metal compound, this is accomplished by diluting the sample with a low-absorbent material such as boron nitride. It is possible to calculate exactly what mixing ratio one should use; however, a good rule of thumb in most cases is that a dilution of one part model compound and ten parts boron nitride will be sufficient for a 1 mm-thick sample. The model compound and boron nitride should be mixed together, ground with a mortar and pestle, and pressed into the sample cell to ensure uniformity. Even a "dilute" powder sample will be ~1,000× concentrated relative to a typical biological sample, so running model compounds with a decent signal-to-noise ratio should not take very long relative to the total experimental time.

9. SixPACK is a good, basic program for beginners, as it is easy to use and can produce normalized absorption data and EXAFS calculations quickly. EXAFSPAK and FEFF have a greater learning curve, but allow more control over the normalization and fitting.

10. Care should be taken when performing the fits, as EXAFS parameters are highly correlated [3]. For example, the number of atoms in a given shell, N_j, is correlated with the Debye-Waller parameter for that shell. If the initial value for N_j is set too high, the corresponding peak in the model radial distribution function will be higher than the corresponding peak in the EXAFS Fourier transform, and the fit will partially compensate by raising the Debye-Waller parameter for that shell to dampen the peak. It is easy to obtain an erroneous result despite a "good" fit quality. One should check the fit results to see if the model is physically reasonable (*see* Subheading 3.4, **step 10**). It is also helpful to reduce the number of floating parameters in the fit by calibrating some of these parameters to fits to known compounds (*see* **Note 8**). For example, many EXAFS fitting programs allow one to adjust the edge energy E0 by an offset dE0 during the fit, to partially account for the effect of the chemical environment on the edge energy. However, it has been noted that a change as low as 1 eV in the edge energy can have a noticeable effect on the fit result [23]. Thus, it is best to take scans of similar, well-characterized model compounds, collect the EXAFS, and perform fits allowing dE0 to float

while keeping the bond lengths and number of nearest neighbors constant. The resulting dE0 can be applied to the entire data set, reducing the number of fit parameters and improving the quality of the fits.

11. For a complete unknown, we usually use a "brute force" method of compiling a list of known metal environments and some simple linear combinations of the same, then performing the fits while keeping the number of scattering atoms N_j constant and allowing the bond lengths and Debye-Waller factors to float. We can then compare the goodness of fit for the best fits for each environment. This effectively reduces the number of fit parameters and ameliorates the effect of correlations between the parameters (*see* **Note 10**). Since the variety of local binding environments in biological samples is typically small, all things considered (binding through tyrosine, glutamate, aspartate, and water oxygens, histidine nitrogens, and cysteine and methionine sulfurs, with between 3-coordinate and 6-coordinate environments typical for biological samples), and since EXAFS has difficulty distinguishing scattering atoms with similar atomic numbers (so oxygen and nitrogen scatterers can effectively be grouped together as "low-Z" ligands), the total number of atom combinations needed for a "brute force" analysis is also typically small—usually fewer than 20.

Acknowledgements

This work was supported by Argonne National Laboratory, operated by UChicago Argonne, LLC, for the US Department of Energy under Contract No. DE-AC02-06CH11357.

References

1. Penner-Hahn JE (2005) Characterization of "spectroscopically quiet" metals in biology. Coord Chem Rev 249(1–2):161–177
2. Penner-Hahn JE (1999) X-ray absorption spectroscopy in coordination chemistry. Coord Chem Rev 190–192:1101–1123
3. Teo BK (1986) EXAFS: basic principles and data analysis. Springer-Verlag, Berlin
4. Knauth P, Schoonman J (2007) Nanocomposites: ionic conducting materials and structural spectroscopies. Springer, New York, NY
5. Glaser T, Hedman B, Hodgson KO, Solomon EI (2000) Ligand K-edge x-ray absorption spectroscopy: a direct probe of ligand–metal covalency. Acc Chem Res 33(12):859–868
6. Stemmler TL, Sossong TM, Goldstein JI, Ash DE, Elgren TE, Kurtz DM, Penner-Hahn JE (1997) EXAFS comparison of the dimanganese core structures of manganese catalase, arginase, and manganese-substituted ribonucleotide reductase and hemerythrin. Biochemistry 36 (32):9847–9858
7. Bianconi A, Garcia J, Benfatto M, Marcelli A, Natoli CR, Ruiz-Lopez MF (1991) Multielectron excitations in the K-edge x-ray-absorption near-edge spectra of V, Cr, and Mn $3d^0$ compounds with tetrahedral coordination. Phys Rev B 43(9):6885–6892
8. Glatzel P, Jacquamet L, Bergmann U, de Groot FMF, Cramer SP (2002) Site-selective EXAFS in mixed-valence compounds using high-resolution fluorescence detection: a study of iron in Prussian blue. Inorg Chem 41 (12):3121–3127

9. Khare T, Chishti Y, Finney LA (2012) Electrophoretic separation and detection of metalloproteins by X-ray fluorescence mapping. In: Kurien BT, Scofield RH (eds) Protein electrophoresis methods and protocols, vol 869, Methods in molecular biology. Humana, Clifton, NJ, pp 533–542
10. Finney L, Chishti Y, Khare T, Giometti C, Levina A, Lay PA, Vogt S (2010) Imaging metals in proteins by combining electrophoresis with rapid X-ray fluorescence mapping. ACS Chem Biol 5(6):577–587
11. Zahler NH (2010) Gel electrophoresis and X-ray fluorescence: a powerful combination for the analysis of protein metal binding. ACS Chem Biol 5(6):541–543
12. Thompson A, Attwood D, Gullikson E, Howells M, Kim K-J, Kirz J, Kortright J, Lindau I, Pianetta P, Robinson A, Scofield J, Underwood J, Vaughan D, Williams G, Winick H (2001) X-ray data booklet. Lawrence Berkeley National Laboratory, Berkeley, CA
13. Goulon J, Goulon-Ginet C, Cortes R, Dubois JM (1982) On experimental attenuation factors of the amplitude of the EXAFS oscillations in absorption, reflectivity and luminescence measurements. J Phys France 43(3):539–548
14. George GN, Pickering IJ (1993) EXAFSPAK: a suite of computer programs for analysis of X-ray absorption data. Stanford Synchrotron Radiaiton Lightsource, Stanford, CA
15. Webb SM (2005) SIXPack: a graphical user interface for XAS analysis using IFEFFIT. Phys Scripta 115:1011–1014
16. Newville M (2001) EXAFS analysis using FEFF and FEFFIT. J Synchrotron Radiat 8(2):96–100
17. Scott RA (1985) Measurement of metal-ligand distances by EXAFS. In: Hirs CHW, Timasheff SN (eds) Methods in enzymology, vol 117. Academic, Waltham, MA, pp 414–459
18. Bunker G (2010) Introduction to XAFS: a practical guide to X-ray absorption fine structure spectroscopy. Cambridge University Press, New York, NY
19. Rehr JJ, Ankudinov AL (2003) New developments in the theory and interpretation of X-ray spectra based on fast parallel calculations. J Synchrotron Radiat 10(1):43–45
20. Rehr JJ, Ankudinov AL (2003) Time-dependent density functional theory calculations of X-ray absorption. Int J Quant Chem 95(4–5):487–492
21. Benfatto M, Congiu-Castellano A, Daniele A, Della Longa S (2001) MXAN : a new software procedure to perform geometrical fitting of experimental XANES spectra. J Synchrotron Radiat 8(2):267–269
22. Vallee BL (1993) The biochemical basis of zinc physiology. Physiol Rev 73(1):79–118
23. Clark-Baldwin K, Tierney DL, Govindaswamy N, Gruff ES, Kim C, Berg J, Koch SA, Penner-Hahn JE (1998) The limitations of X-ray absorption spectroscopy for determining the structure of zinc sites in proteins. When is a tetrathiolate not a tetrathiolate? J Am Chem Soc 120(33):8401–8409
24. Marino S, Hayakawa K, Hatada K, Benfatto M, Rizzello A, Maffia M, Bubacco L (2007) Structural features that govern enzymatic activity in carbonic anhydrase from a low-temperature adapted fish, Chionodraco hamatus. Biophys J 93(8):2781–2790
25. Abrahams IL, Bremner I, Diakun GP, Garner CD, Hasnain SS, Ross I, Vasák M (1986) Structural study of the copper and zinc sites in metallothioneins by using extended X-ray-absorption fine structure. Biochem J 236 (2):585–589
26. Hart PJ, Balbirnie MM, Ogihara NL, Nersissian AM, Weiss MS, Valentine JS, Eisenberg D (1999) A structure-based mechanism for copper–zinc superoxide dismutase. Biochemistry 38(7):2167–2178

Chapter 13

X-Ray Crystallographic Studies of Metalloproteins

Anne Volbeda

Abstract

Many proteins require metals for their physiological function. In combination with spectroscopic characterizations, X-ray crystallography is a very powerful method to correlate the function of protein-bound metal sites with their structure. Due to their special X-ray scattering properties, specific metals may be located in metalloprotein structures and eventually used for phasing the diffracted X-rays by the method of Multi-wavelength Anomalous Dispersion (MAD). How this is done is the principle subject of this chapter. Attention is also given to the crystallographic characterization of different oxidation states of redox active metals and to the complication of structural changes that may be induced by X-ray irradiation of protein crystals.

Key words X-ray diffraction, X-ray fluorescence, Anomalous dispersion, MAD phasing, Redox reactions

1 Introduction

Many metals are essential for life. Incorporated inside proteins, their properties are tuned in specific ways, allowing the combination of protein environment and metal to perform tasks that would be difficult to access otherwise. This chapter focuses on structural studies of such metalloproteins, using X-ray diffraction. Unlike the electrons of light atoms as C, N, and O, those of metals may undergo significant resonance with X-ray photons at the wavelengths that are typically used for a crystal structure determination. The resonance frequencies correspond to absorption energies, which induce an intensity difference between X-ray reflections measured at indices (h,k,l) and $(-h,-k,-l)$, thus breaking the Friedel symmetry [1] of $I(+) = I_{h,k,l}$ and $I(-) = I_{-h,-k,-l}$. The corresponding anomalous scattering depends strongly on the used X-ray wavelength (λ). Each atom type has its own specific absorption energies, of which the *K*-edge (Table 1) corresponds to the energy that is required to remove one core 1s electron. Some of the absorbed X-ray photons are emitted at the same energy but with

Juan C. Fontecilla-Camps and Yvain Nicolet (eds.), *Metalloproteins: Methods and Protocols*,
Methods in Molecular Biology, vol. 1122, DOI 10.1007/978-1-62703-794-5_13,

Table 1
Absorption *K*-edges and KL_3 fluorescence energies of selected elements

Element	*K*-edge (eV)	*K*-edge (Å)	KL_3 emission (eV)
S	2,471	5.018	2,308
Ca	4,038	3.070	3,692
V	5464	2.269	4,952
Mn	6,538	1.896	5,898
Fe	7,111	1.744	6,404
Co	7,709	1.608	6,930
Ni	8,331	1.488	7,478
Cu	8,980	1.381	8,048
Zn	9,661	1.283	8,639
Se	12,656	0.9796	11,223

Source: www.nist.gov/pml/data/xraytrans

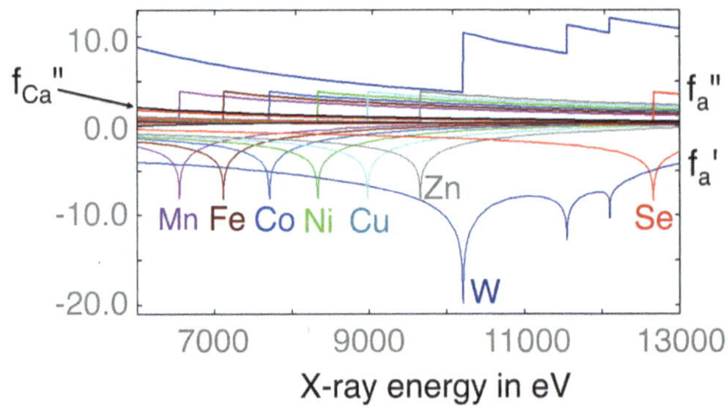

Fig. 1 Calculated anomalous scattering coefficients (f′ and f″) of selected metals. *Source*: www.skuld.bmsc.washington.edu/scatter

a phase delay compared to non-absorbed X-rays that are scattered normally. Other photons are lost for the X-ray diffraction pattern due to permanent absorption and transitions of electrons in higher occupied orbitals to the vacant 1s orbital. The X-ray fluorescence energies corresponding to the latter transitions are also atom-specific (Table 1).

The total X-ray scattering factor of an atom, $f_{at}(\lambda)$, may be described as $f_{at}^{0} + f'_{at}(\lambda) + if'_{at}(\lambda)$. Here f^{0} is wavelength independent, whereas f′ and f″ correspond to the real and imaginary component of the anomalous atomic scattering factor, which is negligible for light atoms. The f′ and f″ components are related by the Kramers–Kronig transformation [2]. Calculated values of f′ and f″ of common metals bound to proteins are shown as a function of X-ray energy in Fig. 1. These values are inaccurate close to absorption edges, where the resonance between the X-ray photons and the electrons depends both on the oxidation state of the metal and on its specific chemical environment. An experimental determination of f′ and f″ is therefore required here.

Because of its wavelength dependence, anomalous scattering may be used to assign specific metals to observed atomic positions in a crystallographic analysis. Such an analysis requires that the phases of the diffracted X-rays have first been determined. If this is the case, a $[F(+,\lambda_{max}) - F(-,\lambda_{max})]$ Bijvoet [3] Fourier difference map calculated with data collected at the maximum of the absorption edge (λ_{max}) will give a significant positive peak at the metal site. The metal identity may also be determined by using data collected at the inflection point (λ_{infl}), where f′ is minimal, and any other wavelength significantly different from λ_{infl} (*see* also Fig. 1), after which a dispersive $[F(\lambda) - F(\lambda_{infl})]$ Fourier difference map should give a positive peak at the same position. Functionally important redox transitions of metal sites may be studied with electron density maps calculated from structure factor data sets of crystals frozen in specific states.

The structure factor amplitudes (F) can be easily obtained from the measured diffracted intensities ($I = F^2$), but it is much more difficult to obtain their phases. However, only properly phased structure factors corresponding to vectors F allow to obtain an interpretable electron density map of the structure by Fourier transformation. The presence of metals allows phasing of X-ray diffraction data from a single crystal by the method of Multiple Anomalous Dispersion (MAD). A large body of literature exists on this method (e.g., 4–7). The basics may be explained with the Harker [8] construction in Fig. 2, which shows that in the absence of other sources of phase information it may be sufficient to collect diffraction data at two different X-ray wavelengths (λ_1 and λ_2), provided that there is significant anomalous scattering for one of them (λ_2 in Fig. 2). The major problem is to locate the metals from the two data sets. As further discussed in the methods section, this is in principle possible with F_a^2 [4], $[F(+,\lambda_2) - F(-,\lambda_2)]^2$ [5] or $[F(\lambda_1) - F(\lambda_2)]^2$ Patterson functions [9], which will show peaks at locations that correspond to vectors between all the metal sites that are present in the structure. From these vectors it is possible to obtain the metal positions and from there to calculate the vector F_a (the total structure factor of all anomalously scattering metals, including its phase ϕ_a) by a Fourier transformation. Alternatively, the metals may be located by direct methods [10], which may also be combined with Patterson methods [11]. Once F_a is known for each reflection, the correct phase for the rest of the protein (ϕ_1) is determined by the point of intersection of the three circles having radii of $F(\lambda_1)$, $F(+,\lambda_2)$ and $F(-,\lambda_2)$, as shown in Fig. 2.

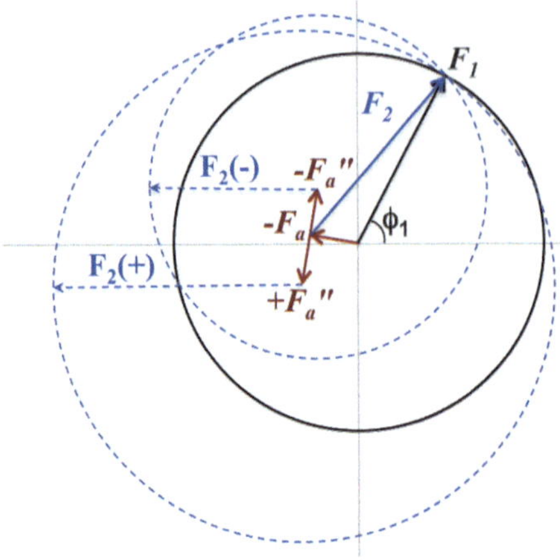

vector addition:

$$F(\lambda 2) = F(\lambda 1) + F_a(\lambda 2)$$

or (in short notation):

$$F_2 = F_1 + F_a$$

using basic trigonometry (and $I=F^2$):

$$I_2 = I_1 + F_a^2 + 2F_1F_a\cos(\phi_1 - \phi_a)$$

$$I_2(\pm) = I_2 \pm 2F_1F_a''\sin(\phi_1 - \phi_a)$$

with $\beta = \cos^{-1}\{[I_2 - I_1 - F_a^2]/2F_1F_a\}$:

$$\phi_1 = \phi_a + \beta \text{ or } \phi_1 = \phi_a - \beta$$

with $\gamma = \sin^{-1}\{[I_2(+) - I_2(-)]/4F_1F_a''\}$:

$$\phi_1 = \phi_a + \gamma \text{ or } \phi_1 = \phi_a + 180 - \gamma$$

the correct solution for ϕ_1 is determined by the point of intersection of the three circles

Fig. 2 Harker construction for the determination of structure factor phases (ϕ), using differences between structure factor intensities ($I = F^2$) measured at two different wavelengths (λ_1 and λ_2) in the simple case when there is no anomalous scattering at λ_1. Phases can be calculated from measured I_1 and $I_2(\pm)$ once the vectors F_a (real component of the anomalous scattering) and F_a'' (imaginary component) have been determined (*see* Subheading 3)

In this chapter, I briefly discuss how to verify the presence of metals in crystals of metalloproteins and choose wavelengths for X-ray data collections, how to collect X-ray data sets and perform the steps that are involved in a successful MAD-phasing experiment, how to prepare specific redox states, and how to limit unwanted effects like structural changes induced by X-rays. Some steps are illustrated with results from personal experience.

2 Materials

The crystallization of metalloproteins, the focus of this chapter, is not fundamentally different from that of any other protein. This crucial step in a crystallographic analysis may nowadays be automated to a large extent and it is not further discussed here. Instead a short list of specialist equipment that is needed for crystallographic studies of metal sites in protein crystals is briefly discussed. These are a tunable synchrotron beamline, an X-ray fluorescence detector, and UV/Vis and/or Raman spectrometers that may eventually be used during X-ray data collection. In addition, significant computer power is needed for the collection and processing of all crystallographic data.

2.1 Tunable Synchrotron Beamline

For exploiting the wavelength dependence of the X-ray diffraction of a metalloprotein a tunable synchrotron source is particularly well suited (*see* **Note 1**). High-energy radiation is produced when the trajectory of electrons moving in a storage ring is changed by a bending magnet [12] or an undulator device [13]. X-ray wavelength selection takes place at a monochromator (*see* **Note 2**). Other optical elements that are not discussed here are included for further controlling the shape and intensity distribution of the monochromatic beam. Depending on the beamline, various detectors may be used for collecting X-ray diffraction data during rotation of the studied protein crystal, which is usually cooled by a cold (100 K) nitrogen stream in order to limit radiation damage. Often a robot is available [14] for the automated transfer of flash-frozen crystals from a storage dewar with liquid nitrogen to the X-ray beam.

2.2 Additional Beamline Equipment

A special X-ray detector should be available to measure X-ray fluorescence and perform absorption edge scans [15]. In addition, advantage may be taken of the fact that many metalloproteins absorb light in the UV/visible range to characterize specific redox states and monitor putative X-ray induced redox changes by adding a UV/Vis spectrometer to the beamline [16]. With a proper setup, both UV/Vis absorption and fluorescence measurements are possible (*see* **Note 3**). Another technique that has already been used online is the use of a Raman spectrometer [17] in order to probe vibrations in the 200–2,000 cm^{-1} range and monitor putative X-ray induced changes near redox active metal sites.

2.3 Crystallographic Software

Many computer programs are involved during the many subsequent stages of a crystallographic structure determination. Although a detailed discussion of them is outside the scope of this chapter, some attention is given in the next section to software that may be specifically used to study metalloproteins.

3 Methods

This section outlines the basic steps that are involved in a crystallographic analysis of a metalloprotein once it has been purified and crystallized. These include crystal preparation, X-ray data collection, MAD phasing, assignment of metals in atomic models, and studies of redox changes.

3.1 Crystal Preparation

1. Proteins with oxygen sensitive metal sites are best crystallized under a controlled atmosphere, using a glove box. Changes of redox state may be induced by exposing a crystal to a different atmosphere or by soaking it in a crystallization solution that is typically enriched with 2 % extra precipitant, includes a cryoprotectant (explained below) and an appropriate cocktail of different redox mediators, with typical concentrations between 2 and 10 mM.
2. In order to increase the lifetime of a crystal and keep it in a fixed state, it is mounted in a cryoloop and flash-frozen in liquid nitrogen after addition of a cryoprotecting agent (*see* **Note 4**). A special procedure is required for crystals of air-sensitive metalloproteins grown inside a glove box, where liquid propane may be used for freezing [18].

3.2 X-Ray Data Collection

1. Following its transfer to the synchrotron beamline, the crystal should be kept in a cold (≈100 K) stream of N_2. This is necessary to limit radiation damage effects.
2. For selecting the proper wavelength for data collection, measurement of an X-ray fluorescence spectrum is very useful (Fig. 3). This takes little time and shows which metals and other heavy atoms are potentially bound to the protein (*see* **Note 5**).
3. The presence of a certain metal should be further verified by performing an edge-scan. In the case of a significant absorption edge, precise experimental values of the imaginary (f'') and real (f') components of the anomalous scattering factor may be obtained as a function of wavelength with the program *CHOOCH* [19], starting from the obtained absorption edge spectrum (Fig. 4a).
4. After the analysis with *CHOOCH*, it should be clear which wavelengths (λs) give maximal Bijvoet $[F(+,\lambda_{peak}) - F(-,\lambda_{peak})]$ and dispersive $[F(\lambda_0) - F(\lambda_{infl})]$ differences (Figs. 1 and 4a). When MAD phasing is needed (*see* **Note 6**), these three λs may next be used for data collection (*see* **Note 7**). As shown in Fig. 2, a minimum of two λs is required (*see* **Note 8**), but in many cases data collection at a single wavelength, chosen at λ_{max}, has been sufficient to solve the phase problem (*see* **Note 9**). However, because this is not always successful, in general it is

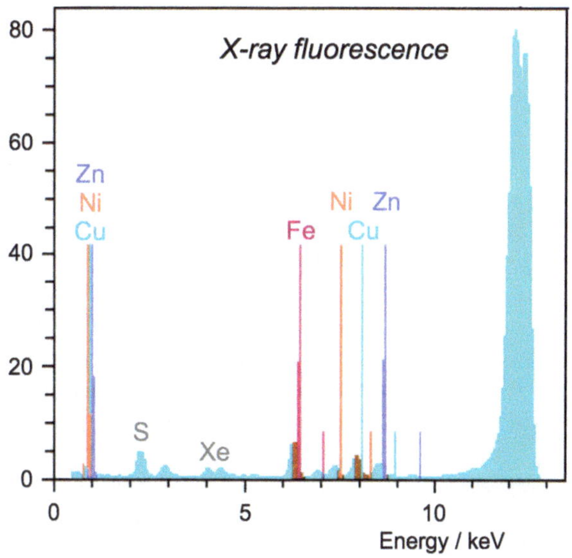

Fig. 3 Fluorescence spectrum of a Xe-treated crystal of CODH/ACS from *Moorella thermoaceticum*, collected with a Röntec detector at the BM30A beamline of the ESRF after excitation with $\approx$ 12.8 keV X-rays. The line-widths of the various peaks arise from non-elastic Compton scattering

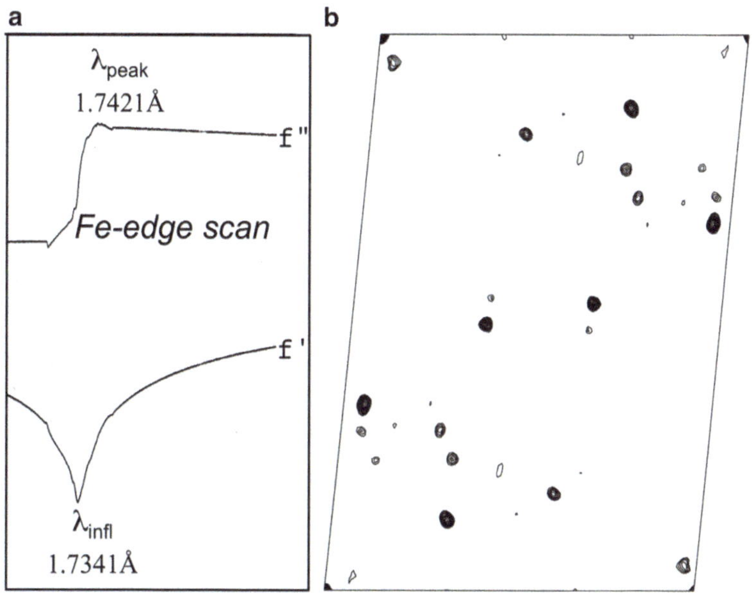

Fig. 4 (**a**) Fe-edge scan and determination of f'_{Fe} from f''_{Fe} with *CHOOCH*. This allows to find the wavelengths where f'' is maximal (λ_{peak}), f' is minimal (λ_{infl}), and f' is much greater than f'_{min} (λ_0, not shown). (**b**) Harker section of a 4.5–15 Å resolution $(|F^+| - |F^-|)_{peak}{}^2$ Bijvoet difference Patterson map used for the location of iron sites

better to have a maximum number of independent observations for each reflection, including at different λs, for phasing the X-ray data.

5. Once a good crystal has been selected for data collection (*see* **Note 10**), important decisions concern the choice of exposure time and resolution limits (*see* **Note 11**). High resolution is needed to observe fine structural details, but this may require long X-ray exposition times that may cause structural changes due to reduction of metal sites (*see* **Note 12**). Continued exposure may lead to severe radiation damage even at 100 K [20]. Unless the goal is to investigate X-ray induced structural changes, it is best to avoid these. For elongated crystals, translations or helical scans may be used to reduce the X-ray dose per crystal volume [20, 21]. If the phase problem has not yet been solved, it is much better to go for data redundancy than for resolution. A crystal rotation range of 360° will maximize both the completeness and the redundancy of the F(+) and F(−) data. If needed, several crystals may be used (*see* **Note 13**). Apart from potentially giving rise to radiation damage, overexposure should be avoided because overloaded strong reflections will cause high noise levels in electron density maps and even more so in Patterson maps (*see* **Note 14**). In order to minimize measurement errors due to differences in absorption from different exposed crystal volumes, it may be important to collect F(+) and F(−) data of the same reflection nearby in time by the application of an inverse beam strategy (*see* **Note 15**).

3.3 MAD Phasing of X-Ray Data

1. First the tutorial of the chosen software should be checked (*see* **Note 6**). When MADSYS is used [4], the F_A amplitudes (Fig. 2) are determined from data collected at the different (at least two) wavelengths and used to find the metal positions. With the CCP4 [22] and Phenix [23] programs these positions are usually determined from [F(+) − F(−)] Bijvoet differences by treating the problem in terms of a SIRAS (single isomorphous replacement and anomalous scattering) experiment [5] (*see* **Note 16**). Important steps are briefly discussed below.
2. It is crucial that all the collected data be properly scaled. All independent observations of I(+) and I(−) should be kept separate until the used procedure, which may include local scaling [6], has converged. At the end of the scaling, correlation factors defined by the agreement between multiple (including symmetry related) observations of the same reflection are good criteria to assess the data quality as a function of resolution (*see* **Note 17**).

3. The next step is to calculate a Patterson map with either F_A^2 or $[F(+) - F(-)]^2$ coefficients (*see* **Note 18**), depending on the used phasing software. All peaks above noise level in these maps correspond to vectors between metal sites. An example of one section of a $[F(+) - F(-)]^2$ map is shown in Fig. 4b. A dispersive difference $[F(\lambda_{max}) - F(\lambda_{infl})]^2$ map using data from the same crystal showed the same peaks with lower heights (*see* **Note 19**). In the used example [24], these peaks are due to the presence of seven FeS clusters (*see* **Note 20**).
4. The interatomic vectors showing up in the Patterson map are used to deduce the distribution of the metal positions in the protein. This may be done by hand, but the available computer programs do the same job much faster (*see* **Note 21**). Direct methods may also be used to locate the metal sites (*see* **Note 22**). The most powerful approach appears to be to combine these with Patterson methods [11]. Multiple potential solutions may be tested and scored by correlating the Patterson map calculated from the obtained atom positions with the experimental one. At the end there will always be two solutions that are equally well compatible with the experimental Patterson map, differing only by hand (*see* **Note 23**).
5. The two possible solutions of the Patterson correspond to a phase of $\pm\phi_a$ in Fig. 2. The correct phase, corresponding to the correct hand, will produce right-handed α-helices in an electron density map (*see* **Note 24**) and give the strongest signal for the anomalous scatterers in a Bijvoet $[F(+) - F(-)]$ difference map, calculated with ϕ_1 (*see* **Note 25**), as determined from the Harker construction in Fig. 2. These maps may also be used to verify the presence of metal sites, predicted from the Patterson map, that are not already included in the calculation of F_a. Improved phases will be obtained until all metal sites have been positioned.

3.4 Assignment of Metals to Atomic Models

1. Once good phases are available for the X-ray data, specific metals may be assigned to specific density features thanks to their unique anomalous scattering characteristics (Fig. 1). The X-ray data collection steps are the same as described under Subheading 3.2. For the example used in Fig. 3, X-ray fluorescence for three other metals beside Fe is observed: Ni, Cu, and Zn. Subsequent absorption edge scans showed significant signals for Ni and Zn (Fig. 5a, b) and a small signal for Cu.
2. Upon choosing a wavelength at the maximum of an absorption edge, a Bijvoet $[F(+) - F(-)]$ difference map should show the location(s) of the corresponding anomalous scatterer when it is bound to the protein (*see* **Note 5**). In the used example [24], two molecules related by non-crystallographic symmetry surprisingly showed a different active site metal composition

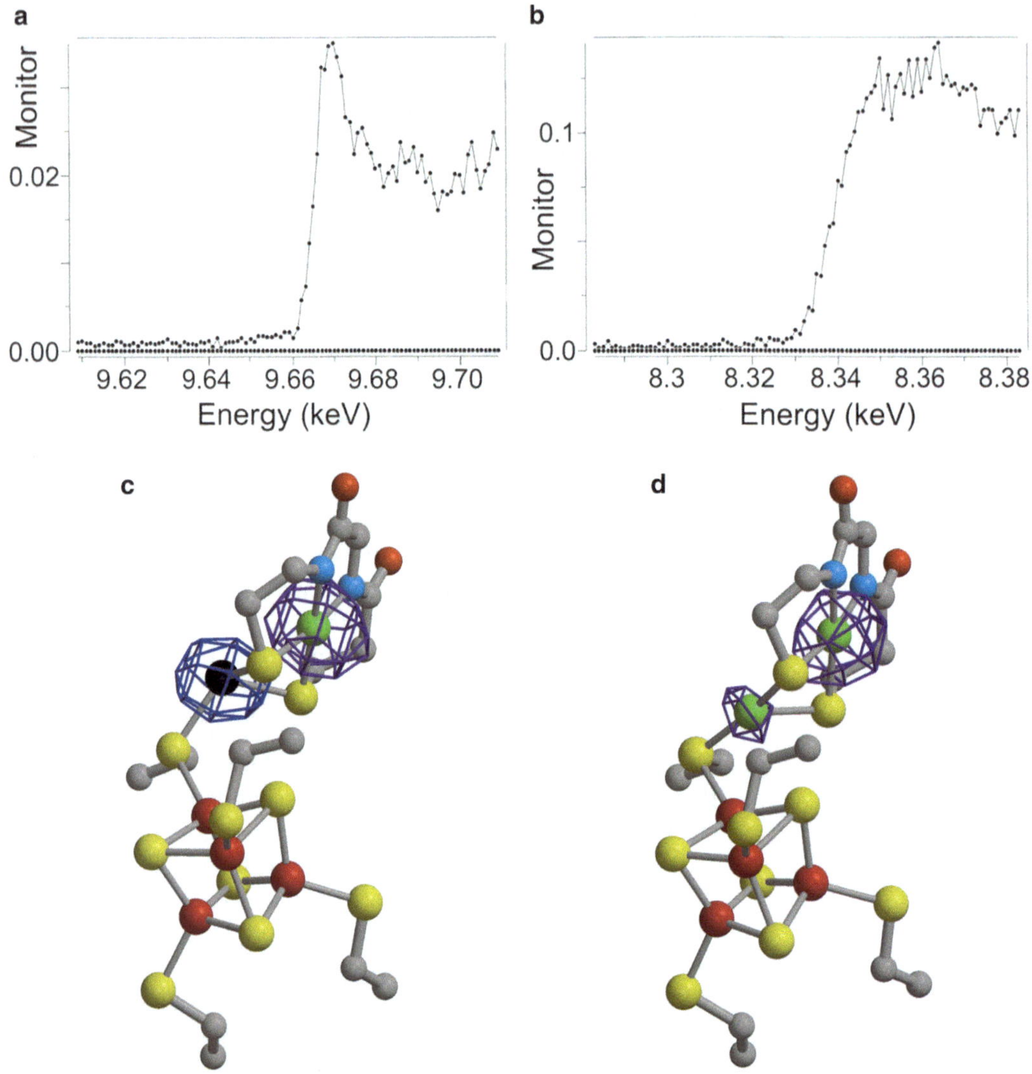

Fig. 5 Measured X-ray absorption spectra of CODH/ACS at (**a**) the Zn-edge, (**b**) the Ni-edge, and anomalous difference maps calculated for the ACS A-cluster with data collected at λ_{max} for the Ni-edge (*violet*) and the Zn-edge (*blue*) in molecules *C* (**c**) and *D* (**d**). Not shown external ligands complete a tetrahedral and a square planar coordination for the central Zn and Ni atoms, respectively [24]. Atoms are colored *black* (Zn), *green* (Ni), *red* (Fe), *yellow* (S), *orange* (O), *light*-blue (N) and *gray* (C)

(Fig. 5c, d). All metal assignments may be further confirmed with $[F(\lambda_2) - F(\lambda_1)]$ difference maps when λ_1 is chosen sufficiently close to λ_{inf} (*see* **Note 26**).

3.5 Studying Redox Changes

1. High-resolution data from crystals frozen in specific states, as explained under Subheading 3.1, may allow to localize structural changes that are due to metal based redox reactions. In order to obtain unbiased results, phase information is usually obtained from a refined atomic model of the protein (*see* **Note 27**) from which all or part of the atoms making up

the metal site are removed (*see* **Note 28**). In the final refinement cycles to be made upon including all the atoms, bias introduced by wrong restraints for metal–ligand interactions should be minimized (*see* **Note 29**). As explained above under Subheading 3.2 (**step 5**), it is important to avoid complications arising from X-ray induced redox changes during data collection (*see* **Note 30**).

2. Many metalloproteins change color after reduction/oxidation. For such favorable cases, it may be possible to use UV/Vis spectrometry to identify redox states and monitor putative changes during data collection (*see* **Note 31**). In several cases it has been found that X-ray induced reductions of metal sites appear much more rapidly than other signs of radiation induced changes, including damage [16, 25, 26].
3. Some redox changes at a metal site may significantly change the absorption edge (*see* **Note 32**). This may be checked by performing edge scans at the start and the end of data collection [26], or as a function of exposure time.
4. In certain cases structural changes induced by X-rays have also been detected by following Resonance Raman spectra during X-ray data collection [17, 27].

4 Notes

1. See http://en.wikipedia.org/wiki/List_of_synchrotron_radiation_facilities for available sources.
2. A typical monochromator consists of a liquid nitrogen cooled Si crystal. Its orientation (θ) with respect to the reflected beam determines the selected wavelength according to Bragg's [28] law, $n\lambda = 2d \sin \theta$ (d denoting a specific reflection). Before selecting a synchrotron beamline for a particular experiment, care should be taken to check whether the absorption edge of the metal to be studied is accessible for the beamline optics. If this is not the case, useful data may still be collected, provided that an anomalous signal is detectable, but the assignment of this signal to a specific metal may not be possible.
3. All additional beamline equipment should be oriented perpendicular to the X-ray beam in order to prevent interference with X-ray data collection. As a result, experimental setups may become quite crowded around the studied crystal, not leaving much space for manual interventions. Ideally, it should be possible to add or remove additional detectors and light sources from the beamline in a flexible user-friendly way.
4. A cryoprotecting agent is needed for the crystal to survive the flash-freezing procedure. A typical cryoprotectant is glycerol,

e.g., at 20 %, but it may be necessary (or worthwhile) to test other molecules in order to find optimal conditions for keeping the crystal in a good state.

5. The atoms showing up in a fluorescence spectrum could also be present in the solvent regions of the crystal. Toxic metals like Cd and Hg, which are not included in Table 1 and Fig. 1, have their own specific f' and f'' curves and may also be detected this way.
6. Popular software that may specifically be used for MAD phasing of metalloproteins includes MADSYS [4], CCP4 [22] and Phenix [23]. Clear tutorials are available for each of these (*see* xtl.cumc.columbia.edu/comp/madsys/madsys.html, www.ccp4.ac.uk and www.phenix-online.org/ for more information).
7. X-ray diffraction depends on crystal orientation. During a diffraction experiment the crystal is rotated about an axis that is oriented perpendicular to the X-ray beam. Different software is available for the indexation and integration of all resulting reflections. My personal preference is XDS [29].
8. In Fig. 2 it is assumed that there are no errors in the measured reflection intensities. In reality experimental errors must be taken into account. The treatment of these is included in the MAD phasing programs already mentioned and not further discussed here.
9. This is called a SAD (Single-wavelength Anomalous Dispersion) experiment. Omitting the circle corresponding to F_1 in Fig. 2 shows two possible phases for F_2, each with equal probability, that are defined by the points of intersection of the dashed circles corresponding to $F_2(+)$ and $F_2(-)$. Additional sources of phase information like solvent flattening or non-crystallographic symmetry averaging are now needed to obtain an interpretable electron density map. Such sources provide unimodal phase probability distributions that may be combined with the bimodal SAD probabilities, using Hendrickson–Lattman coefficients [30], to obtain improved phases. A detailed treatment of all this is beyond the scope of this chapter.
10. The speed of data collection that is nowadays possible at high intensity synchrotron beamlines allows the rapid testing of many crystals. Often, each crystal is put in the beam and, after proper centering, 180° of data are collected in a few minutes (the "shoot and think later" strategy). A rotation range of 180° is in most cases sufficient for data completeness. Alternatively, using an automated transfer system, many crystals may be tested by collection of two images separated by 90°, before the best diffracting ones are selected for collecting complete datasets. In that case, the two images that have already been

collected may be indexed and used for the determination of crystal cell parameters, orientation and space group, from which an optimal strategy for data collection may be obtained. Whatever the chosen strategy, before starting data collection it is *crucial* that the crystal is correctly centered. For small crystals, centering may take even more time than data collection. However, when the beam intensity is low and much longer times are necessary for data collection, it is a good idea to plan this such that the data are completed before refill of the synchrotron beam. This is because beam intensity and center may significantly change after a refill.

11. In order to prevent loss of high resolution reflections the X-ray detector should be positioned close enough to the crystal. However, at high resolution there may be an increasing problem of overlap between neighboring reflections, especially for crystals with large unit cell dimensions. For reducing the number of overlapped reflections it is advantageous to collect diffraction images over small $\Delta\phi$ ranges like 0.1°, although this means that many images are needed for a complete dataset (3,600 for 360° of data).
12. Upon their interaction with a protein crystal, X-rays produce photoelectrons. These may either reduce redox active sites like metals or form radical species that, even at 100 K, may give rise to further irreversible reactions. In order to limit such radiation damage, it is often necessary to significantly reduce the beam intensity at high-intensity beamlines. This is possible by the application of suitable filters. More information on beam transmission and X-ray exposure times to be used may be found at beamline Web sites. For a proper choice of X-ray dose programs like *RADDOSE* and *RADDOSE-3D* may be used [20]. Furthermore, for a MAD experiment it may be advantageous to start data collection at the shortest wavelength, where there is less absorption of X-rays and fewer photoelectrons are produced.
13. Even weak anomalous scatterers as S have been successfully used for phasing purposes, using data collected far from the absorption edge. However, in such cases high data redundancy from the use of multiple crystals and optimized data collection strategies may be crucial for obtaining sufficient phasing power [7]. Tutorials for special data collection strategies are usually available at the beamlines.
14. Such maps are very sensitive to strong reflections. Accordingly, these should be correctly measured or, if this is not the case, be excluded from the map calculation.
15. This strategy for maximizing anomalous [F(+) − F(−)] signals consists of applying a 180° rotation between successively

collected wedges of data. For example, after collecting a first 30° wedge (0–30°), a second one is collected from 180° to 210°, a third one from 30° to 60°, a fourth one from 210° to 240°, etc., until diffraction data from all crystal orientations between 0 and 360° (or less, depending on crystal symmetry) have been recorded.

16. For a SIRAS experiment two crystals are required (one for the native protein and one for the heavy atom derivative), whereas a two-wavelength MAD experiment may be performed with a single crystal. The great advantage of the latter is that there are no complications arising from putative differences in crystal shape and cell dimensions between data sets. In addition there is no need for the search of a heavy atom derivative in a MAD experiment.

17. Such correlation criteria may also be used for optimal data selection from each crystal when multiple crystals are needed for data completion.

18. Patterson maps are calculated with the function:

$$P(uvw) = [\Sigma_{(hkl)} A_{hkl}^2 \cos 2\pi(hu + kv + lw)]/V.$$

Here V is the volume of diffraction space (also called reciprocal space), the summation (Σ) goes over all reflection indices (hkl) and uvw denotes a position in real space. The terms A_{hkl} are either F_A, $[F(+) - F(-)]$ or $[F(\lambda_1) - F(\lambda_2)]$. $P(uvw)$ is the Fourier transformation of the map coefficients A_{hkl}^2.

19. It is important to optimize the signal-to-noise level of the Patterson map. Apart from the chosen wavelengths such a map may be quite sensitive to the used resolution limits. A proper choice of these may be guided by the obtained correlation statistics after data scaling. As already noted, incorrectly measured strong reflections should be excluded for the map calculation. These show up as outliers after a statistical analysis of the data quality. Because dispersive and Bijvoet differences are completely independent, a comparison of the corresponding Patterson maps may be used to distinguish noise peaks from real ones: the latter should show up in both maps.

20. The individual atoms of an iron–sulfur cluster are only resolved at high resolution. Such clusters behave as a single superatom at 4.5 Å or lower resolution (Fig. 4b).

21. For N metal sites there are $N(N-1)$ interatomic vectors. Many additional peaks are caused by crystallographic symmetry. In the case of many sites, manual interpretation of Patterson maps becomes very difficult.

22. Conventional direct methods use the tangent formula, derived by Karle and Hauptman [10]:

$$\tan(\phi_H) = \Sigma_{(K)} E_K E_{H-K} \sin(\phi_K + \phi_{H-K}) / \Sigma_{(K)} E_K E_{H-K} \cos(\phi_K + \phi_{H-K}).$$

Here the indices H and K denote reflections (h1,k1,l1) and (h2,k2,l2), E_K and E_{H-K} are normalized structure factor amplitudes and the summations Σ go over all reflections K. The equation shows how to calculate the unknown phase (ϕ_H) of E_H from known phases of E_K and E_{H-K}. It has been incorporated in the popular *Shake-and-Bake* algorithm for the determination of structures containing up to a few hundred atoms [31]. Often a 3.5 Å high resolution limit is enough to solve heavy atom distributions in a protein structure by direct methods [11].

23. For each set of positions x1,y1,z1 and x2,y2,z2 that explain a peak in the Patterson map, there is always an alternative possibility defined by −x1,−y1,−z1 and −x2,−y2,−z2.

24. The electron density map $\rho(xyz)$ is obtained by the Fourier transformation of the phased structure factors F. The best maps are obtained when phase errors are taken into account by a weighting factor called figure of merit m that lies between 0 and 1 and is calculated from the obtained phase probability distribution (see also **Note 8**). Taking Bijvoet pairs together (see also **Note 18**):

$$\rho(xyz) = 2\Sigma_{(hkl)} m_{hkl} F_{hkl} \cos[2\pi(hx + ky + lz) - \phi_{hkl}]/V.$$

25. Difference maps correspond to the Fourier transformation of phased ΔF terms, including figure of merit weighting. For anomalous (Bijvoet) differences the phases should be retarded by 90°, giving (using $\cos[\alpha - 90] = \cos[90 - \alpha] = \sin \alpha$):

$$\Delta\rho(xyz)_{\text{anom}} = \Sigma_{(hkl)} m_{hkl}[F(+)_{hkl} - F(-)_{hkl}]\sin[2\pi(hx + ky + lz) - \phi_{hkl}]/V.$$

Such maps may be very helpful to check if phasing by other methods like molecular replacement has been successful.

26. Near the inflection point of anomalous scattering, λ_{inf}, f' is very sensitive to small changes of wavelength (*see* Fig. 1 and 4a). A small error in the determination of λ_{inf} may therefore have a large effect on the signal-to-noise ratio of dispersive difference maps.

27. Errors in the model generate phase errors that, in turn, may generate severe bias in electron density maps calculated with model phases. It has been shown that the map coefficients to be used for Fourier transformation that result in an electron density map with the least model bias are $[2mF_{\text{obs}} - DF_{\text{calc}}]$. Here F_{calc}, the calculated structure factors, are obtained by an inverse

Fourier transformation of the electron density map that is directly calculated from the atomic model, F_{obs} are the measured structure factor amplitudes, with phase information included from the atomic model, *m* the figure of merit of the calculated model phases and *D* depends on the overall coordinate error (being 1 if there is no error), as described by Read [32].

28. Omit maps calculated by Fourier transformation of $[mF_{obs} - F_{calc}]$ map coefficients show positive electron density only for the atoms that are absent from the used protein model.
29. Except at very high resolution, where the ratio of observations to refined parameters is sufficiently high, crystallographic refinement normally requires the use of geometric restraints in order to produce stereochemically sensible protein models. For metal sites such restraints are often not available: metal–ligand bond lengths may depend on unknown redox and spin states of the metal and on the exact identity of the various ligands. In addition, electron-poor ligands like hydride will normally not be detectable in the electron density map. In order to prevent unwanted van der Waals repulsions between metals and ligands, either weak (but probably inaccurate) bond restraints may be used or the van der Waals radius of the metal may be reduced by adapting its Lennard-Jones parameters.
30. The highest resolution structures are likely to have reduced metal sites. Shorter X-ray exposures giving lower resolution may be needed for the characterization of more oxidized states. What counts is the X-ray exposure per crystal volume. Helical scans or crystal translations may be employed to reduce this [20, 21]. For very radiation-sensitive metal sites, multiple crystals may be needed for obtaining complete data sets of oxidized states.
31. If there are several metal sites, correlating changes of color with a specific metal may be difficult. Still, for a proper interpretation of the obtained results it is important to be able to define if these depend or not on the used X-ray exposure dose. This dose may be quantified by using programs like *RADDOSE* and *RADDOSE-3D* [20].
32. After oxidation, the absorption edge is expected to shift to a higher energy as further electron removal from a more positively charged metal should be more difficult. However, changes in metal ligation resulting from its oxidation may significantly reduce such edge shifts or even cancel them completely.

Acknowledgements

This work was supported by institutional funding from the Commissariat à l'Energie Atomique et aux Energies Alternatives (CEA) and the Centre National de la Recherche Scientifique (CNRS). The author wishes to thank David Cobessi and Jean-Luc Ferrer of the Synchrotron Group of the IBS for helpful discussions.

References

1. Friedel G (1913) Sur les symétries cristallines que peut révéler la diffraction des rayons Röntgen. CR Hebd Acad Sci 157:1533–1536
2. Kronig R d L, Kramers HA (1928) Zur Theorie der absorption und dispersion in den Röntgenspektren. Z Phys 48:174–179
3. Bijvoet JM (1949) Phase determination in direct Fourier synthesis of crystal structures. P K Ned Akad Wetensc 52:313–314
4. Hendrickson WA (1991) Determination of macromolecular structures from anomalous diffraction of synchrotron radiation. Science 254:51–58
5. Ramakrishnan V, Biou V (1997) Treatment of multiwavelength anomalous diffraction data as a special case of multiple isomorphous replacement. Method Enzymol 276:538–557
6. Terwilliger TC, Berendzen J (1999) Automated MAD and MIR structure solution. Acta Crystallogr D55:849–861
7. Liu Q, Dahmane T, Zhang Z et al (2012) Structures from anomalous diffraction of native biological macromolecules. Science 336:1033–1037
8. Harker D (1956) The determination of the phases of the structure factors of non-centrosymmetric crystals by the method of double isomorphous replacement. Acta Crystallogr 9:1–9
9. Patterson AL (1934) A Fourier series method for the determination of the components of interatomic distances in crystals. Phys Rev 46:372–376
10. Karle J, Hauptman H (1956) A theory of phase determination for the 4 types of non-centrosymmetric space groups 1P222, 2P22, 3P12, 3P22. Acta Crystallogr 9:635–651
11. Schneider TR, Sheldrick GM (2002) Substructure solution with SHELXD. Acta Crystallogr D58:1772–1779
12. Roth M, Carpentier P, Kaïkati O et al (2002) FIP: a highly automated beamline for multiwavelength anomalous diffraction experiments. Acta Crystallogr D58:805–814
13. De Sanctis D, Beteva A, Caserotto H (2012) ID29: a high-intensity highly automated ESRF beamline for macromolecular crystallography experiments exploiting anomalous scattering. J Synchrotron Radiat 19:455–461
14. Ohana J, Jacquamet L, Joly J et al (2004) CATS: a cryogenic automated transfer system installed on the beamline FIP at ESRF. J Appl Crystallogr 37:72–77
15. Leonard GA, Solé A, Beteva A et al (2009) Online collection and analysis of X-ray fluorescence spectra on the macromolecular crystallography beamlines of the ESRF. J Appl Crystallogr 42:333–335
16. Ronda L, Bruno S, Bettati S et al (2011) Protein crystal microspectrophotometry. Biochim Biophys Acta 1814:734–741
17. McGeehan JE, Bourgeois D, Royant A et al (2011) Raman-assisted crystallography of biomolecules at the synchrotron: instrumentation, methods and applications. Biochim Biophys Acta 1814:750–759
18. Vernède X, Fontecilla-Camps JC (1999) A method to stabilize reduced and or gas-treated protein crystals by flash-cooling under a controlled atmosphere. J Appl Crystallogr 32:505–509
19. Evans G, Pettifer RF (2001) CHOOCH: a program for deriving anomalous scattering factors from X-ray fluorescence spectra. J Appl Crystallogr 34:82–86
20. Zeldin OB, Gerstel M, Garman EF (2013) Optimizing the spatial distribution of dose in X-ray macromolecular crystallography. J Synchrotron Radiat 20:49–57
21. Flot D, Mairs T, Giraud T et al (2010) The ID23-2 structural biology microfocus beamline at the ESRF. J Synchrotron Radiat 17:107–118
22. Winn MD, Ballard CC, Cowtan KD et al (2011) Overview of the CCP4 suite and current developments. Acta Crystallogr D67:235–242

23. Adams PD, Afonine PV, Bunkóczy G et al (2011) The Phenix software for automated determination of macromolecular structures. Methods 55:94–106
24. Darnault C, Volbeda A, Kim EJ et al (2003) Ni-Zn-[Fe_4-S_4] and Ni-Ni-[Fe_4-S_4] clusters in closed and open α subunits of acetyl-CoA synthase/carbon monoxide dehydrogenase. Nat Struct Biol 10:271–279
25. Meharenna YT, Doukov T, Li H et al (2010) Crystallographic and single-crystal spectral analysis of the peroxidase ferryl intermediate. Biochemistry 49:2984–2986
26. Antonyuk SV, Hough MA (2011) Monitoring and validating active site redox states in protein crystals. Biochim Biophys Acta 1814:778–784
27. Merlino A, Fuchs MR, Pica A et al (2013) Selective X-ray-induced NO photodissociation in haemoglobin crystals: evidence from a Raman-assisted crystallographic study. Acta Crystallogr D69:137–140
28. Bragg WL (1913) The diffraction of short electromagnetic waves by a crystal. Proc Camb Philos Soc 17:43–57
29. Kabsch W (2010) XDS. Acta Crystallogr D66:125–132
30. Hendrickson WA, Lattman EE (1970) Representation of phase probability distributions for simplified combination of independent phase information. Acta Crystallogr B26:136–143
31. Chang C-S, Weeks CM, Miller S et al (1997) Incorporating tangent refinement in the Shake-and-Bake formalism. Acta Crystallogr A53:436–444
32. Read RJ (1986) Improved Fourier coefficients for maps using phases from partial structures with errors. Acta Crystallogr A42:140–149

Chapter 14

Quantum Mechanical Methods for the Investigation of Metalloproteins and Related Bioinorganic Compounds

Luca Bertini, Maurizio Bruschi, Ugo Cosentino, Claudio Greco, Giorgio Moro, Giuseppe Zampella, and Luca De Gioia

Abstract

It is well known that transition metal ions are often bound to proteins, conveying very specific functional properties. In fact, metalloproteins play crucial biological roles in the transport and activation of small molecules such as H_2, O_2, and N_2, as well as in several other biochemical processes. However, even if the presence of transition metals in the active site of proteins allows a very rich biochemistry, the experimental disclosure of structure–activity relationships in metalloproteins is generally difficult exactly because of the presence of transition metals, which are intrinsically characterized by a very versatile and often elusive chemistry. For this reason, computational methods are becoming very popular tools in the characterization of metalloproteins. In particular, since computing power is becoming less and less expensive, due to the continuous technological development of CPUs, the computational tools suited to investigate metalloproteins are becoming more accessible and therefore more commonly used also in molecular biology and biochemistry laboratories. Here, we present the main procedures and computational methods based on quantum mechanics, which are commonly used to study the structural, electronic, and reactivity properties of metalloproteins and related bioinspired compounds, with a specific focus on the practical and technical aspects that must be generally tackled to properly study such biomolecular systems.

Key words Metalloproteins, Metalloenzymes, Bioinorganic chemistry, Computational chemistry, DFT, QM

1 Introduction

About 40 % of known proteins contain metal ions, which can play a structural role, such as Ca and Zn, or a catalytic role. In fact, the catalytic site of several enzymes contains transition metal ions such as Fe, Mn, Cu, Ni, V, Mo, and others, which allow to catalyze many reactions under mild conditions and often with high selectivity. Relevant examples include nitrogen fixation, hydrocarbon oxidation, hydrogen activation, generation of radicals, dismutation of superoxide and hydrogen peroxide, photosynthesis, electron

Juan C. Fontecilla-Camps and Yvain Nicolet (eds.), *Metalloproteins: Methods and Protocols*, Methods in Molecular Biology, vol. 1122, DOI 10.1007/978-1-62703-794-5_14,

transfer in the respiratory chain, DNA synthesis, and many other processes. However, the presence of transition metal ions in the active site of metalloproteins makes the disclosure of structure–activity relationships more difficult than for canonical proteins, because the electronic properties (and consequently the reactivity) of transition metals are intrinsically complex (due to the presence of partially filled *d* orbitals) and strongly affected by the environment (i.e., by the amino acids directly coordinated to the metal ions, but also by long-range interactions with other amino acids of the protein, as well as with solvent molecules). As a consequence, experimental approaches, such as X-ray crystallography, UV–Vis, NMR, EPR, IR, Mössbauer and Raman spectroscopy, as well as electrochemical techniques, must be often combined with computational experiments, to fully characterize metalloproteins.

The computational methods used to study biological processes at the molecular level can be roughly divided into two families. One includes a variety of methods based on classical physics, such as Molecular Mechanics (MM) and Molecular Dynamics (MD). Classical approaches are well suited to study properties such as the dynamics of proteins in solution, the interaction between proteins, as well as between protein and DNA fragments or between proteins and small molecules (as in drug discovery or enzyme engineering studies), but cannot be used to study chemical properties that explicitly depends on electrons, such as most spectroscopic properties and reaction pathways involving covalent bond cleavage/formation. Instead, Quantum Mechanical (QM) methods allow, at least in principle, also the quantitative evaluation of reactions and activation energies, as well as of spectroscopic properties of molecules. Even if methods based on classical physics are orders of magnitude faster than corresponding QM methods, the development of quantum chemical theories, coupled to the present efficiency of CPUs, has recently allowed the application of QM methods to the investigation of large molecular systems, such as the active site of enzymes and related bioinspired synthetic compounds [1, 2]. In particular, efforts in the QM modeling of transition metal complexes (see ref. [3] for a recent review), including metalloproteins [4–8], have led to an ever increasing area of applicability of QM methods, together with a consistent reduction in the computer time needed for such calculations. However, the QM investigation of metalloproteins is still limited by the fact that metalloproteins are often too large to be studied in a reasonable time frame using presently available CPUs. Recently, the development and critical evaluation of hybrid Quantum Mechanics/Molecular Mechanics (QM/MM) methods, where the reactive portion of a protein is described at the quantum mechanical level of theory, whereas its molecular environment, including solvent molecules, is described using a molecular mechanics theoretical description, has allowed the investigation of whole proteins, and consequently to shed light on the interplay between the chemical properties of the active site and the effects on its properties due to the molecular environment [9].

In this contribution, the strategies and QM methods most commonly used to study the structural, electronic, and reactivity properties of metalloproteins and related bioinspired compounds will be presented, focusing on practical and technical issues. In particular, the first two sections deal with the choice of the proper level of theory to study metalloproteins, as well as with different strategies that can be used to design proper models of the molecular system under investigation (cluster models vs. QM/MM models). The third section deals with the peculiarities associated to the computation of the electronic properties of metalloproteins and bioinorganic molecules, which are often characterized by unpaired electrons (i.e., partially filled molecular orbitals). In the fourth section, the methods used to characterize reaction profiles (i.e., the characterization of intermediate species and transition states formed in reactions and catalytic cycles) of metalloenzymes and related compounds are discussed. Finally, the fifth section deals with QM methodologies suited for the investigation of excited electronic states, which are relevant to study the photochemical processes in which several metalloproteins, as well as related bioinspired catalysts, are involved.

2 Materials

Adequate computer power and programs to be described below.

3 Methods

3.1 The Choice of the Level of Theory in the Investigation of Metalloproteins and Bioinorganic Molecules

Among all available QM methods, those based on the Density Functional Theory (DFT) have become the preferred ones for modeling complex chemical systems like metalloproteins, due to their positive trade-off between accuracy and computational efficiency.

The electronic structure of molecules within DFT is represented in terms of the three dimensional electronic density of the system [10], which is a great simplification over wave function theory (WFT), which would imply to use a 3N-dimensional antisymmetric wave function to describe a system with N electrons. As a consequence, DFT computational costs scale more favorably with system size than the corresponding cost of other methods, nevertheless maintaining a comparable accuracy. The advantages of DFT are still greater when dealing with transition metal-containing biomolecules, where correlation effects (i.e., the effects due to interaction between electrons in a molecule) on the ground-state structure, on the electron distribution, and on the energy can be very large. In particular, while the inclusion of correlation effects in a well-balanced way in WFT calculations can be sometimes very difficult, the DFT approach remains simple and is often surprisingly accurate. These empirical evidences, together with the advantages associated to computational efficiency, simply make DFT the most used method to study metalloproteins.

In DFT, the electronic energy of a molecule is expressed by a summation over four energy terms: T_n, ε_{ne}, ε_{ee}, and ε_{xc}. T_n is the kinetic energy of an ensemble of non-interacting electrons with the same spin densities of the real system; ε_{ne} is the interaction energy of the electron distribution with the nuclei; ε_{ee} is the classical Coulomb energy between interacting spin densities; and ε_{xc} is the so-called exchange-correlation energy, which is written as a functional of the spin densities (the so-called spin-density functional). Usually, the density functional is divided into two contributions: the exchange and the correlation functionals. The Hohenberg–Kohn theorem [11] shows that the density functional exists, but the exact spin-density functional does not present a closed-form expression; moreover, no systematic routes to improving an approximate functional can be outlined. Nevertheless, useful approximations have been proposed, and they keep getting better. The first approximation to a density functional has been the Dirac–Slater approximation to exchange [12, 13], which has been derived from the exact exchange energy of a uniform electron gas in which the constant electron density is neutralized by a continuous distribution of positive charges (rather than by discrete nuclear charges). In the DFT theory, this approximation is usually referred to as the local spin density approximation (LSDA), since it depends only on spin densities (and not on their derivatives). The next development in density functionals was the addition of a dependence of the spin densities on their gradients: the so-called generalized gradient approximations (GGAs). GGA functionals may combine an exchange functional from one source with a correlation functional from another. As an example, BP86 combines the Becke's 1988 exchange functional (usually abbreviated as B) [14] and the Perdew's 1986 correlation functional P86 [15], whereas BLYP includes the Lee–Yang–Parr correlation functional, LYP [16]. PW91 and PBE refer to the Perdew and Wang's 1991 exchange and correlation functional [17]; and to the Perdew, Burke, and Ernzerhof functional [18], respectively. mPWPW, which is a modified Perdew–Wang functional by Adamo and Barone [19], is similar to PBE.

The density functional theory with LSDA or GGA functionals includes self-exchange and self-correlation contributions, both of which are unphysical. As a consequence, such functionals tend to predict too small energy gaps between the highest occupied and the lowest unoccupied molecular orbitals (HOMO and LUMO, respectively). By including partial or full Hartree–Fock exchange contributions in the exchange functional, it is possible to reduce some of the inaccuracies of the LSDA and GGA functionals, while retaining many of the computational advantages [20, 21]. In fact, the inclusion of Hartree–Fock exchange leads to the so-called hybrid functionals that are often used to study models of biomolecules, because of their generally good predictions of reaction energies. The most straightforward hybrid functionals result from simply replacing a percentage (x, from 15 to 50 %) of local density functional exchange by Hartree–Fock exchange. By far the most popular hybrid functional used to study metalloproteins is B3LYP ($x = 20$ %) [22], where the correlation functional is based on LYP, but it is optimized specifically for its use in a hybrid functional. Other examples of hybrid functionals include B1LYP [23] (BLYP with $x = 25$ %), mPW1PW [19] (mPWPW with $x = 25$ %), PBE1PBE, PBE0 [23, 24], (PBE with $x = 25$ %), MPW1K [25] (mPWPW with $x = 42.8$ %), and BH&HLYP (with $x = 50$ %).

3.1.1 The Choice of the Basis Set

One of the approximations adopted in QM methods, both at the WFT and DFT level, is the introduction of a basis set, consisting in expanding the unknown molecular orbitals as a finite linear combination of atomic orbitals (AOs). Because the computational effort of ab initio methods scales formally with M^4, where M is the number of basis functions, it is important to use the smallest basis set for describing the molecular system without compromising the accuracy of the results. Two types of basis functions are generally used in QM calculations: the Slater Type Orbitals (STOs) and the Gaussian Type Orbitals (GTOs) [26]. Even if the former best describe the exponential dependence on the distance of the electron from the nucleus, they involve serious complication in the calculation of electron interactions (that is, in the calculations of two-electron integrals). This problem is overcome by using GTOs that allow to easily calculate two-electron integrals. However, GTOs do not describe accurately the behavior of electrons near and far from the nucleus. To address this issue, additional GTOs are generally used to describe each AO. The increase in the number of basis functions is more than compensated for by the faster calculation of the required integrals.

Once decided the type of function for the AOs, the most important next decision concerns the number of functions to be used. In principle, one could use one function to describe one atomic orbital (this choice is known as *minimum basis set*). In fact, this choice is generally not suitable to obtain reliable results. A better strategy implies the use of at least two basis functions for each AO (the *Double Zeta*, DZ, basis set). However, because the "chemistry" of molecules is in the "valence," that is, the valence electrons are those involved in chemical bonding and reactivity, a better choice consists in doubling only the number of valence orbitals, thus providing a *split valence basis set*. Further increasing in the valence orbitals provides the *Triple Zeta* (TZ), and *Quadruple Zeta* (QZ) *split valence basis sets*.

An even better description of the valence electrons can be achieved including also functions with higher angular momentum, which are known as *polarization functions*. For instance, in the case of a C–H bond, the electron distribution along the bond will be different than the distribution perpendicular to the bond. If only *s*-type functions are present on the hydrogen atom, this behavior cannot be correctly described. For this reason, one set of *p*-orbitals is added to the hydrogen atom, allowing a polarization of the *s*-orbital. Similarly, *d*-orbitals can be used for polarizing *p*-orbitals, *f*-orbitals for polarizing *d*-orbitals, and so on. The long experience collected on molecular calculations makes the *polarized split valence basis sets* a reasonable entry-level for medium–large molecular system calculations [27]. To even better describe the electron distribution far from the nucleus, *diffuse functions* can be added to the basis set, allowing a better description of loosely bound electrons (which is important for anions or excited states) or of properties dependent on the wave function tail (for example polarizability). On the other hand, when properties depending on electron distribution very close to the nucleus (as in the case of hyperfine coupling constant of EPR spectra) are of interest, standard basis sets must be replaced by purposely optimized basis sets [28].

3.1.2 DFT Investigation of Biomolecules Containing Elements from the Third or Higher Rows in the Periodic Table

The study of systems including elements from the third or higher rows in the periodic table rises additional problems concerning the definition of appropriate basis sets. In particular, heavy elements present a large number of core electrons, which are generally less relevant from the chemical point of view. Moreover, relativistic effects assume a relevant role in heavy elements. Both problems can be addressed by using an *Effective Core Potential* (ECP), also called *pseudopotential*, to represent the core electrons. That is, the core electrons are modeled by a suitable function, and only the valence electrons are treated explicitly in the calculations. In this way, the computational efforts can be focused on the most chemically relevant part of the system (the valence shell) and part of the relativistic effects may be also taken (implicitly) into account without performing full relativistic calculations (*see* **Notes 1** and **2**).

3.1.3 The Choice of the QM Software

Several QM software can perform DFT calculations using different density functionals. The Gaussian program [29], based on Gaussian basis sets, is one of the most widely used program to study isolated or solvated molecules. It is very flexible in the choice of basis sets, effective core potentials, and density functionals, and it has excellent geometry optimizers. Gaussian supports analytic Hessians (the matrices containing second-order partial derivatives) for all functionals and it features an interface to external programs. The Amsterdam Density Functional (ADF) program [30–32], somewhat uniquely, uses Slater-type orbitals rather than Gaussians. In addition, ADF uses density fitting to reduce computational costs. For LSDA and GGA functionals, analytic Hessians are available. The General Atomic and Molecular Electronic Structure System (GAMESS) package [33, 34] has a broad range of Gaussian-based electronic structure capabilities, many of which are parallelized. Solvent effects can be modeled both by effective fragment models and a continuum solvation model. Jaguar is the electronic-structure program in the suite of computational chemistry and drug design tools offered by Schrodinger, Inc., and can carry out calculations using Gaussians for both isolated and solvated molecules, using a continuum solvent model. MOLCAS [35] is another Gaussian-based electronic structure program for studying molecules. Methods are included for modeling the solvent effects, as well as the embedding in ionic solids or macromolecular environments. TURBOMOLE [36] is another Gaussian-based program in which an auxiliary Gaussian basis set is used to fit the electron density in addition to the primary Gaussian basis set used to describe the orbitals. In some cases this provides large savings in computer time for calculations on large molecules. DALTON 2.0 [37] is a Gaussian-based program including linear, quadratic, and cubic response functions for both singlet and triplet perturbing operators; the properties section is especially complete. Last but not least, ORCA is an efficient and easy-to-use quantum chemistry software, with a specific emphasis on the calculation of spectroscopic properties [38]. It features a wide variety of standard quantum chemical methods ranging from semiempirical methods to DFT to single and multireference correlated ab initio methods.

3.2 The Definition of the Computational Models of Metalloproteins and Related Bioinorganic Molecules

In metalloproteins, the architecture of the protein matrix surrounding the active site can play an important role in determining the geometry properties of the active site, the regiochemistry of substrates binding and the interaction with inhibitors. Therefore, explicit consideration of the protein environment when studying the structure and reactivity of metalloproteins is usually important. Clearly, a high-level quantum mechanical treatment of an entire protein molecule would provide the most reliable approach, and some studies of this kind are appearing in the literature. However, the latter are limited to small proteins for which only the computation of the electronic structure was possible [39]. On the other hand, geometry optimization and evaluation of dynamic behavior are essentially unaffordable within a fully quantum mechanical setup, given the constraints imposed by the computational power of currently available computers.

Two approaches are generally used in the definition of the computational model of a metalloprotein; the so-called cluster approach, in which the metal containing active site and a limited number of atoms in its surrounding are "isolated" from the protein and all treated quantum mechanically, and the QM/MM approach in which a small number of atoms of the active site and its surrounding are described quantum mechanically, whereas all other atoms of the protein are described using a classical force field in the framework of molecular mechanics methods. Both approaches are extensively discussed in recent reviews [1, 40–43].

3.2.1 Cluster Models

In the cluster approach, only a portion of the protein surrounding the active site is explicitly included in the model. In particular, the cluster approach has proven fruitful in the modeling of metalloenzymes, since generally all relevant chemical steps involve the metal centers and nearby residues [44–47].

Several practical issues should be addressed for a correct definition of the cluster model, and among them the most important ones concern: the selection of the atoms to be included in the model in order to describe adequately the active site properties and reactivity, how to treat the peripheral atoms of the model (i.e., where covalent bonds are cut), how to model the effects of the surrounding part of the protein not included in the model, and finally, the choice of the computational scheme (level of theory and basis set).

Considering the size of the model, one would generally like to include as many atoms as possible, in order to take into account the effects due to long-range interactions, which can be very important for an accurate description of a biochemical process. However, it is important to stress that the inclusion of many atoms in the model may increase the occurrence of artifacts at the interface between the QM part and the rest of the protein, which is not included in the model. Therefore, a meticulous inspection of the environment of the active site is always recommended to properly design the model. A minimal model of a metalloenzyme should include all residues and exogenous ligands coordinated to the metal atoms of the active site, and all other residues, which are assumed to be involved in the catalytic mechanism. These residues are generally polar or charged species, which in the protein are often constrained to a functional position through the formation of H-bonds or ionic interactions with other nearby residues. These interactions may also affect the acidity of the residues involved in the catalytic mechanism, tuning their pK_a to values, which are functional for catalysis. Other polar or charged residues, which are not involved in the catalytic mechanism, but are close to the metallic active site, may affect the reduction potential of the metal site. Therefore, the inclusion of fragments of these residues in the cluster model can be crucial in the modeling of redox reactions. Hydrophobic residues that may be important for the catalytic mechanism, for example to properly orient the substrate in the active site, should also be included in cluster models. They are much less affected by nearby polar or charged residues, and the orientation of the side chains of these residues is generally regulated by nonspecific interactions with surrounding atoms. When they are at the interface of the QM model, geometry optimizations can lead to large movements of their side chains, in regions that, in the protein, should be occupied by other residues. Therefore, the behavior of such residues should be carefully monitored during geometry optimizations, and if necessary selected atoms of the side chains should be constrained to the X-ray positions to avoid "unrealistic" movements of these fragments.

Another important issue to be addressed is related to the choice of the peripheral atoms of the model. Generally, the side chains of peripheral amino acids included in the model are truncated at the C_β or C_α atoms, which are saturated with hydrogen atoms. However, carbonyl groups or amide groups of the backbone could be involved in H-bonds with other residues included in the model. In this case, it is a good practice to include also selected atoms of the backbone in the model. The peripheral atoms, i.e., the terminal atoms in the model, are generally fixed at the X-ray position in order to avoid large "unrealistic" displacements of the molecular fragments during geometry optimizations. This means that constrained geometry optimizations are performed maintaining the Cartesian coordinates of the selected peripheral atoms at the original position. Clearly, constraints can give rise to artifacts, in particular for models with a small number of atoms, since they could not allow a proper relaxation of the system. The inclusion in the model of longer side chain fragments and more atoms of the backbone should minimize this problem, since the fixed atoms should be far from the active site, making the system more flexible. However, it is important to underline that an increase in model size also increases the area of the interface between the QM portion and the surrounding protein, which is not included in the model, and therefore the number of interactions at the interface, which are ignored in the calculation. Therefore, a small model including only the metal active site and a few surrounding residues is "rigid," since it does not take into account long range interactions, but it has the advantage that large rearrangements of the surrounding residues should not occur. On the other hand, in larger models which include a large number of residue side chains and atoms of the backbone, large rearrangements can occur if selected atoms are not properly fixed at the X-ray position. In this context a detailed analysis of the interface between the QM portion and the surrounding protein is always necessary in order to identify all interactions, which are ignored. In particular, if elements of secondary structures such as α-helices or β-sheets are truncated at the interface, relevant H-bonds or polar interactions, which may increase local structural rigidity, will be ignored. Atoms participating to form these bonds may be fixed at the X-ray position to model the constraints imposed by these interactions.

As a general protocol, in order to evaluate possible problems and artifacts in the definition of the QM model, it is recommended to start with a minimal set of atoms and to increase gradually the size of the model, by checking the effects of the additional included fragments. This approach also allows to check the convergence of the properties investigated as a function of the size of the model.

As outlined in a recent review by Rulisek et al. [1], the selected cluster model should be validated by checking the correlation between calculated spectroscopic and thermodynamic properties with their experimental counterparts.

It should be noted that the direct application of the cluster model approach only allows the rigorous determination of potential energy differences. For the evaluation of free energy differences vibrational frequency calculations are necessary (*vide infra*). However, when selected atoms are constrained to the X-ray positions, geometry optimizations do not lead to genuine minima or first order saddle points for transition states (*vide infra*). In particular, several imaginary frequencies appear for the reactants and products, and more than one imaginary frequency appear for transition states. However, these imaginary frequencies generally correspond to soft vibrational modes and are almost constant for the reactants, products, and transition states. For transition states, these imaginary frequencies are also generally well separated in magnitude from the one associated to the reaction coordinate. Therefore, entropic corrections based on such calculations offer normally a relatively safe estimate of free energy differences between stationary states.

A further important issue is related to the description of the surrounding protein matrix, which is not included in the model. It can be expected that the protein matrix has a polarizing effect, which can be modeled by soaking the QM portion in a continuum dielectric (i.e., using a continuum solvation model). This is particularly important for metal-containing systems, which often possess non-zero molecular charge. Indeed the presence of a net charge strongly affects the value of the energy of the orbitals localized on the region in which the charge is concentrated (generally the metal site). This "unbalanced" charge distribution can produce unrealistic electron transfers within the model. The continuum dielectric has a "leveling effect" on the orbital energies, minimizing the occurrence of unrealistic electron transfer. However, it is always a good practice to evaluate the computed charge of the separate groups of atoms in the model, in order to check the charge distribution in different regions in the model. As a continuum dielectric, several solvation models like the Conductor-like screening model (COSMO) [48–51], or the Polarizable continuum model (PCM), in various formulations, are commonly used [52–57]. Several values of the dielectric constant varying between 2 and 20 are used to represent the protein environment, even if in most studies a value of 4 is used [41, 58–61].

The effect of model size on computed energies has been recently reviewed by Himo et al. for the enzymatic reaction mechanisms of 4-oxalocrotonate tautomerase (4-OT) and haloalcohol dehalogenase HheC (HheC) [40, 62, 63]. The effects of the protein environment were also studied, using the conductor like polarizable continuum model CPCM [52]. For both enzymes it was shown that small models of the active site, containing 77 and 83 atoms, respectively, do not accurately reproduce reaction energies and barriers. It was also shown that solvation of the model in a continuum medium significantly affects the energy profile of the reactions, increasing the accuracy of the results. Calculations performed on larger models, containing 177 and 161 atoms for 4-OT and HheC, respectively, gave energy profiles fully consistent with experimental observations. In addition, solvation effects practically vanished for these large models (i.e., the energy profile is almost identical with and without addition of homogeneous solvation).

3.2.2 QM/MM Models

A complementary approach to study metalloproteins is represented by the so-called hybrid quantum mechanical/molecular mechanical (QM/MM) modeling, which was pioneered by Warshel and Levitt [64], and has become increasingly popular in the last 20 years. In a standard QM/MM model, the protein matrix surrounding the metal-containing cofactor is described by means of a Molecular Mechanics representation, whereas the active site is represented using a suitable high-level quantum chemical method. As for the latter, DFT is the most popular choice. However, alternative computational strategies based on semiempirical methods [65–69] can be also applied, but their description is beyond the scope of this chapter.

The selection of the most appropriate functional and basis sets for the quantum chemical portion of the model is based on the same considerations exposed above for pure DFT models. Therefore, in the following sections we will mainly focus on other key issues of QM/MM modeling: after a brief discussion of the nature of molecular force fields, we will provide hints on the criteria to be adopted in order to set up the model and to properly choose the atomic composition of the QM and MM systems. Finally, we will discuss the details of possible couplings between the QM and the MM systems, and mention the strengths and limitation of QM/MM approaches as compared to purely QM modeling. Before proceeding with such details, it is crucial to illustrate the "gross features" of a QM/MM model, taking Fig. 1 as reference. The computationally expensive QM portion (System 1) is normally chosen to be much smaller than the surrounding MM part

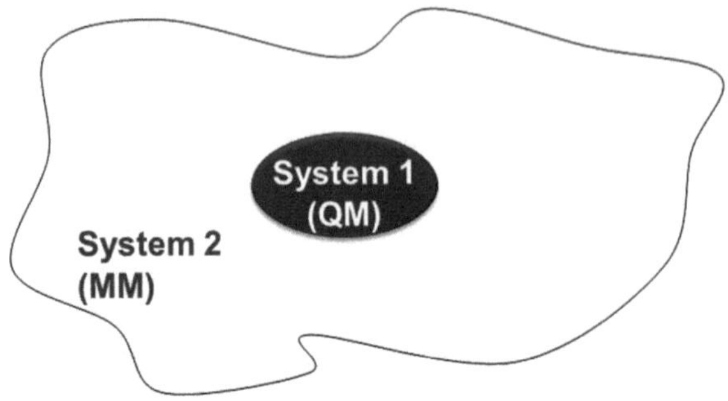

Fig. 1 Schematic representation of a QM/MM model

(System 2); the latter can include not only most of the atoms that compose the protein, but also a suitable representation of the solvent. The total energy of QM/MM model can be expressed as follows:

$$E_{\mathrm{QM/MM}} = E_{\mathrm{MM}}^{\mathrm{System1+2}} + E_{\mathrm{QM}}^{\mathrm{System1}} - E_{\mathrm{MM}}^{\mathrm{System1}} \tag{1}$$

where $E_{\mathrm{MM}}{}^{\mathrm{System1+2}}$ is the energy of the whole metalloprotein (possibly with solvent) computed at MM level, $E_{\mathrm{QM}}{}^{\mathrm{System1}}$ is the QM energy of the quantum mechanical portion, and $E_{\mathrm{MM}}{}^{\mathrm{System1}}$ is again the energy of System 1, but computed at MM level. The rationale behind such equation will become clearer after a detailed discussion of the composition of the quantum region (*vide infra*).

In general, the energy of an MM system is computed using a simple summation of classical terms—usually referred to as "force field" in the MM terminology—that depend on bond lengths, bond angles, dihedral angles, and include also nonbonding contributions. Atoms are modeled as points in space that possess a given charge and are at the center of a van der Waals sphere of given radius; the electrostatic and van der Waals non-bonding energies between pairs of atoms depend on such parameters. Several force fields are available; AMBER [70, 71], CHARMM [72], GROMOS [73], and OPLS [74] are very popular choices.

In the definition of the MM portion of a QM/MM model it is fundamental that the atomic composition of the whole protein is described correctly. In this context, a crucial practical issue is represented by the protonation state of titratable side chains, such as histidine residues. In fact, the pK_{a} of histidine is 6.5; therefore, at physiological pH values its protonation state can vary from the neutral form (only one of the two N atoms in the imidazole side chain is bound to an H atom) to the cationic form (in which both the N atoms in the ring are bound to H). The cationic state of histidine is usually referred to as "HIP," while in the case of the neutral form two isomers are possible: the so-called HID and HIE forms that feature H-bound N^{δ} and N^{ε} atoms, respectively (*see* Fig. 2).

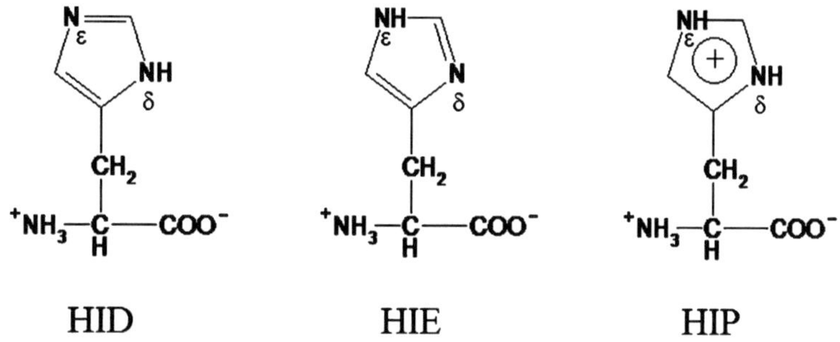

Fig. 2 Possible isomers for neutral and protonated histidine residues

What follows is a simple protocol to set the protonation state of His residues:

- When a histidine residue is buried within the protein matrix, the presence of possible hydrogen bond donors or acceptors in the neighborhood is the first aspect to be taken into account. In particular, when inspection of the crystal structure shows that one of the N atoms is a hydrogen bond acceptor, then the other N atom of the imidazole ring must be protonated. On the other hand, strong hydrogen bonds (characterized by very short inter-residue distances) usually involve the negatively charged carboxylic groups of aspartate and glutamate moieties, and this is a strong indication for the presence of HIP.
- When the imidazole ring is solvent exposed, it is still important to carefully look at the environment: the presence of a strong hydrogen bond to one of the N atoms is again an indication for the presence of HIP. Another indication may come from the overall charge of the protein before assigning the specific protonation states of the histidine residues: if the overall charge is very negative, this may suggest the presence of a relatively large number of HIP residues. Clearly, solvent-exposed histidine residues that do not establish hydrogen bonds with other residues are the first ones to be considered for an increase in the number of HIP residues in the model.

The choice of the subregion of a metalloprotein to be included in the quantum mechanical portion of a QM/MM model is another crucial step in the setup of the computational model. In particular, two key issues are: (1) the treatment of hydrogen bonding at the border between QM and MM regions; (2) the extension of the QM region in the case of metalloproteins containing several metal atoms. As far as hydrogen bonding is concerned, when the metal-containing active site (i.e., the QM subsystem) establishes hydrogen bonds with surrounding residues, an MM representation of the latter residues usually provides satisfactory results in terms of reproduction of donor–acceptor distances, as well as in terms of interaction energies. As for the extension of the QM region in the case of metalloproteins containing several metal ions, it is important to remark that when ancillary metal clusters are thought to be able to establish electronic communication with the active site, their inclusion in the MM portion may lead to possible artifacts. In fact, the resulting QM/MM model will be unable to reproduce intramolecular redox processes that might occur among the various metal ions. In such cases, extension of the QM system to some or all metal clusters may represent a better choice.

When a hybrid QM/MM model of a protein is set up, a rule must be defined to couple the MM potential with the quantum mechanical potential, keeping in mind that System 1 and System 2 always establish non-bonded interactions and are usually connected also by covalent bond(s). Various ways to implement the coupling between the two regions of a QM/MM model have been reported. In the following section, we will specifically focus on the "electrostatic embedding" approach for the Coulombian terms [75], and on the "hydrogen link" scheme for the treatment of covalent interactions between the QM and MM regions. Brief mention of alternative techniques for QM/MM coupling will be also provided. The reader interested in thorough and authoritative overviews of all aspects of hybrid modeling of protein systems is referred to reviews recently published by Guallar [76], Tateno [77], Salahub [78], Thiel [66], Morokuma [79], and their respective coworkers.

The electrostatic embedding approach is based on the introduction of the point charges of the MM region directly in the Hamiltonian of the quantum mechanical system. Most of the available quantum chemical codes allow to efficiently perform DFT calculations characterized by this kind of embedding, which lets the quantum mechanical system be polarized by the surrounding protein matrix in a self-consistent way. Electrostatic embedding allows to directly exploit a highly accurate quantum chemical Hamiltonian to compute the long-range Coulombian interactions between System 1 and System 2 [75]. As far as the covalent connection between the QM and MM system is concerned, it is usually convenient to place the border at the level of C–C bonds. For example, if a protein-embedded Fe_4S_4 cluster has to be included in the QM system of a QM/MM model, a reasonable choice is to impose that the border between System 1 and 2 lays between the α and β carbons of the four cysteine residues that coordinate the Fe ions (*see* Fig. 3). However, it is clearly not reasonable to describe the four terminal groups as CH_2 radicals. The nonpolar nature of the C–C bond suggests that a feasible option would be to saturate the QM system with hydrogen atoms, in consideration of the fact that the C–H bond is similarly nonpolar. Such H atoms are usually referred to as link hydrogen atoms. However, the substitution of carbon with hydrogen raises important issues: first of all, the spatial localization of the link hydrogen atom has to be defined; secondly, the energy

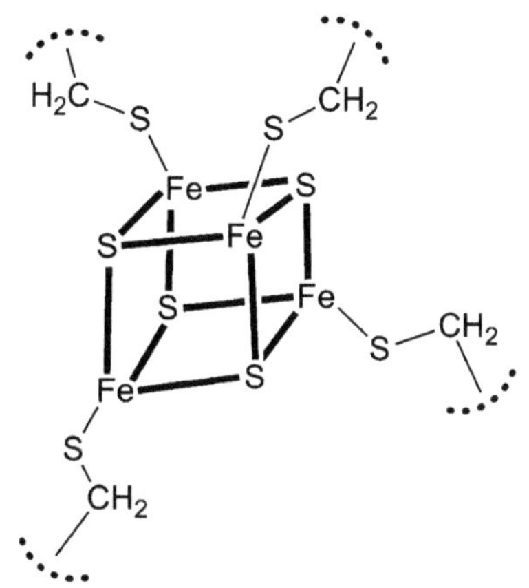

Fig. 3 The choice of the QM system of a QM/MM model. Possible definition of the borders between System 1 and 2, between the α and *β* carbons of the four cysteine residues that coordinate the Fe ions in a protein-embedded Fe_4S_4 cluster

contribution of these atoms—which are fictitious, i.e., nonexistent in the real system—has to be canceled out when computing the total energy of the QM/MM system. As far as the first point is concerned, it is possible to establish a simple linear relation between the C–C bond lengths in the original system and the C–H bonds which saturate System 1 (see for example [80]). As for the exclusion of the hydrogen link atom when the total QM/MM energy is computed, Eq. 1 represents a convenient solution. In fact, the first term on the right hand side is the classical energy of the whole system without hydrogen links; the second term is the QM energy, which includes the capping H atoms, while the third term is the MM energy of System 1 again featuring capping H atoms. The subtractive nature of the last term has the effect of formally eliminating the contribution of hydrogen links from the computed total energy. Clearly, a preliminary step before evaluating the $E_{MM}{}^{System1}$ term is the development of MM parameters for the fictitious C–H bonds, to be included in the energy expression of System 1 when it is treated classically. Such parameters are usually obtained by means of high-level quantum mechanical calculations on the truncated amino acid [81]. Finally, for the sake of completeness, let us mention that alternative techniques to cut out the QM system from the protein have been proposed, based on the inclusion of special local orbitals at the level of the QM/MM borders [64]; for further details and a comparison between the latter approach and the use of link hydrogen atoms, see Karplus et al. [82].

After having set the atomic composition of System 1 and System 2, as well as the functional form of the QM/MM total energy, it is possible to carry out the geometry optimization of the model (Scheme 1). The simplest protocol implies to initially relax only the

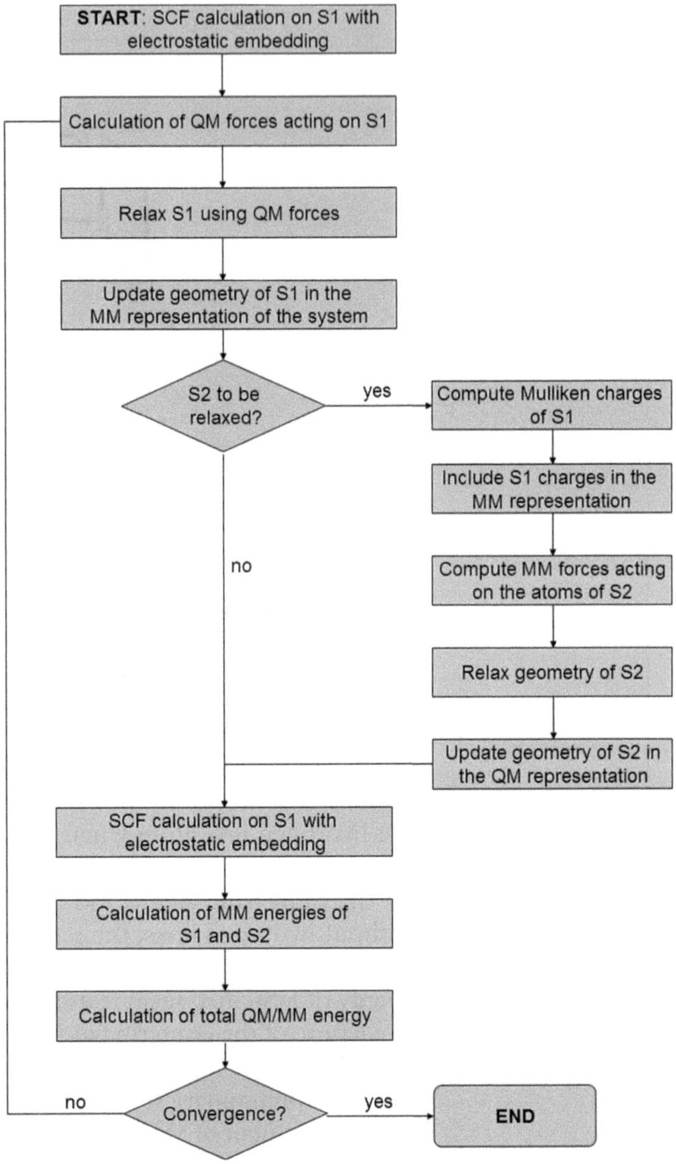

Scheme 1 General protocol for QM/MM calculations

QM system, while keeping the structure of System 2 frozen; in particular, when an X-ray protein structure is used as the starting geometry, freezing System 2 means that its atoms are kept at their respective crystallographic positions. Geometry optimization requires knowledge of the forces acting on atoms, which can be obtained as energy derivatives. To optimize System 1, one has to compute the quantum mechanical forces within the QM portion of the model, and the MM forces from System 2 onto the atoms of System 1 (*see* **Note 3**). The coordinates of the atoms of System 1 are then modified based on the composition of such forces, leading to relaxation of the QM/MM model. Calculation of both QM and MM forces, as well as the relaxation steps, are efficiently performed by many DFT and molecular mechanics packages currently available. It is also possible to let System 2 (or a portion of the latter) be free to relax as well. In this case, it is first necessary to obtain atomic charges for the atoms of System 1, which can be easily derived from DFT calculations. Subsequently, one can use these atomic charges to define an MM representation of the whole protein, and compute the MM forces acting on the atoms of System 2 (*see* **Note 4**). The information coming from computed forces can then be exploited to relax the latter system (or just a portion of it). The updated geometries of Systems 1 and 2 can then be taken as starting structures for new optimization rounds, that are terminated when the change in energy and geometry becomes lower than an arbitrarily low threshold (Scheme 1).

The reader might have noticed that all the discussion about QM/MM models lacked any reference to time and temperature; however, introduction of the latter two variables in hybrid quantum/classical modeling is actually possible, and normally allows obtaining better convergence in QM/MM energy calculations. An approach that is becoming increasingly popular is the application of classical molecular dynamics (MD) to the atoms of System 2. The application of MD in a QM/MM model requires defining a set of initial velocities for the atoms in the MM system, which enables the model to account for the vibrations taking place within the protein matrix. Basically, MD simulations proceed iteratively by alternatively calculating forces and solving the Newtonian equations of motion based on the accelerations obtained from the new forces. The trajectory thus obtained, which can be propagated for an arbitrary long amount of time (typically in the order of nanoseconds), defines a thermodynamic ensemble. The latter includes many possible states which, as a result of molecular motion, a protein system is rapidly passing through. From a practical perspective, MD allows sampling many (and virtually all) minima that characterize the QM/MM potential energy surface, giving a more realistic representation of the protein, which can be exploited to obtain highly accurate theoretical reaction energies. Examples of methods which couple QM/MM and MD have been described by several authors [83, 84].

3.3 QM Description of the Electronic Properties of the Ground State of Metalloproteins

The most important output of any DFT calculation is the electron density, which is computed from the wave function within the Kohn–Sham approach. Although this latter is not the best wave function from the point of view of the variational principle (in DFT one minimizes the energy optimizing the electron density), it allows to shed light on the electronic structure of the molecular system under investigation. In the DFT Kohn–Sham approach (DFT-KS) electron density is computed from a wave function expressed as an antisymmetric product (Slater determinant) of molecular orbitals (MOs), each of which is expanded as a linear combination of a fixed sets of atom centered functions (the basis set). Both in wave function and DFT methods, MOs are computed using an iterative procedure, starting form an initial guess. Typically, for each electron the orbital is optimized in the field of all the other electrons. This process is repeated until convergence is achieved and each orbital is optimal in the self-consistent field (SCF) of all other electrons. Because each MO is generally expressed as a linear combination of atomic orbitals (AO), the SCF procedure provides the AO coefficients of the final MOs. Current quantum-mechanical software use very efficient methods to perform the SCF procedure: one of the most used method is the *Direct Inversion in the Iterative Subspace* (DIIS) that is fast and overcomes most of the convergence problems of the SCF procedure (*see* **Notes 5** and **6**).

In analogy with the ab initio Hartree–Fock mono-determinant method, three "types" of DFT-KS calculations can be performed depending on the restraints imposed on the wave function. These are the restricted closed shell (RKS), the restricted open shell (ROKS), and the unrestricted (UKS) schemes. The restricted scheme is applied to closed shell molecules (i.e., those molecules in which all electrons are paired to give a system with zero total spin). In this case, MOs are doubly occupied with electrons of opposite spin, and one-electron spin-orbitals are obtained by multiplying the same spatial part of each MO with α and β spin functions (*see* Fig. 4a and **Note 7**). ROKS is an extension of RKS to open shell molecules (i.e., those molecules with unpaired electrons) in which doubly occupied MOs are still used for paired electrons, and singly occupied MOs are used for the unpaired electrons (*see* Fig. 4b). UKS is also generally applied to open shell molecules, but in this case different singly occupied orbitals are generated for electrons of opposite spin (i.e., all α and β orbitals have different spatial parts; *see* Fig. 4c). The UKS scheme, when applied to closed shell molecules, should lead to the RKS wave function, even if in some specific cases unrestricted calculations converge to a lower energy solution than RKS. The UKS scheme has several advantages with respect to the ROKS one: it generally gives lower energy than the corresponding ROKS scheme, it is capable of providing a qualitative correct description of bond dissociation, and it accounts for spin polarization, which can be very important for the correct description of the electronic structure of metalloproteins. The major disadvantage of the UKS scheme is that, contrary to RKS and ROKS, the calculated wave function is not an eigenfunction of the $\widehat{\mathbf{S}}^2$ operator, giving rise to the problem of spin contamination (*see* **Note 8**).

For a spin-restricted determinant wave function of N electrons (hereafter indicated with the label ψ_i), the electron density is:

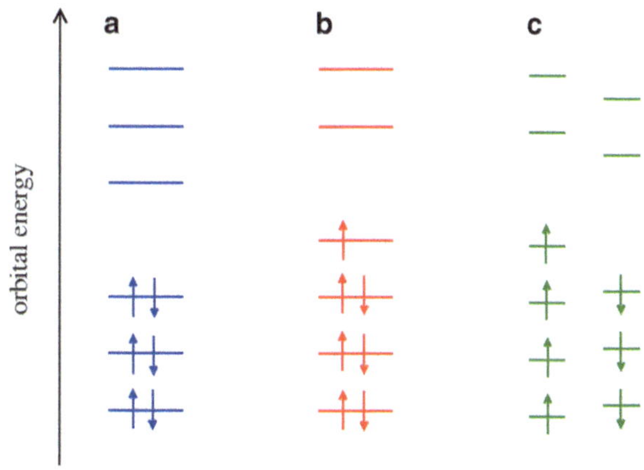

Fig. 4 Schematic representation of (**a**) restricted closed shell (RKS), (**b**) restricted open shell (ROKS), and (**c**) unrestricted (UKS) schemes

$$\rho(r) = 2\sum_{i}^{N/2} |\psi_i(r)|^2 \tag{2}$$

with $\rho(r)dr$ corresponding to the probability of finding one of the N electrons in dr at a distance r. The integral of $\rho(r)$ over the volume is the total number of electrons N.

For a spin-unrestricted wave function, in which electrons with spin α and β occupy MOs with different spatial distributions (hereafter labeled ψ_a^α and ψ_a^β, respectively) the electron density can be partitioned in the α and β contributions:

$$\begin{aligned} \rho^\alpha(r) &= \sum_{a}^{N\alpha} |\psi_a^\alpha(r)|^2 \\ \rho^\beta(r) &= \sum_{a}^{N\alpha} |\psi_a^\beta(r)|^2 \end{aligned} \tag{3}$$

where $N\alpha$ and $N\beta$ are the number of α and β electrons, and the total charge density is given by the sum of the two contributions:

$$\rho(r) = \rho^\alpha(r) + \rho^\beta(r) \tag{4}$$

In this case, a spin density function can also be defined as:

$$\rho^S(r) = \rho^\alpha(r) - \rho^\beta(r) \tag{5}$$

so that $\rho^S(r)dr$, if positive, corresponds to the probability of finding an "excess" of α electrons in dr at distance r, while if negative, it corresponds to the probability of finding an "excess" of β electrons in dr at distance r. Therefore, $\rho^S(r)$ describes the spatial distribution of spin in a molecule with $N_S = N\alpha - N\beta$ unpaired electrons. The integral of $\rho^S(r)$ over the volume corresponds to the number of unpaired electrons N_S.

Population analysis of charge and spin densities are generally carried out in order to calculate atomic partial charges and atomic spin densities, which can be very useful for a qualitative description of the charge and spin distributions in a metalloprotein. In this context, it is important to underline that atomic partial charges are not physical observables (they cannot be measured), and there is no unique mathematical method to calculate them. The problem basically resides in the definition of the electron density "belonging" to an atom in a molecule. Many different schemes have been proposed to partition the electron density among the atoms in a molecule, giving rise the different methods to calculate atomic partial charges. The charges calculated using different schemes can vary significantly, and clearly, they should never be discussed in absolute values, but rather as relative values to compare different protein states and/or redox states. The simplest scheme for the calculation of atomic partial charges, implemented in all QM software, is the Mulliken population analysis. In fact, the number of electrons on each atom is obtained summing over the population of all basis functions centered on that atom. Then, the atomic charge is calculated as the difference between nuclear and electron charges. This method can give unreasonable results when the basis set is not "well balanced." In particular, the presence of diffuse basis functions in large basis sets can also account for electron density far from the atom where they are centered, increasing incorrectly the net charge on that atom.

Another common step in the investigation of the electronic structure of a protein is the analysis of the computed Frontier Molecular Orbitals (FMOs) and of their energies (or eigenvalues). In proteins containing transition metal atoms, the FMOs can be usually qualitatively described using the ligand field theory, where each MO can be classified as essentially localized on a metal center or a ligand. In fact, FMOs can be conveniently displayed using one of the many available visualization programs, which allow to display the isosurface of a given MO computed on a regularly spaced grid of points. However, it is important to underline that often the FMOs can be delocalized on a large portion of the molecule and therefore this approach provides only a qualitative description of the electronic structure. A more quantitative picture can be obtained from population analysis (vide infra). In any case, analysis of the highest occupied (HOMO) and the lowest unoccupied (LUMO) molecular orbitals might allow to disclose and to rationalize basic reactivity properties. For example, the portion of the system where the HOMO is localized should be susceptible of electrophilic attack, whereas nucleophilic attacks are more likely where the LUMO is localized.

The HOMO and LUMO orbital energies and the HOMO/LUMO energy difference (also referred to as the HOMO/LUMO gap) can allow the evaluation of ionization potentials, electron affinities, electronegativity, as well as hardness and softness properties, which are important to rationalize several properties of biomolecules. Ionization potentials and electron affinities can be computed from the HOMO and LUMO orbital energies in a rigorous way (although approximated) when using Hartree–Fock (HF) methods. In DFT the meaning of orbital energies (in particular unoccupied MOs) is less clear. In addition, self-interaction errors (SIE) due to the spurious Coulomb interaction on an electron with itself could affect the energies and the ordering of FMOs. Therefore, properties which depend on orbital energies obtained by DFT calculations should be discussed cautiously.

3.3.1 Magnetic Properties

Several metalloproteins contain metal ions with unpaired electrons, which require for their proper description unrestricted spin polarized methods, where α and β electrons are treated with a different potential, and therefore singly occupied orbitals are generated for electrons of different spin. The corresponding one-electron orbital energies are also spin dependent, due to the different exchange interactions within α and β electrons.

In the unrestricted Hartree–Fock (HF) method the calculated wave function is not an eigenfunction of the $\widehat{\mathbf{S}}^2$ operator, and therefore, it does not represent a pure spin state (*see* **Note 8**). In this case, the $<S^2>$ value is higher than that of the pure spin state, meaning that the wave function is mixed with higher spin multiplicity states. The spin contamination of the wave function is as large as the difference between the value of $<S^2>$ and that of the pure spin state. This concept has been exported in the unrestricted DFT formalism, even if the meaning of $<S^2>$ in this case is still a question of debate.

The spin multiplicity of metal ions in proteins is generally determined by experimental techniques such as EPR and Mössbauer spectroscopies, and it is therefore given as an input parameter in DFT calculations. However, for elusive states which are not characterized experimentally, or for which the data are ambiguous, the spin multiplicity of the metal cluster must be predicted by comparing the energy of different spin states. In this respect, it is known that DFT functionals are often not well suited to describe the correct multiplicity of the ground state. As an example, different density functionals yield results differing by up to 1 eV for transition-metal complexes with sulfur-rich ligands [85]. As mentioned above, hybrid functionals include a fraction of HF exchange, which can improve descriptions of many properties [86]. In particular, the widely used B3LYP, with 20 % HF exchange, performs well for several structures and chemical reactions [10, 87, 88]. However, the exchange contribution tends to stabilize those states with higher multiplicities owing to the explicit consideration of Fermi correlation, and for systems with near-degenerate ground states, such as many transition metal complexes, the 20 % HF exchange in B3LYP is excessive, favoring high spin with respect to low spin states. On the other hand, "pure" functionals are expected to favor low spin states. In this context, Reiher found that relative multiplet energies depend almost linearly on the amount of HF exchange, and therefore reducing the HF exchange contribution in B3LYP improves accuracy [85, 89]. A good agreement with spectroscopic data is obtained using the B3LYP* functional in which the HF exchange is decreased to 15 %, or the BP86 hybrid with 10 % HF exchange. In a recent review on iron spin-crossover compounds, B3LYP* is still reported as the most suitable DFT functional for the calculation of spin state splitting [90] (*see* **Note 9**).

The metal atoms contained in enzymes can interact weakly, such as in 4Fe4S clusters, generating antiferromagnetic coupling between spins localized on different atoms. These interactions can be described using a phenomenological Hamiltonian, as proposed by Heisenberg, Dirac, and Van Vleck [91–93]. For a molecular system composed by two spin centers the phenomenological Heisenberg Hamiltonian takes the form:

$$\widehat{\mathbf{H}} = J_{AB}\left(\widehat{\mathbf{S}}_A \cdot \widehat{\mathbf{S}}_B\right) \tag{6}$$

where $\widehat{\mathbf{S}}_A$ and $\widehat{\mathbf{S}}_B$ are the spin operators on centers A and B, respectively, and J_{AB} is the exchange coupling constant, which represents the strength of the exchange interaction between the two spin centers. By convention, J_{AB} is lower than 0 for antiferromagnetic interactions, and greater than 0 for ferromagnetic interactions. The eigenvalues of the $\widehat{\mathbf{S}}_A \cdot \widehat{\mathbf{S}}_B$ operator for different total spin states can be derived from the $\widehat{\mathbf{S}}^2$ operator using the relationship:

$$\widehat{\mathbf{S}}^2 = \left(\widehat{\mathbf{S}}_A + \widehat{\mathbf{S}}_B\right)^2 = \widehat{\mathbf{S}}_A^{\ 2} + \widehat{\mathbf{S}}_B^{\ 2} + 2\widehat{\mathbf{S}}_A \cdot \widehat{\mathbf{S}}_B \tag{7}$$

In Eq. 7, the eigenvalue of $\widehat{\mathbf{S}}^2$ for the low spin singlet state (where unpaired electrons are localized on the two centers with opposite spins) is equal to 0, the eigenvalues of $\widehat{\mathbf{S}}_A^2$ and $\widehat{\mathbf{S}}_B^2$ are equal to 3/4, and therefore, the eigenvalue of $\widehat{\mathbf{S}}_A \cdot \widehat{\mathbf{S}}_B$ must be equal to $-3/4$. Using this approach, for spin centers with more than two electrons the Heisenberg Hamiltonian gives the spin state energies:

$$E(S_{\text{tot}}) = 1/2 J_{AB} S_{\text{tot}}(S_{\text{tot}} + 1) \tag{8}$$

where S_{tot} is the total spin ranging from $|S_A - S_B|$ to $|S_A + S_B|$.

The difference between the energies of two successive spin states is:

$$E(S_{\text{tot}}) - E(S_{\text{tot}} - 1) = J_{AB} S_{\text{tot}} \tag{9}$$

allowing to construct, given the J_{AB} value, the full ladder of pure spin states energies.

Such spin-coupled systems are difficult to treat with DFT because their ground (low spin) state wave functions typically correspond to linear combinations of multiple determinants. However, to discuss energies and properties of these metal clusters, it is important to describe the weak antiferromagnetic coupling at the same level of theory of strong metal–metal and metal–ligand bonds. In the framework of the unrestricted formalism, these exchange interactions can be treated reasonably well by employing the broken symmetry (BS) approach developed by Noodleman et al. [94, 95] (*Editors' note: This approach is discussed in detail by J.M. Muesca in Chapter 14*). The BS approach consists in the localization of the opposite spins in different parts of the molecule to give a mono-determinant representation of the spin exchange interactions within the molecule. For example, in a system composed by two spin centers a BS wave function can be calculated by localizing unpaired electrons with spin α on center A and with spin β on center B. Clearly, if symmetry is used, the BS state is determined by reducing the symmetry of the electronic wave function relative to the nuclear geometry; a condition from which the term broken symmetry is derived. The BS wave function is not a pure spin state of the molecule (it is an eigenfunction of S_z, but not of S^2) and therefore it represents an unphysical situation. In this context, the BS wave function can be considered as a low-spin wave function affected by a large spin contamination. In spite of this restriction, Noodleman [94, 95] showed that the BS state is a weighted average of pure-spin wave functions, where the weights are the Clebsch–Gordan coefficients relating the spin quantum numbers for oppositely aligned spin vectors of the subunits, and that correspondingly the BS state energy is a weighted average of pure spin state energies.

In the case of weakly interacting systems, for which unpaired electrons are well localized onto the metal centers, the BS state energy can be correlated to the spin state energies derived from the Heisenberg Hamiltonian, from which the value of the exchange coupling constant can be evaluated.

In particular, for a two spin centers system the J_{AB} value can be determined from the BS wave function and the pure high spin wave function, which is easily calculated by aligning ferromagnetically the spins of the metal sites.

The expectation values of the spin operator product in Eq. 6 for the HS and BS state are:

$$\langle S_A \cdot S_B \rangle_{\mathrm{BS}} = -S_A S_B \quad (10)$$

$$\langle S_A \cdot S_B \rangle_{\mathrm{HS}} = +S_A S_B \quad (11)$$

from which the difference between HS and BS state energies can be written as:

$$E(\mathrm{HS}; S_{\max}) - E(\mathrm{BS}; M_S = |S_A - S_B|) = +2J_{AB} S_A S_B \quad (12)$$

and used to determine J_{AB}:

$$J_{AB} = \frac{E(\mathrm{HS}; S_{\max}) - E(\mathrm{BS}; M_S = |S_A - S_B|)}{2S_A S_B} \tag{13}$$

The calculated value of J_{AB} can then be used to project the BS state energy to any pure spin state energy. In particular, the energy difference between the high spin and the low spin ground state ($E(S_{\max}) - E(S_{\min})$) can be derived from Eq. 8 as:

$$E(S_{\max}) - E(S_{\min}) = 1/2 J_{AB} S_{\max}(S_{\max} + 1) - 1/2 J_{AB} S_{\min} (S_{\min} + 1) \tag{14}$$

and the difference between this equation and Eq. 10 gives $\Delta E_{\mathrm{spin}} = E(\mathrm{BS}) - E(\mathrm{GS})$, which represents the energy correction of the BS state energy to the pure spin ground state energy.

The description of the BS approach given above for a two-spin centers system can be easily generalized to a system with multiple spin centers by using the Heisenberg Hamiltonian:

$$\hat{\mathrm{H}} = \sum_{ij} J_{ij}\left(\hat{\mathbf{S}}_i \cdot \hat{\mathbf{S}}_j\right) \tag{15}$$

where the summation is extended over all the pairs of spin centers i and j bearing spin S_i and S_j, and J_{ij} is the exchange coupling constant between them. The J_{ij} parameters can be determined from the HS state and a consistent number of different BS states calculated by aligning ferromagnetically or antiferromagnetically the spins of different centers, and applying the spin algebra of Eqs. 10 and 11. In particular, for N spin centers, $2^N/2$ BS states can be defined by non-equivalent combinations of spins. In order to reduce the number of coupling constants to be determined, the summation in Eq. 15 is generally restricted to adjacent spin centers.

From a practical viewpoint, different protocols can be adopted to generate BS wave functions, depending on the complexity of the molecular system and the spin coupling scheme. A fast way to generate a BS wave function implies to start from a high spin (HS) wave function as a guess for the BS state. Consider for example a two spin centers model in which two unpaired electrons, each localized on one spin center, are coupled antiferromagnetically. In this case, one can first generate the high spin (triplet) wave function in which the two unpaired electrons with the same spin are localized on the two spin centers. The MOs generated from this calculation can then be used as a starting guess in order to obtain the BS state by: (1) replacing the β MOs with the α ones, and (2) restarting the calculation with the multiplicity of the molecular system equal to 1 (singlet; no unpaired electrons) rather than 3 (triplet; two unpaired electrons). Starting from this guess, the calculation may converge to the BS wave function in which the two electrons remain localized on the two metal atoms with α and β spin, modeling the antiferromagnetic interaction between the two spin centers. It should be noted that, even starting from a proper guess, the SCF convergence to the BS state can be challenging. Population analysis of the BS wave function should reveal an atomic spin density equal to about 1 on one metal center and about −1 on the second metal center. Clearly the sum of the spin density for all of the atoms must be equal to 0.

The protocol discussed above cannot be adopted for molecular systems with more than two antiferromagnetic spin centers. In addition, it often fails for molecular systems with several unpaired electrons on each atom, or with highly asymmetric environments of the spin centers. In these cases, BS wave functions can be generated starting from a guess of MOs in which the system is partitioned into different fragments with the proper number of α or β unpaired electrons. When using the Turbomole QM package the following protocol can be followed: (1) the molecular system is divided into fragments containing each spin center. (2) A high spin calculation is then performed for each fragment with the number of unpaired electrons coupled antiferromagnetically to those of the other spin centers. The atoms of the fragments must be frozen at the corresponding coordinates of the entire molecular system. This step generates MOs for each fragment with the proper number of unpaired α electron localized on the spin center. (3) If a BS wave function with unpaired β electrons on such a fragment is needed, α and β MOs should be exchanged. (4) The MOs of two adjacent fragments are "merged" in order to generate the MOs for the system composed by the two fragments. In our laboratory this step is carried out using a home-made program, which simply reads the MOs of the two fragments and generates new MOs for the composite system, by setting to zero the MO coefficients of the AO basis functions centered on the other fragment. In other words, as shown in Scheme 2, starting from $n \times n$ and $m \times m$ matrices, containing the MOs of the two fragments, a $(m + n) \times (m + n)$ matrix is produced in which the missing coefficients are set to zero. The new MOs are then ranked according to orbital energies. This new set of MOs represents a system composed by the two noninteracting fragments, in which unpaired electrons are localized on the different spin centers with the proper spin. This procedure is repeated by merging the MOs of the other fragments, one at a time, with those obtained from a previous merging-process, in order to generate a set of MOs for the entire molecular system. In these MOs, electrons of opposite spins are localized on each spin center, and therefore, they can be used as a guess to generate the selected BS solution. Several QM software have recently implemented procedures to routinely generate a guess for a BS calculation, which are also based on the partitioning of the molecular system into fragments. For example in the popular QM package Gaussian,

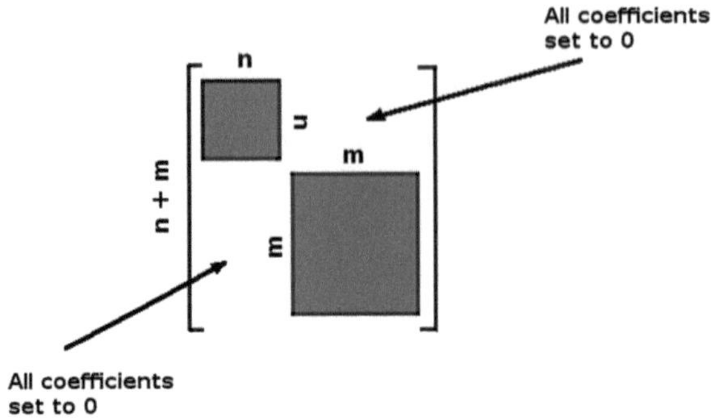

Scheme 2 $(m + n) \times (m + n)$ matrix containing MOs of fragments 1 and 2

the different fragments in which the molecule is partitioned, as well as charge and spin multiplicity of each fragment, can be specified in the input file.

As mentioned above, even if the guess of MOs has been properly generated, the convergence of the wave function in the SCF procedure to the BS state may be a difficult task. Indeed, it is common to observe that after a few SCF cycles the energy value begins to oscillate, with an amplitude of several hartree. In this context, damping and level shifting techniques can be very useful (*see* **Note 5**). In fact, to avoid energy oscillations it is often necessary to set the level shifting to very large values (e.g., 0.4–0.5 hartree). Damping should also be set to large values in order to have large mixing of the new MOs with those of the previous SCF cycle. Clearly, both damping and level shifting should be reduced during the calculation, when the wave function starts to converge properly.

A biologically relevant case that can be used as reference example to show how to compute the properties of metal atoms that are coupled antiferromagnetically is the 4Fe4S cluster found in many metalloproteins. This cluster is generally anchored to the protein by the sulfur atoms of four cysteine residues which are coordinated to the Fe atoms, and it is often involved in electron transfer from the active site to the surface of the protein, or it is an integral part of the active site, as in [FeFe]-hydrogenases, in Acetyl Coenzyme A synthases and in nitrite and sulfite reductases. The electronic structure of the $[Fe_4S_4(SCH_3)_4]^{-x}$ model cluster has been extensively investigated by Noodleman et al. [96–100]. Spectroscopic studies on 4Fe4S clusters in synthetic complexes and in proteins [101–107] have shown that the four Fe atoms generally occur in equivalent pairs of metal centers coupled ferromagnetically, forming two Fe_2S_2 layers coupled antiferromagnetically to give the overall low spin state. This coupling scheme supports the use of a spin Hamiltonian defined in terms of pairs of centers. For the highest ($x = 0$) and lowest oxidation state ($x = -4$), the two $Fe^{2+}Fe^{2+}$ or $Fe^{3+}Fe^{3+}$ layers coupled antiferromagnetically are treated as single spin centers with $S = 4$ and $S = 5$, respectively, and the correction for the energy of the diamagnetic pure spin state is readily obtained by applying the equations given above. In the case of the $[Fe_4S_4\ (SCH_3)_4]^{-2}$ model cluster, which is composed by the two mixed valence $Fe^{2+}Fe^{3+}$ dimers, Mössbauer studies have shown that the four Fe atoms are equivalent, indicating an $Fe^{2.5+}Fe^{2.5+}$ oxidation state for each layer. This intralayer delocalization of the valence electrons can be described by introducing an additional term into the Heisenberg Hamiltonian operator which produces a splitting of the spin states [108–110]. Several authors have shown that the spin state of the system is generally the result of the competition between the Heisenberg and this resonance delocalization terms of the Hamiltonian [108, 111, 112].

The $[Fe_4S_4(SCH_3)_4]^{-1/-3}$ clusters with charge −1 or −3 are composed by a mixed-valence $Fe^{2.5+}Fe^{2.5+}$ layer and by a $Fe^{3+}Fe^{3+}$or $Fe^{2+}Fe^{2+}$ layer, respectively. In these cases a third parameter, which describes the possible interlayer resonance delocalization between the ferrous or ferric layers and the mixed-valence layer, has to be added to the spin Hamiltonian operator.

The values of J calculated for the 4Fe4S cluster in different redox states range from about 100 cm^{-1} in the all ferrous $[Fe_4S_4(SCH_3)_4]^{-4}$ model cluster to about 900 cm^{-1} in the all ferric $[Fe_4S_4(SCH_3)_4]^0$ model cluster [98]. For a detailed explanation of the approaches used for the calculation of the exchange parameters the reader is referred to the extensive work of Noodleman et al. [97].

A slightly more complicated example of the application of the BS approach can be found in studies dealing with the H-cluster contained in the [FeFe]-hydrogenases, in which the 4Fe4S cluster is an integral part of the active site. The application of the BS approach to the H-cluster should account not only for exchange interactions within the 4Fe4S cluster but also for the exchange interaction between the 4Fe4S and $[2Fe]_H$ clusters, a requisite that further complicates the analysis of this large system. Mössbauer studies carried out on the isolated forms of [FeFe] hydrogenases have shown that the 4Fe4S cluster is always in the formally diamagnetic oxidation state +2, indicating that the changes in the oxidation state occur only at the $[2Fe]_H$ cluster [113, 114]. However, Fe atoms of the 4Fe4S cluster acquire a paramagnetic character due to the exchange coupling between the two subunits of the H-cluster. The molecular and electronic structure of the H-cluster has been investigated using DFT by Bruschi et al. [115] and Brunold et al. [116] considering the model complex $[(CH_3S)_4(Fe_4S_4)(\mu\text{-PDT})(Fe_2)(CO)_3(CN)_2]^{3-}$ which refers to the H_{ox} redox state of the enzyme [117]. In H_{ox} the 4Fe4S cluster was found to feature the formal Fe^{3+}–Fe^{3+}–Fe^{2+}–Fe^{2+} redox state and the $[2Fe]_H$ cluster the formal Fe^{1+}–Fe^{2+} redox state. The two Fe_2S_2 layers coupled antiferromagnetically can be treated as single spin centers formally composed by Fe^{2+} and Fe^{3+} high spin ions, with $S = S_1(2) + S_2(5/2) = 9/2$. Therefore, H_{ox}, in which an unpaired electron is also localized on the $[2Fe]_H$ cluster, can be represented by a three-spins model Hamiltonian of the form:

$$\widehat{\mathbf{H}} = J_{12}\left(\widehat{\mathbf{S}}_1 \cdot \widehat{\mathbf{S}}_2\right) + J_{23}\left(\widehat{\mathbf{S}}_2 \cdot \widehat{\mathbf{S}}_3\right) + J_{13}\left(\widehat{\mathbf{S}}_1 \cdot \widehat{\mathbf{S}}_3\right) \quad (16)$$

which, as suggested by Brunold et al. [116], neglecting the interaction between the nonadjacent spin centers 1 and 3, may be written as:

$$\widehat{\mathbf{H}} = J_{\text{cube}}\left(\widehat{\mathbf{S}}_A \cdot \widehat{\mathbf{S}}_B\right) + J_{\text{H}}\left(\widehat{\mathbf{S}}_A \cdot \widehat{\mathbf{S}}_{\text{H}}\right) \quad (17)$$

where J_{cube} represents the exchange interaction between the two Fe_2S_2 layers of the diamagnetic 4Fe4S cluster with $S_A = S_B = 9/2$, and J_H corresponds to the exchange interaction between the vicinal Fe_2S_2 layer and the $[2Fe]_H$ cluster with $S = 1/2$.

The vicinal mixed-valence Fe_2S_2 layer A can be coupled to the unpaired electron of the $[2Fe]_H$ cluster in a ferromagnetic (F) or antiferromagnetic (AF) way generating the two BS solutions BS_1 and BS_2, respectively. The exchange coupling constant J_{cube} and J_H can then be obtained from the energies of the HS and the two BS states, using the Heisenberg Hamiltonian:

$$E(\text{HS}) = J_{\text{cube}}\left(\frac{9}{2} \cdot \frac{9}{2}\right) + J_{\text{H}}\left(\frac{9}{2} \cdot \frac{1}{2}\right) \quad (18)$$

$$E(\text{BS}_1) = -J_{\text{cube}}\left(\frac{9}{2} \cdot \frac{9}{2}\right) + J_{\text{H}}\left(\frac{9}{2} \cdot \frac{1}{2}\right) \quad (19)$$

$$E(\text{BS}_2) = -J_{\text{cube}}\left(\frac{9}{2} \cdot \frac{9}{2}\right) - J_{\text{H}}\left(\frac{9}{2} \cdot \frac{1}{2}\right) \quad (20)$$

from which:

$$\begin{aligned} E(\text{F}; M_{\text{BS}} = 5) - E(\text{AF}; M_{\text{BS}} = 4) &= 2J_{\text{H}}(9/2)(1/2) \\ &= (9/2)J_{\text{H}} \end{aligned} \quad (21)$$

and

$$\begin{aligned} E(\text{F}; M_{\text{HS}} = 19/2) - E(\text{AF}; M_{\text{BS}} = 17/2) &= 2J_{\text{H}}(9)(1/2) \\ &= (9)J_{\text{H}} \end{aligned} \quad (22)$$

As noted by Brunold [116], the Fe atoms in the 4Fe4S cluster are not equivalent, and therefore up to three different solutions of localized spins could be considered, each of them being characterized by a different definition of the Fe_2S_2 layers. For each of the three schemes of spin localization, BS_1 and BS_2 can been calculated, from which three distinct J_H values can be obtained.

3.4 QM Investigation of the Reactivity of Metalloproteins and Related Bioinorganic Molecules

The computational dissection of reaction pathways of metalloproteins and related bioinorganic models necessarily implies the characterization of the structure of reactants, products, intermediate species and corresponding transition states, as well as their energy differences. In general, when the QM models are limited in size (up to about 200 atoms), full geometry optimizations of both reaction intermediates (which correspond to energy minima) and transition states (which correspond to saddle points) can be performed. If the characterization of the reaction pathway is limited to intermediates and reactants/products, systems up to about 400 atoms can presently be handled. A recent DFT study in which very large full-QM models have been studied regards the evaluation of how secondary bonding interactions between metal lone pair electrons and peptide atoms can affect the binding mode of toxic ions (As(III) and Pb(II)) to proteins [118]. Other recent examples are provided by the DFT investigation of Mo and W ions within the active site of pterin cofactor containing enzymes, such as aldehyde oxidoreductases, acetylene hydratase, formate dehydrogenase, and xanthine oxidase [119].

3.4.1 Peculiarities in the QM Investigation of the Reactivity of Metalloproteins and Related Bioinorganic Molecules

The computation of structures (and their associated energies) of reactants, products and intermediates is generally a straightforward task to accomplish when investigating most organic species, whereas specific issues often arise studying metalloproteins and/or bio-inspired transition metal compounds. In general, the greater complexity emerging when considering the structure–energy problem when dealing with transition metals, compared to organic chemistry, can be mainly ascribed to the involvement of d electrons in the valence shell. As an example, several isomers can coexist in solution, especially when dealing with Me-L_n complexes with $n = 5$. In such cases all possible permutations of the various L_n ligands in the metal coordination environment should be evaluated. If two or more configuration isomers are close in energy, the picture becomes even more complicated, especially if all low-energy isomers might be important to understand the reactivity of the metal. Another example illustrating the high degree of complexity in the theoretical modeling of metalloproteins and related models regards the computational characterization of reaction energy profiles associated with seemingly trivial processes, such as unimolecular rearrangements. As an example, the investigation of the intramolecular isomerization of octahedral complexes is an intrinsically complicated task and requires to compute different types of transition state structures (referred to as Ray–Dutt and Bailar twists), besides all the plausible intermediates species [120, 121]. The task complicates further if dissociative and/or associative mechanisms can be operative, leading to the transient formation of five- and/or seven-coordinated species.

Another tricky aspect which is typically associated with metalloproteins regards the characterization of the electronic state configuration of the metal atoms (and therefore the definition of the oxidation states) which are involved in a specific biochemical process. Unless the reaction involves ions such as Zn^{2+}, which has a very stable ($4s^0 3d^{10}$) ground state electronic configuration, associated with a formal +2 oxidation state, the electronic configuration of transition metal ions commonly found in living organisms (such as V, Mn, Ni, Fe, and Cu) can vary in different states and systems. In this context, a typical task in the quantum chemical investigation of metalloenzymes regards the elucidation of the oxidation state, which is involved in the binding and activation of a given substrate. As an example, within the hydrogenase research area, one relevant issue has been the evaluation of which metal oxidation state could be more likely to bind/split dihydrogen [122].

3.5 Computation of Reaction Energies

To obtain equilibrium and rate constants of a reaction, one should compute standard free energies (G^0). However, DFT calculations (or any wave function based method) provides directly only the so-called E_{SCF}, i.e., the pure electronic energy, computed assuming that the system is at the temperature of 0 K. Indeed, even at 0 K, at least another contribution should be included to obtain a more realistic picture of the reaction energies. This extra contribution is the vibrational zero-point energy (ZPE, namely, the residual energy that any quantum system has at 0 K, arising directly from Heisenberg uncertainty principle, and which for molecules corresponds to atomic vibrations around their equilibrium positions), which is a significant energy term to consider when investigating several biochemical reactions. As an example, the ZPE cannot be neglected when accurate deprotonation energy values are needed [123], but also to evaluate proton-transfer energies and barriers, proton tunneling, and kinetic isotope effects both for protons and for heavier atoms [124, 125].

ZPE is defined as:

$$\mathrm{ZPE} = \sum_i^N \frac{1}{2} h\nu_i \tag{23}$$

where the sum runs over all the possible N vibration normal modes. Assuming that only the lowest vibrational level ν_0 is populated, one can see that the residual energy of molecules, even at 0 K, is $1/2h\nu_0$ (*see* **Note 10**).

At temperatures higher than 0 K, i.e., in biologically relevant conditions, thermal and entropic contributions should also be estimated. In fact, entropic contributions can be calculated by evaluating the roto-translation partition function of the system, at given T and P conditions. In all modern QM software, however, only approximated partition functions can be computed, since for real thermodynamic (many particles) ensembles the true partition function $Q(N, V, T)$ is extraordinarily complicated, due to its many-body nature and to the very large number of energy levels to be considered (*see* **Note 11**). Entropic corrections, even though approximated, play a major role in adjusting pure electronic energies (i.e., those resulting from a SCF-DFT energy minimization) when associative and/or dissociative elementary reaction steps are to be modeled computationally. This is the case when ligand substitution (CO, NO, H_2O) or oxidative addition of small molecules (H_2, HX) to transition metal centers take place.

The choice of the appropriate exchange-correlation functional is another critical issue that one must consider when energy profiles and related kinetic-thermodynamic parameters must be computed to study models of metalloproteins. The literature about such topic is wide and many reviews [126–128] and excellent books [26, 129, 130] have been devoted to benchmarking different kinds of DFT functionals. Nonetheless, the performance level of various functionals to study transition metal compounds is somewhat controversial (*see* **Notes 12** and **13**), since some authors reported in favor of GGA (i.e., Generalized Gradient Approximation; with BP86 and PBE being usually the most popular choices), whereas others prefer hybrid functional [131, 132] (amongst which B3LYP and PBE0 are certainly the most commonly employed). It must be noted that often the "better" performance of a single functional (or a family of) over others is simply due to cancellation of errors [130]. Therefore, even though new-generation functionals now exist which improve performances, the lesson to be kept in mind is that no generally and universally valid functional is presently available for a complete covering of all chemistry. As an example, M06 and M06-L functionals are well suited for transition metal thermochemistry, the former especially when bond cleavage/formation reactivity is the target of the investigation, but the latter performs worse than the former with main-groups elements [126]. According to our experience, the statement that within DFT no "best" functional exist aprioristically is particularly true for the investigation of metalloproteins and related models. Even worse, one might expect in principle that a given functional/basis set scheme, which had previously worked out with a given reaction occurring at a given metal center, could perform well with a system containing the same metal atom but involved in a different reaction (e.g., featuring new substrate). However, this is not guaranteed a priori in DFT practical usage. It is therefore strongly suggested to adopt a systematic case-by-case approach whenever one is at the very beginning of an investigation of metalloproteins and/or related bio-inspired compounds. This implies the necessity, especially when species not previously subjected to systematic benchmarking are present, to carry out accurate calibration of different DFT methods, i.e., to test which functional(s) proves to be able to offer the best performance in terms of reproducibility of experimental data.

3.6 Practical Considerations Related to the Localization and Characterization of Stationary Points

The following section is dedicated to more technical and practical details related to the localization and characterization of stationary points (i.e., where gradients are zero and therefore the forces are null) on the potential energy hypersurface (PES) associated to a molecule. This is a critical issue in the investigation of the reactivity of molecules because reactants, intermediates, products and transition state structures of a given reaction profile are all stationary points on such PES. In particular, the computation of thermodynamic and kinetic parameters associated with a reaction implies the structure optimization of relevant intermediates and transition states along a specific reaction coordinate. According to mathematic criteria, reactants, products and possible intermediates correspond to energy minima on the PES, whereas first order saddle points (characterized by only one negative eigenvalue in the Hessian matrix, which is formed with the second derivatives of the energy respect to the nuclear coordinates) correspond to transition state structures.

First of all it should be noted that the Born–Oppenheimer approximation is generally used in the investigation of the reactivity of bioinorganic molecules, since without this assumption the PES conceptual model is not valid. In a few words, owing to the extremely fast electron velocity compared to that of nuclei, which arises from their relative mass ratio, the total wave function of a molecule is broken into electronic and nuclear equations, solved separately. This assumption corresponds to a situation in which electrons follows instantaneously—adiabatically—all moving nuclei, meaning that in general (i.e., without specifying which kind of QM method is adopted for energy-structure determination) nuclei are thought of as moving on a "potential energy hypersurface," which is a solution exclusively of the *electronic* Schrödinger equation.

The evaluation of free energy differences between energy minima and transition states should allow to calculate the kinetic constant of each elementary step, by means of the Eyring–Polanyi equation [133, 134]:

$$k = \frac{k_B T}{h} e^{\frac{-\Delta G^{0\neq}}{RT}} \tag{24}$$

with k_B, Boltzmann constant 1.3806×10^{-23} J/K; T, temperature (K); h, Planck's constant $= 6.6261 \times 10^{-34}$ J s; R, universal gas constant $= 8.3143$ J/mol/K.

The term $\Delta G^{0\neq}$ is the standard free energy of activation, directly evaluated from the barrier height, i.e., from the energy difference existing between the transition state structure (with energy G^0_{TS}) and either the preceding intermediate G^0_{I1} (if the forward reaction is being considered) or the following one G^0_{I2} (if the reverse reaction is being considered). Therefore, rate constants are evaluated through the exponential term $\Delta G^{0\neq}$, which is computed as $G^0_{TS} - G^0_{Ii}$. Note that the experimental way to proceed is the opposite, since k values are first determined (from kinetic data) and then $\Delta G^{0\neq}$ are calculated (*see* **Note 14**).

Most DFT programs can calculate energy gradients analytically, thus it is quite trivial (at least from a technical standpoint, but not always from a chemical one) to determine the structure and energy of reactants, products and intermediates. Hence, with the energy of two given minimum structures in hand, one can easily deduce their relative equilibrium distribution (or population) in solution using the Boltzmann statistics. Quite often, one may be simply interested in computing a given energy difference between two minima (configuration isomers), with the sole purpose to exclude the existence in solution of a given isomer: since the energy difference appears at the exponent of the Boltzmann distribution, even an energy difference of few kcal/mol implies that the higher energy isomer does not exist in solution. In general, when dealing with metal–ligand complexes, which are often studied as models of the active site of metalloenzymes, the number of minima to locate is not huge, provided that: (1) metal nuclei are in limited number; (2) the ligands structure does not contain long and flexible substituents. The first step in the computational optimization of a molecular structure (i.e., finding the structure corresponding to an energy minimum on the PES) always implies the definition, using computational graphics tools, of a guess structure, which ideally must be as close as possible to the structure corresponding to the energy minimum (*see* **Note 15**).

The electronic structure of QM models including both transition metal compounds and non-metal components should be investigated at least with a triple-ζ basis on the valence electronic shell of the metal center and on all atoms forming its first coordination sphere [36]. If substrates or other molecules come to play a relevant role in the reactivity associated with the metal atom, they should also be treated at the highest possible level of theory, also to avoid artifacts such as the Basis Set Superposition Error (BSSE). A lower quality basis set can be employed on molecular regions away from the metal or, however, from the reaction center.

Differently from the characterization of energy minima, the characterization of transition state structures requires computing, storing, inverting, and diagonalizing the Hessian matrix, a very demanding step from a computational standpoint, especially with large size systems. Therefore, strategies have been designed that use an approximate inverse of the Hessian matrix, which is first constructed in an approximate way and then updated (i.e., made better and better at each subsequent point along the search path, see below). Among different popular techniques, it must be cited the BFGS (Broyden–Fletcher–Goldfarb–Shanno) algorithm, which is useful in the localization of energy minimum structures. The Powell updating scheme is instead suited when transition state structures are computed, since this technique allows for negative eigenvalues in the Hessian update scheme [26]. Besides being demanding in terms of computational costs, the location of transition state structure is all but a trivial task. The strategies typically employed for determining TS can be generally grouped in two families:

1. Interpolation between two minima, requiring that both reactants and products structures (or two adjacent intermediates along a multi-step reaction coordinate) are known.
2. Local methods, requiring a guess of the true TS, in which the initial structure is propagated along a direction based upon gradient and Hessian information.

For a survey on the first group of methods, specialized textbooks are recommended [26]. In local methods, such as the Newton method, a quadratic function (a Taylor expansion truncated at the second order term) approximating the potential energy surface is used to extrapolate the point where the gradient vector will be zero. Because the quadratic model surface is only an approximation of the actual surface, and because the Newton method requires analytic second derivatives, more flexible pseudo-Newton (sometimes referred to as quasi-Newton) algorithms are normally employed [26]. Among them, the Newton-Raphson method is the most commonly used, for its flexibility and rapid convergence, provided that a good guess TS structure and the correct degrees of freedom are used to describe the evolving coordinate reaction.

A protocol for locating transition state structures by local methods can be summarized as follows:

1. Definition of the reaction coordinate (i.e., which bonds, angles, torsions vary during a given reaction).
2. Design and constrained (vide infra) structure optimization of a guess TS.
3. Computation of the Hessian to guarantee (at least qualitatively) that the curvature has the proper structure.
4. Newton-Raphson (or pseudo-NR) search of the true TS along the direction defined by the lowest eigenvalue in the Hessian.

The first two points deal with chemical intuition and aprioristic knowledge of the system under investigation, and they are the basic reasons, which make the use of local methods somehow less naïve than others. The constrained structure optimization of a guess TS implies that all the conformational degrees of freedom associated with the reaction coordinate must be frozen. This is because one has to be sure that the TS search is confined along the Minimum Energy Path (MEP) connecting two consecutive intermediates, or reactant and products in an elementary reaction event. Thus, by means of such constrained energy minimization, all molecular coordinates are adjusted iteratively according to a DFT-SCF procedure [129, 130] which leads (provided that the reaction coordinate has been represented by the proper set of degrees of freedom) to the "valley" path connecting two minima. Essentially, the saddle point on the PES in the peculiar region associated with the reaction is just the maximum along the MEP in such region.

Once the constrained minimum has been reached, computation of the Hessian (H) is carried out. This step represents normally the most time/resource consuming one. However, the analytical computation of H (i.e., without using numerical interpolation techniques) is crucial for two reasons. First, as aforementioned, one has to characterize the nature of the zero-gradient point located, checking that the curvature at that point has the correct structure (one and only one negative eigenvalue of the H matrix, with all other eigenvalues being positive). Second, even though in principle NR methods, associated with an eigenvector following algorithm (*vide infra*), could locate first order saddle points even starting from points with wrong curvature, a guess TS featuring the proper Hessian structure is extremely advantageous in terms of rapid convergence of calculations. The negative eigenvalue will, in fact, feature the correct eigenvector to be followed in the final step of the procedure, i.e., the pseudo-NR search for the TS (*see* **Note 16**).

The final step of the TS search consists in the real NR step. If the molecular coordinates have been brought very close to the true TS, the NR method minimizes the energy along all directions but that indicated by the eigenvector of the Hessian associated with the negative eigenvalue, which corresponds to the transition mode vibrating with imaginary frequency. The pure NR method approximates the true function (energy in our case) by a Taylor polynomial truncated at the second order around the point of interest r_0. In matrix notation, we have:

$$E(r) \approx E(r_0) + \mathbf{g}^t(r - r_0) + 1/2(r - r_0)^t\mathbf{H}(r - r_0) \qquad (25)$$

where $\mathbf{g}$ is the gradient vector and $\mathbf{H}$ is the Hessian matrix. The superscript "t" indicates matrix transposition.

In proximity of a stationary point $\mathbf{g} = 0$, which leads, after differentiation with respect to the generic coordinate (i) of the vector r and some rearrangements, to

$$r = r_0 - \mathbf{H}^{-1}\mathbf{g} \qquad (26)$$

which provides a prescription on how to generate a trajectory to follow in order to locate a stationary point, once a previous arbitrary point, as well as the first and second derivatives calculated at that point are known. Note that a matrix inversion is also required.

If we select a new coordinate system $\mathbf{r}'$ in which $\mathbf{H}$ is diagonal (these are referred to as normal coordinates),

$$(\boldsymbol{r}-\boldsymbol{r}_0)' = \left[(r_1-r_0)',(r_2-r_0)',(r_N-r_0)'\right]^t \tag{27}$$

The generic step can be written as:

$$(r_i-r_0)' = \frac{g_i}{\varepsilon_i} \tag{28}$$

Where g_i is the projection of the gradient along the particular $\mathbf{H}$ eigenvector which is associated with the ε_i eigenvalue. This formulation makes evident the versatility of the NR method: if the Hessian has only positive ε_i values, i.e., close to a minimum, the direction of the trajectory will be opposite to that of the gradient, as it is for example for other minimization techniques exploiting only gradient information (such the Steepest Descent algorithm). But if, otherwise, a TS search has to be performed, the NR step will be taken in the same direction where the gradient points, which is uphill directed (*see* **Note 17**). This is why, when a single eigenvalue in the Hessian is present, NR leads to maxima (in just one direction).

The step control issue is not the only potential problem with NR methods; especially when large QM systems are treated, also the computation and even the simple storing of the Hessian can become troublesome. These are the basic reasons why the pure NR step equation is ordinarily corrected introducing so-called shift-parameters (usually denoted as λ) which serve to scale down and modulate the denominator of the pure NR methods. One of the techniques employed to such aim is the Eigenvector Following procedure [26]. In general, all methods employing shift parameters are also referred to as *augmented*-Hessian NR techniques. Among them, two methods have proven useful in TS structure optimization, since their definition of λ simultaneously alleviates the double-faceted issue of guarantying that the step size does not exceed the Trust Radius and also that the line search is kept always along gradient, which implies that a maximization of the energy along a single direction is being performed. The two implementations are referred to as Partition Rational Function Optimization, which adopts two different types of λ and the Quadratic Approximation, which uses a single λ. We show explicitly only the equation of the latter:

$$\begin{aligned} \left|(\boldsymbol{r}-\boldsymbol{r}_0)'\right|^2 &= \textstyle\sum_{i\neq \mathrm{TS}} (g_i/\varepsilon_i-\lambda)^2 \\ &\quad + \left(g_{\mathrm{TS}}/(\varepsilon_{\mathrm{TS}}-\lambda)^2 = R^2\right) \end{aligned} \tag{29}$$

with R being the Trust Radius. If the actual energy change between two subsequent steps is in line with the corresponding variation of the Taylor polynomial approximation of the energy, the R value can also be adjusted to larger values (to speed up convergence).

As discussed previously, the eigenvector following (or augmented Hessian) procedure could allow the TS search starting from an energy minimum structure (for example the optimized reactant structure of a given reaction), instead of from a structure very close to the real TS. Nonetheless, such strategy fails unless the eigenvector leading to TS is strongly uncoupled to all other Hessian eigenvectors, which mathematically means that the derivatives of order higher than the second are close to zero. The choice of the set of the various degrees of freedom which compose the reaction coordinate in the TS search is really appropriate also if it prevents higher order derivatives from assuming too high values.

As for the problem arising with Hessian size and computation (and inversion), updating schemes are employed to approximate the true Hessian by using gradient information. The Powell procedure guarantees, among others, that negative eigenvalues can be obtained in the approximate Hessian matrix used along the search path. The usage of updating schemes for simplifying the Hessian structure during the iterative process defines the so-called pseudo-NR methods. The level of approximation of the initial Hessian (i.e., that calculated after the constrained minimization) is crucial when updating schemes are adopted. Whereas updating schemes can be used in the initial step of a simple energy minimum search, in TS search this is not possible because the correct direction to follow is needed, which can only be achieved by computing the second derivative matrix analytically. After that, the updating scheme may be switched on (and this is actually the case unless very complicated instances occur, requiring more than a single analytical computation of **H**).

The computation of vibrational frequency should always be carried out to confirm the nature of the stationary points located (i.e., distinguish between energy minima and transition states). Because the Hessian (converted to the equivalent mass-weighted coordinate force constant matrix $\boldsymbol{F}$) is positively definite for energy minima (after diagonalization), all eigenvalues λ from the equation

$$\boldsymbol{F} - \lambda \boldsymbol{I} = 0 \tag{30}$$

must be positive [135], and the normal-mode frequencies ν_i (also positive) are given by

$$\nu_i = \frac{\sqrt{\lambda_i}}{2\pi} \tag{31}$$

When dealing with transition state structures (corresponding to a first order saddle point) this matrix equation has strictly only one negative eigenvalue λ_k, associated with one imaginary normal-mode frequency ν_i for the normal mode having $i = k$ along the reaction coordinate, whereas all the other frequencies are positive (*see* **Notes 18** and **19**).

3.7 Computation of the Redox Potentials of Bioinorganic Molecules

Metalloproteins are often involved in redox reactions, and therefore the possibility to reliably compute redox potentials is crucial to better understanding structure activity relationships in several biochemical processes. The accurate prediction of redox potentials of proteins and related models containing transition metal complexes necessitates a detailed understanding of the electronic structure of the molecule under investigation. Also in this case DFT proved to give a reasonable trade-off between accuracy and computational costs. In fact, as recently summarized by Hughes and Friesner [136], there are several examples about the investigation of the redox chemistry of transition metal complexes using DFT [26, 27, 137, 138], showing different levels of success. In general, errors can range from reasonable values of about 100–200 mV, to very large errors up to 900 mV when dealing with coordination compounds involving fourth-, fifth-, and sixth-row metal centers.

From a practical perspective, the computation of standard reduction potentials implies the following steps. The free energy change associated to the redox reaction can be represented by a thermodynamic (Born–Haber) cycle (Scheme 3), in terms of the free energy change in the gas phase, $\Delta G^0_{(g)}$, and the solvation free energies of reactants and products. Therefore, the standard Gibbs free energy in solution $\Delta G^0_{(\mathrm{soln})}$ can be derived as follow:

$$\Delta G^0_{(\mathrm{soln})} = \Delta G^0_{(\mathrm{g})} - \Delta G^0_{(R,\mathrm{solv})} + \Delta G^0_{(P,\mathrm{solv})} \tag{32}$$

As discussed previously, the free energy change in the gas phase, $\Delta G^0_{(g)}$, can be derived according to the classical equation:

$$\Delta G^0_{(\mathrm{g})} = \Delta H^0_{(\mathrm{g})} - T\Delta S^0_{(g)}, \quad \text{where } \Delta H^0_{(\mathrm{g})} = \Delta E_{\mathrm{SCF}} + \Delta E_{\mathrm{TRANS}} + \Delta E_{\mathrm{ROT}} + \Delta E_{\mathrm{VIB}} + \Delta E_{\mathrm{ZPE}} \tag{33}$$

The solvation free energies $\Delta G^0_{R,\mathrm{solv}}$ and $\Delta G^0_{P,\mathrm{solv}}$ can be usually evaluated by the self-consistent reaction field (SCRF) approach based on PCM modeling (where the solvent is represented by a polarized dielectric medium and a set of optimized radii are used to build a cavity occupied by the solute in the

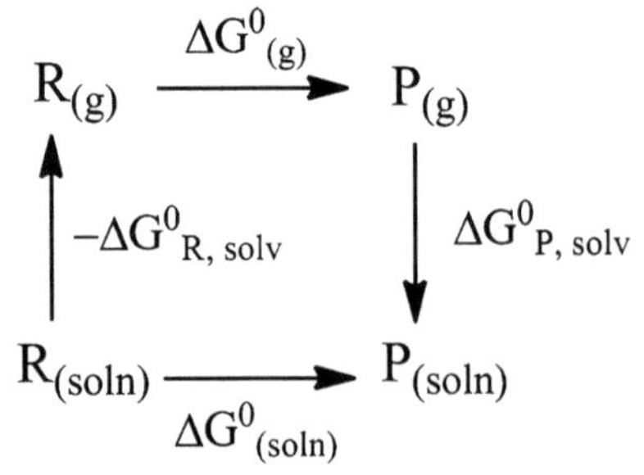

Scheme 3 Representation of the free energy change in a redox reaction using a thermodynamic (Born–Haber) cycle

solvent), using the geometries optimized in gas-phase calculations. Indeed, the solvation free energy, $\Delta G^0_{(soln)}$, is often partitioned into two contributions: $\Delta G^0_{(soln)} = \Delta G^0_{EE} + \Delta G^0_{CDR}$, where the first term accounts for the bulk electrostatic effects (EE) including the energy difference between the solute polarized by the reaction field of the solvent minus the energy of the solute in vacuum, and the second term accounts for the sum of the cavitation and the dispersion-repulsion energies (CDR). Then, the standard one-electron redox potentials, E^0 (in V), can be easily calculated from the Nernst equation:

$$\Delta G^0_{(soln)} = -FE^0 \tag{34}$$

where F is the Faraday constant (23.06 kcal/mol/V).

It must be noted that the computed absolute potential, E^0, is not referenced to any standard electrode. Therefore, in order to compare computed values to experimental data, it is necessary to subtract 4.44 ± 0.02 V (at 298.15 K) to the absolute reduction potential, i.e., to refer this latter value to the reference electrode *par excellence*, the standard hydrogen electrode (SHE; *see* **Note 20**).

Another factor which should be taken into account in the computation of reduction potentials is the liquid junction potential. If one has to deal with the computation of standard redox potentials in nonaqueous solutions (such as acetonitrile), a contribution keeping into account the liquid junction potential should be also taken into account. In fact, the tabulated potential for the interphase of interest should be subtracted from the original value of the reference electrode. As an example, the value of 2.44 V versus SHE, associated with the saturated calomel electrode in aqueous medium, should be decreased by 0.093 V (liquid junction potential acetonitrile-water) [139] to obtain the standard redox potential in acetonitrile. Finally, it is worth noting that to have a correct representation of the change from gas phase to solution state, a correction of RTln 24.46 must be included, to take into account the change from the different standard states (1 atm in gas phase, 1 mol/L in solution) [140]. Also, some authors adopt the electron convention (correcting the $\Delta G^0_{(g)}$ term above) in which the free electron is treated as an element having the thermal energy of an ideal monoatomic gas, i.e., $3/2RT$, which has to be subtracted from $\Delta G^0_{(g)}$ [141, 142].

3.8 The QM Investigation of Excited Electronic States and Photochemical Processes in Bioinorganic Systems

Bioinorganic photochemistry deals with all light-assisted or light-induced processes that take place in living organism and involve metalloproteins or metalloenzymes [143]. The prototypical bioinorganic photochemistry process is photosynthesis. Other processes that involve the interaction of metalloproteins or biologically relevant inorganic molecules with light are NO photolysis in nitrile hydratase [144], CO photolysis in hemoglobin [145] or in

(FeFe)-hydrogenases [146], as well as photodynamic processes in Vitamin B_{12} derivatives [147]. In many photobiological processes the solar light energy conversion is assisted by specific molecules that work as antenna or *photosensitizers*, and this scheme has been proved useful also to design bio-inspired light induced processes, in which part of the chemical energy needed to catalyze a process is supplied to the system by light absorption. In these and other examples of photochemical process, the rationalization of the phenomena observed experimentally often requires their investigation using quantum chemical methods. In fact, the experimental characterization of photochemical reactions and processes is often hampered by the short life of relevant species.

Before presenting a general protocol for quantum chemical calculations of photochemical processes in bioinorganic systems, it is important to briefly introduce the Time-Dependent Density Functional Theory (TDDFT) approach, which is the most popular and extensively used method for the calculation of electronic excited state energies and properties [148]. Based on the extension of the ground state Hohenberg-Kohn theorem, TDDFT is suitable for computing electronic excited state transition energies and properties of medium to large size organic and inorganic compounds (up to several hundreds of atoms). The role of TDDFT in the computation of excited state properties can be considered equivalent to that of DFT for ground state properties, and indeed it suffers of the same problems and limitations of DFT (in particular the dependence of the computed properties on the approximated exchange-correlation functional adopted). Generally, in TDDFT computational studies the environment of the active site is not explicitly considered, since it is assumed to not influence significantly the process of interest. Therefore, after removing partially or totally the active site environment, the system to be studied is generally reduced to the metal cofactor plus the chemical groups that directly interact with the cofactor. The role of the environment can be taken into account using implicit continuous models, which have been recently implemented within the TDDFT approach.

The first step of any excited state investigation is the computation of the absorption electronic spectrum of the system. In TDDFT this implies the computation of the vertical excitation energies (Frank-Condon region) at the ground state minimum geometry, using the set of MOs (ϕ) previously computed for the ground state. Each excitation is identified by the corresponding oscillator strength, which gives an estimation of the intensity of each absorption, and by a specific set of coefficients P_n associated with the single mono-electronic transitions $\phi_i \rightarrow \phi_a$ between occupied and virtual MOs that contribute to a given excited state in order to have $\Sigma_n \mathrm{P}_n^{\ 2}(\phi_i \rightarrow \phi_a) = 1$ (Fig. 5).

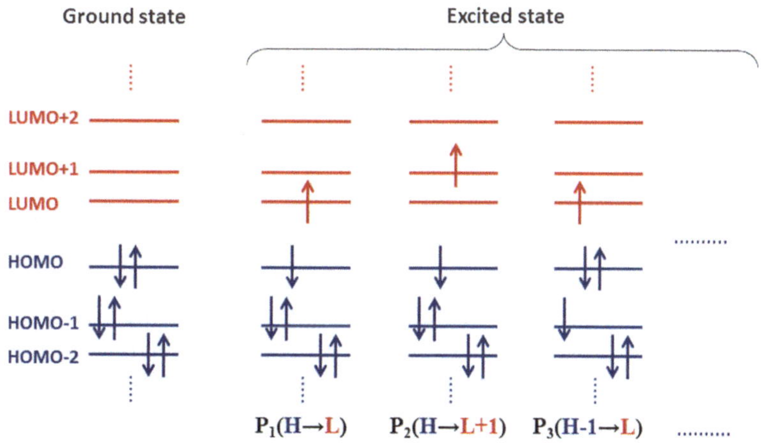

Fig. 5 Example of hypothetical MO occupations of the singlet ground state and the TDDFT description of an excited state in terms of mono-electronic transitions between occupied (*blue*) and virtual (*red*) ground state MOs, each characterized by a coefficient P_i(occ. → virt.)

Starting from these coefficients it is possible to characterize the type of excitation by computing the differential population of the MOs involved or the transition density (as an example, an electronic density shift between a transition metal center and a ligand suggests a metal-to-ligand charge transfer (MLCT) character of the excitation). The number of excited states to be considered depends on the range of the experimental spectrum that should be covered and by the number of atoms of the system, and its value can be as high as several tens, in particular when more than one metal atom is involved. Once the excitation energies have been computed, the spectrum is simulated as a simple sum of Gaussian functions centered on a computed transition with a full width at half-maximum value of about 50 nm and weighted by the oscillator strengths (*see* **Note 21**).

Together with oscillator strengths, it is also possible to easily compute the rotatory strengths, to simulate circular-dichroism (CD) spectra [149], which give very useful information in the investigation of optically active molecules and to evaluate the fraction of secondary structure in proteins.

In principle, when studying photochemical processes, one should consider only excited states with the same spin multiplicity of the ground state. However, when dealing with transition metals, the spin–orbit coupling can mix states with different spin multiplicity, allowing transition between states otherwise forbidden. In fact, the transition between excited states of different spin-multiplicity is quite common in transition metal complexes and it is referred to as *inter-system crossing* (ISC). As an example, one of the most important features of a photosensitizer like $(Ru(bpy)_3)^{2+}$ is its ability, due to ISC, to populate a long-lived Ru → bpy MLCT low energy triplet state, that can undergo oxidation or reduction much more easily than the ground state.

Using quantum chemical methods, it is also possible to compute fluorescence and phosphorescence spectra, i.e., the emission spectra for singlet and triplet states (considering as reference a system with a singlet ground state). Because an emission spectrum is due to the spontaneous emission from the vibrationally relaxed excited state to the ground state, in order to compute these transitions one needs to optimize the structure of the molecule in the excited state, and compute the TDDFT excited state energy first derivatives with respect to the nuclear positions. In the first step of this type of investigation only the first excited singlet and triplet states (usually refer as S_1 and T_1) are considered, investigating the so-called 0-0 vibronic transitions (also known as well bottom to well bottom transitions), in which the emission transition energy ($S_1 \rightarrow S_0$ and $T_1 \rightarrow S_0$ for fluorescence and phosphorescence, respectively) is computed considering the TDDFT energy of the excited state optimized structure and the DFT minimum ground state energy. A more detailed and accurate simulation of the emission spectra consists in the vibrational resolution of each transition by computing the overlap among the vibrational states of the ground state and the excited state obtained from a normal mode frequency calculation [150].

Beside the simulations of the spectroscopic properties (absorption or emission spectra), it is often interesting to understand how excited states behave as a function of time, since the photo-reactivity properties of a system depend on the excited state dynamic that takes place after photon absorption. From the experimental point of view, these properties can be investigated by means of time-resolved infrared spectroscopy on a wide range of time scales, from femtosecond to nanosecond, allowing to evaluate the time-scale of a process and the lifetime of the excited state species involved. From the theoretical point of view, the investigation of the photo-reactivity entails the study of the vibrational relaxation of the system or, more generally, the exploration of its excited state PES. Here it is important to underline that the investigation of photochemistry processes using TDDFT generally implies the detailed characterization of the initial steps of the photodynamic process, whereas the characterization of the final fate of the excited state is generally more complicated.

A simple computational protocol to study how excited states behave as a function of time implies the exploration of the excited state along a given reaction coordinate computed on the ground state. As a first step, a reaction pathway on the ground state PES must be determined. As an example, let us consider the photodissociation of a ligand (L) bound to a metal atom (M). The M–L dissociation coordinate can be easily computed starting from the ground state structure by elongating the M–L bond distance in a

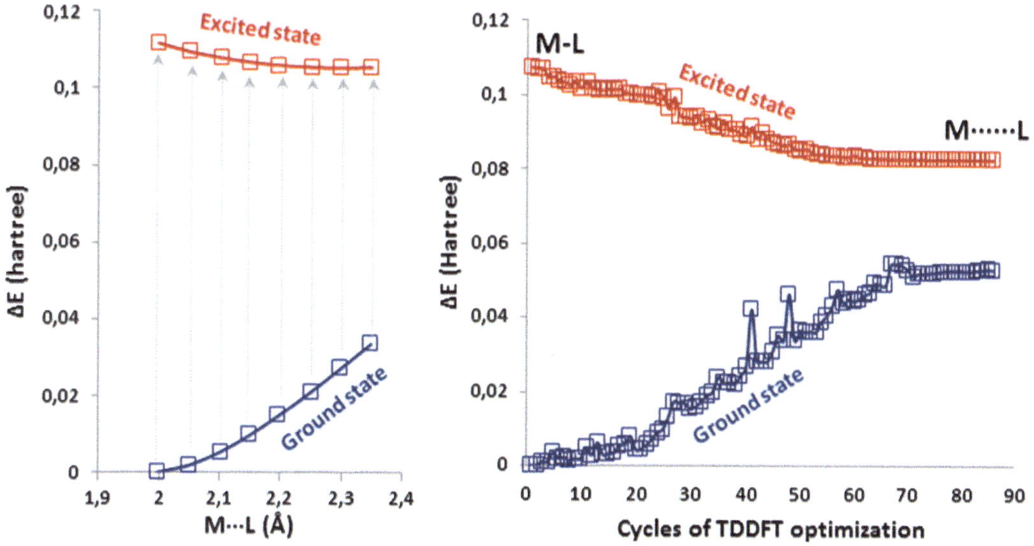

Fig. 6 Two approaches in the TDDFT exploration of the excited state PES. On the *left*, the exploration of the excited surface along a coordinate computed on the ground state PES; on the *right*, the TDDFT geometry optimization (ΔE is computed with respect to the ground state minimum energy; in this example we show the data obtained investigating CO photolysis on the first singlet excited state of $Fe_2(CO)_9$ using the BP86 functional)

number of steps with constant increment; for each step, the geometry parameters are optimized imposing the constraint on the M–L distance. Finally for each structure, the TDDFT electronic spectrum is computed and the energies of the excited state and the ground state are reported as a function of the M–L distance, as sketched in Fig. 6.

Indeed, the photo-reactivity of a molecule may depend on an higher energy excited state which is populated upon light absorption, and then decays to a lower energy state through ISC or a *internal conversion* (IC) mechanism, i.e., the transition from higher to lower excited states with same spin-multiplicity. This latter decay mechanism is much faster (10^{-14} to 10^{-11} s) than ISC (10^{-11} to 10^{-6} s) and takes place in many photobiological processes. It is worth noting that the molecular geometry at which the IC occurs (conical intersection) is particularly relevant for understanding photo-reactivity mechanisms.

The next step in the investigation of an excited state is generally the optimization of its structure using TDDFT [151] (Fig. 6). Using this approach it is possible to identify the stationary point on the excited PES closer to the ground state structure (note that TDDFT second energy derivatives cannot be presently computed to calculate normal modes and consequently determine the nature of a stationary point on the excited PES). As a final practical remark regarding photodynamic investigations, the choice between the two approaches presented for the exploration of the excited state

(PES scans and TDDFT geometry optimizations) depends on the available information about the experimental time scale of the photochemical process. In general, the first approach is appropriate if photodissociation takes place on a very fast time scale, whereas for longer time scales the redistribution of the vibrational energy and consequently the geometry relaxation of the excited state structure could play an important role, and therefore, the second approach becomes more appropriate.

4 Notes

1. Relativistic effects in chemistry manifest themselves in several ways [152, 153]. First of all, core electrons in heavy metals move at a substantial fraction of the speed of light, thus increasing their mass over the rest mass. The mass-velocity correction, together with the Darwin correction (describing the change in an electron from a point particle to a finite charge distribution, hence reducing nuclear–electron attraction and electron–electron repulsion), are often referred to as scalar corrections. The mass-velocity correction has the effect that the 1*s*-orbital, as well as the higher *s*-orbitals, contract in size. The more effective shielding of the nuclear charge by core electrons causes the increase in size of higher angular momentum orbitals: valence *p*-orbitals are relatively unaffected (due to spin–orbit interactions that counterbalance this effect), while *d*- and *f*-orbitals become larger. Thus, relativistic effects generally result in a shortening of chemical bonds. Relativistic effects for geometries and energetics are normally negligible for the elements in the first three rows in the periodic table (up to Kr), become non-negligible for the fourth row, and relativistic corrections are necessary for the fifth and sixth rows, as well as for lanthanide and actinides metals. The other main relativistic effect consists in the spin–orbit interaction, which is the interaction of the intrinsic magnetic moment of the electron (spin) with the magnetic field generated by the movement of the electron (this is also known as vector corrections). This interaction is present in all the elements. Relativistic effects can be directly considered in QM calculations by solving the Dirac–Fock equation, which is the equivalent of the Hartree–Fock equation (in the wave function theory) or the Kohn–Sham equation (in the Density Functional Theory). However, calculations are computationally very demanding and presently can be routinely applied only to small molecular systems. Another approach consists in the inclusion in the Schrodinger equation of additional operators describing corrections to the non-relativistic wave function.

2. While for transition metals it is quite clear that the outer $(n + 1)s$-, $(n + 1)p$- and $(n)d$-orbitals constitute the valence space, the definition of the core/valence shells in the case of Lanthanides is less obvious. In general, the $5s$ and $5p$ shells can be considered core electrons, as well as $4f$ orbitals given their extremely contracted nature. Several ECP have been developed for Lanthanides, differing for the core dimension: 28-electron core ((Ar)$3d^{10}$), [154] 46-electron core ((Kr)$4d^{10}$), [155], and 54-electron (Xe) [156] core have been developed. Moreover, although $4f$-orbitals form an open shell, their core-like character prompted the development of ECPs in which $4f$-orbitals are attributed to the core, whereas 5 s- and $5p$-orbitals are included in the valence, so that the atoms are modeled as 11-valence electron systems. The resulting large-core ECPs are thus $46 + 4f^n$ electron core ((Kr)$4d^{10}$ plus $4f^n$) [157]. Inclusion of the $4f$ shell into the core prevents calculation of lanthanide properties such as $4f$–$4f$ spectroscopic transitions. However, these ECPs present several advantages when used for determining structures and energetics of medium-large molecules containing lanthanides. In particular, inclusion of $4f$ electrons in the core allows to treat the system formally as a closed shell system, neglecting the huge amount of low lying electronic states characterizing these systems. These large-core ECPs have been used in the calculations of several lanthanides molecular systems, which find applications in many areas of technology, from medical diagnostics to energy. As an example, the electronic properties of the gadolinium ion, that formally presents seven unpaired electrons, make Gd(III) complexes useful in clinical practice as contrast agents in magnetic resonance imaging (MRI) techniques [158].
3. Overpolarization of the QM region is usually observed if no action is taken to modify in a convenient way the MM charges of the atoms that are bound to atoms in System 1. Let us consider for example the simple case of the presence of a junction at the level of the bond between the alpha and beta carbon atoms of a cysteinate residue ($^{-}SH_2C$-CHRR' bond). If the MM charges of the atoms directly bound to the alpha carbon are not adjusted in some reasonable way, then some MM point charges will be found in very close proximity to one of the atoms of the QM region. To avoid this, a popular approach is to zero the charges of neighboring atoms in the MM region, with consequent redistribution of the deleted charges on other MM atoms relatively more distant from the QM region. Such technique is usually referred to as the charge shift scheme.
4. When the relative stabilities of intermediates along a reaction path are computed by QM/MM geometry optimizations with System 2 free to relax, the accuracy of the energy values

obtained is typically in the order of 7–10 kcal/mol. Factorizing the error in QM/MM energy differences is an intrinsically difficult task, given the large size and the complexity of the system. Models including link hydrogen atoms have an obvious source of errors in the fictitious nature of the latter; moreover, the result of a QM/MM optimization depends on the starting structure adopted for energy minimization. In fact, the QM/MM potential energy function features many local minima, and algorithms devoted to energy minimization normally optimize the system by bringing it to the stationary point closer to the starting geometry. When the energy difference between intermediates has to be computed, a better sampling of the potential energy landscape can be obtained by following a recursive approach in geometry optimization, as follows. Initially, QM/MM models of the two intermediates are optimized, usually starting from experimental atomic coordinates; after convergence of the latter calculation, the relaxed geometries of System 2 is exchanged between the two models, and the coordinates thus obtained are used as starting geometries for a new couple of QM/MM optimizations. This procedure can be repeated cyclically, until the computed energy difference between the two intermediates reaches a stable value.

5. In fact, there is no guarantee that the iterative SCF scheme will converge. For starting geometries near equilibrium, when using small basis size sets, the SCF procedure generally converges. However, distorted geometries (such as transition states) and large basis sets containing diffuse functions can present convergence problems. This is especially true for transition-metal complexes with open metal d shells, where several states with similar energies are possible. A fairly common problem encountered is the so-called "SCF oscillations": the SCF procedure bounces back and forth from the density matrix calculated at one step (D_n) to the density matrix calculated at the next step (D_{n+1}) and never converges. In more pathological cases the SCF behaves showing large and apparently random changes occurring in the density matrix at every step. When the SCF procedure does not converge, several techniques for controlling convergence can be used: (1) *Damping*. The current density matrix is replaced with a weighted average of the previous and new matrices, $D_{n+1} = \alpha D_n + (1 - \alpha)D_{n+1}$ where the weighting factor α can be chosen as constant or changed dynamically during the SCF procedure. Damping has the effect of reducing the size of the step in the iteration and limits the tendency for oscillations. On the other hand, it slows down the rate of convergence. (2) *Level shifting*. It can be shown that the iterative SCF procedure can be also formulated in terms of orbital rotations, involving the mixing of occupied

and virtual MOs. During the iterative process these mixings can be large, causing oscillations or making the total energy increase. The degree of mixing can be reduced by artificially increasing the energy of the virtual orbitals. The more the virtual orbitals are raised in energy the more robust is the convergence; however, convergence becomes slower. Also the level shift can be fixed or adjusted dynamically. (3) *Quadratical Convergent SCF* (*QC-CF*). The basic strategy of the QC-SCF procedure consists in expressing the electronic energy in terms of orbital rotation parameters, i.e., the parameters describing the occupied-virtual mixing of the orbitals. The energy hypersurface in the space of these parameters is then approximated by a quadratic surface and the minimum of this surface is sought by a Newton-Raphson procedure involving calculation of the gradient and of the Hessian matrix of the energy with respect to the orbital rotation parameters. This procedure is quite time consuming (twice a DIIS calculation) and the QC-SCF technique is usually adopted as "last resource" when damping and level shifting failed in converging SCF. (4) *Initial guess.* The quality of the initial guess can play a crucial role in the convergence of the SCF procedure. In general it is easier to converge a small basis set calculation than a larger one; thus, it can be convenient to achieve SCF convergence using a small basis set, and then project the result onto a larger basis set to yield a good initial guess for the desired level. Alternatively, the molecular charge can be modified: it is easier to converge a calculation with fewer electrons (but with the same number of unpaired electrons) and if one is interested in an anion, one can first try to converge a large basis set for the neutral (or the cation) species to get a good estimate of the more "compact" MOs, and then import that result as a guess for the anionic system.

6. The SCF procedure produces a solution where the variation of the energy is "stationary" with respect to all orbital variations, i.e., the first derivatives of the energy with respect to the MO coefficients are zero. To be sure that this solution corresponds to an energy minimum, second derivatives should be also calculated. The eigenvalues of the corresponding matrix should all be positive for an energy minimum. A negative eigenvalue means that it is possible to get a lower energy state by "exciting" an electron from an occupied to an unoccupied orbital, indicating that the solution is unstable.
7. The output of a DFT calculation is the KS DFT wave function written as product of doubly occupied MOs. For closed shell systems, the total number of MOs is equal to the number of atomic orbitals included in the basis set, whereas the occupied MOs are equal to the number of electrons divided by two.

Because the MO orbital energy is related to the amount of energy needed to take an electron from the molecule to infinity, occupied MO orbital energy should always be negative; for the same reason, unoccupied MO will have positive orbital energies. In particular cases it could happen that the orbital energy of an occupied MO turns out to be positive, or that a low energy MO is unoccupied. FMOs with positive orbital energies are often found when negatively charged systems are considered, and this behavior is due to the self-interaction error. The presence of an unoccupied low energy MO could be due to the molecular symmetry of the system, but could also be a hint that the DFT calculation has converged to the wrong electronic state. In the latter case, one can change the occupation of this MO and successively run a new DFT calculation.

8. From basic theorems of Quantum Mechanics the "exact" wave function Ψ, obtained from the solution of the Schrodinger equation, should also be an eigenfunction of the $\widehat{\mathbf{S}}^2$ and $\widehat{\mathbf{S}}_z$ spin operators with eigenvalues $S(S+1)$ and M_S, respectively:

$$\widehat{\mathbf{S}}^2\Psi = S(S+1)\Psi \tag{35}$$

$$\widehat{\mathbf{S}}_z\Psi = M_S\Psi \tag{36}$$

where S is the spin quantum number representing the total spin of the system and M_S is a quantum number which can take the values ranging from +S to −S. For a single unpaired electron $S = 1/2$ and M_S can have values $+1/2$ (α spin) or $-1/2$ (β spin). In the case of a system with N unpaired electrons, the total spin quantum number S is given by the sum of the spin of each unpaired electron: $S = S_1 + S_2 + \ldots + S_N$. For example, in a system with two unpaired electrons $S = 1/2 + 1/2 = 1$, the system is said to be a triplet since M_S can have values +1, 0, 1 which corresponds to three degenerate (in the absence of a magnetic field) spin states. In general the spin multiplicity for a molecule of spin S is equal to $2S + 1$.The statement that the exact wave function is an eigenfunction of $\widehat{\mathbf{S}}^2$ and $\widehat{\mathbf{S}}_z$, and therefore, it is a pure spin state, is not necessarily true for approximate wave functions. In particular, it can be shown that a wave function expressed as a restricted Slater determinant is still an eigenfunction of both $\widehat{\mathbf{S}}^2$ and $\widehat{\mathbf{S}}_z$, whereas an unrestricted Slater determinant is not an eigenfunction of the $\widehat{\mathbf{S}}^2$ operator (i.e., it does not represent a pure spin state). This means that the computed wave function can be mixed with spin states of higher multiplicity. The error introduced by such mixing is called spin contamination. For example, the unrestricted determinant calculated for a radical ($S = 1/2$; doublet) can be mixed with a higher energy determinant representing three unpaired electrons ($S = 3/2$; quadruplet). The calculated energy and the electronic structure can be strongly affected by spin contamination, and therefore spin-contaminated wave functions should be considered with caution. The amount of spin contamination of the wave function can be checked considering the expectation value of the spin operator $<S^2>$, which is printed out in most ab initio and dft programs at the end of the wave function calculation. The expectation value depends on the total spin and it is equal to $S(S + 1)$ for a wave function representing a pure spin state. For example, in the case of a radical $S = 1/2$ and $<S^2> = (1/2(1/2 + 1)) = 0.75$, whereas for a molecule in a triplet state $S = 1$ and $<S^2> = (1(1 + 1)) = 2.00$. In the presence of spin contamination the calculated value of $<S^2>$ is larger than that predicted for a pure spin state. One rule of thumb is that the spin contamination is negligible if the calculated value of $<S^2>$ differs from $S(S + 1)$ by less than 10 %.It is important to note that the concept of spin contamination is well established for ab initio methods, like HF, where the unrestricted determinant is indeed an approximate wave function of the Schrodinger equation. In the KS-DFT method the unrestricted determinant is not the true wave function of the molecule, and the physical meaning of $<S^2>$ applied to this wave function is questionable. Beyond these considerations, it is worth noting that unrestricted KS-DFT methods with both GGA and hybrid functionals are generally much less affected by spin contamination than the Hartree–Fock method.

9. An example highlighting the difficulties to predict correctly the ground state of metalloproteins using DFT is provided by the active site of [NiFe]-hydrogenases. These enzymes, which catalyze the reversible oxidation of dihydrogen, are characterized by a very peculiar active site, in which one Ni and one Fe atom are bridged by two cysteine residues; two other cysteines coordinate the Ni ion and the coordination environment of Fe is completed by biologically atypical ligands such as CN and CO groups. A form of the enzyme labeled as Ni-SI has been characterized as containing a four-coordinated Ni(II) and a five-coordinated Fe(II) ions [159–161]. A still open question is related to the ground state of the Ni(II) ion in the Ni-SI form. Most spectroscopic data are compatible with diamagnetic ($S = 0$) Ni(II) [162, 163] and, correspondingly, models of these forms of the enzyme have been theoretically investigated assuming a singlet state, resulting, in the case of Ni-SI models, in a slightly distorted square-planar coordination arrangement [160, 161, 164–166]. However, according to the X-ray data [167], this arrangement of Ni ligands in the active site is unlikely, and nickel L-edge soft X-ray data [168]

have suggested the involvement of high-spin ($S = 1$) Ni(II) species in the catalytic cycle. Geometry optimizations and energy calculations of a simple model of the Ni-SI form, in the high-spin and low-spin states, have been carried out using the hybrid functional B3LYP, its modified version B3LYP*, as well as the pure BP86 functional. The computed energy differences between the fully optimized high- and low-spin Ni-SI states are strongly dependent on the adopted functional. The BP86 functional favors the low-spin state by about 6.0 kcal/mol, whereas the B3LYP functional favors the high-spin state by about 1.5 kcal/mol. The high-spin state is predicted to be more stable than the corresponding singlet species also by the B3LYP* functional, even though the energy difference becomes very small (0.70 kcal/mol).

10. Therefore any stationary point, previously located on a pure electronic energy PES, should be rigorously augmented by such ZPE (or ZPVE, as some authors report) additional contribution, which actually makes possible to construct the "ZPE-including or corrected-PES." A useful application of ZPE is in the theoretical determination of the kinetic isotope effect, which is important when dealing with hydrogen/deuterium-X bond cleavage or formation, and corresponds to the ratio of rate constants associated to the same reaction events occurring with two different isotopes of the same element. Since when dealing with hydrogen/deuterium the two isotopes differ in their mass by 100 %, the ZPE difference becomes significant, because $\nu = \frac{1}{2\pi}\sqrt{\frac{k}{\mu}}$ (k is the bond force constant—the Hessian in multi-dimensional systems- and μ is the reduced mass of a diatomic system). The kinetic isotope effect is important when the H/D-X bond cleavage or formation is the rate determining step in a reaction mechanism, which for metalloenzymes occurs, as an example, in proton/hydride transfer from/to metal ions. Such process was studied, amongst others, in the Mo-dependent formate dehydrogenase enzyme, both experimentally [169] and theoretically [170]. When associated with atoms directly involved in bond cleavage-formation, the effect is referred to as primary isotope effect, while the term secondary is used for all other cases. In this context it is worth recalling that the pure electronic energy (E_{SCF}) does not depend on the atomic masses, whose values, instead, modulate the ZPE term (and other energy contributions associated with the molecular partition function, vide infra). If we consider the energy profile for an hypothetic H/D-X bond breaking process, the ZPE difference between the light and the heavy isotope system is in practice all confined in the reactants, since by definition the reaction mode (vibrating with imaginary frequency) does not contribute significantly to the energetic

level of the TS. But if the E_{TS} is identical for the two isotope systems and the energy of the reactants including ZPE is different ($ZPE_{light} > ZPE_{heavy}$), the two energy barriers will be different and in particular the H–X bond will break faster (lower barrier) than the corresponding D–X bond (higher barrier). Thus, with the two ZPE values in hand one can predict, at least qualitatively, the kinetic isotope effect observed experimentally.

11. The main approximations introduced in the calculations of the roto-translation partition function of a system are:
    (a) Ideal gas model, which allows to simplify the true partition function $Q(N,V,T)$ as a sum of individual contributions $q(V,T)$, each associated with a single molecule of the whole system. If the ideal gas approximation is assumed valid, the PV term in H = U + PV can be substituted by RT (if only one mole of molecules is considered), which is trivially computed.
    (b) Separability of energy terms. It is assumed that the partition function of a single molecule within the system $q(V,T)$, which is a function of the total molecular energy, can be written as the product of individual (and independent) contributions, each being a function of a single energy contribution. This implies to assume that the total energy of a single molecule can be written as a sum of separate components: roto-translational, electronic, and vibrational.
    (c) QM harmonic oscillator approximation of the true potential energy including vibrational levels. This roughly corresponds to approximate the true potential function by a Taylor expansion truncated at the second order, which in close proximity of a minimum energy point is quite an acceptable simplification. Geometrically, this means to replace the actual potential function describing any single chemical bond with a parabola featuring equally spaced vibrational levels, instead of a Morse function with the vibrational state distribution which narrows when going toward bond distances farther from equilibrium. A strong implication of using such approximation is that the resulting harmonic vibration frequencies are always higher than the corresponding real frequencies, because except at the very minimum point (where the parabolic function and the Morse potential are identical) the harmonic potential overcomes the Morse potential. Scaling factors are therefore employed for different DFT functionals to account for such issue [130].
12. As an example, pure functionals perform better than hybrid functionals in reproducing the correct equilibrium population

of different coordination modes of oxo-peroxo dicopper compounds [171]. Truhlar et al. reported explicitly that B3LYP (hybrid) is less accurate than its nonhybrid counterpart, BLYP, both regard to metal–metal and metal–ligand bonding [172]. They also asserted that too premature obituaries of pure functionals had been previously prognosticated, essentially because of too much reliance on databases, originally devised mainly for organic and main-group chemistry.

13. An advantageous property of GGA (which as discussed previously are often referred to as "pure" functionals, since no Hartree–Fock exchange contribution is included) is the possibility to adopt the so-called Resolution-of-Identity (RI) approximation, which allows to approximate the very demanding computation of four-center integrals describing both the Coulomb repulsion and the exchange-correlation, with simpler two or three-center integrals, by using an expansion of the original density in terms of an orthonormalized and atom-centered basis set, known as the "auxiliary basis set" [173–175].

14. Since QM computational modeling deals with microscopic events, it is more appropriate to use the notation "transition state structure," which is just a stationary point on the PES, instead of "transition state." The "transition state" is related to a macroscopic ensemble, namely, to a population of molecules, whose reactive fluxes leading from reactants to products are maximized along a surface of $3N-7$ degrees of freedom (i.e., one less than minima, actually corresponding to the imaginary frequency).

15. When one is only interested in obtaining a computational speciation, which consists in the determination of all possible isomeric structures of a given species, a global minimum search should be carried out. Such a global minimum search can be generally carried out in a reasonable time only with molecular species not too large and without soft (or loosely harmonic) vibration modes (such as, for example, torsions in peptide loops).

16. It is also possible to compute the Hessian at a lower level of theory (i.e., by using a smaller basis set) than that employed for structure optimization, since the structure of the Hessian is qualitatively independent from basis set quality.

17. Problems with NR methods may arise if some H eigenvalues become very small, as it happens, for example, when soft vibrational modes are present. When considering metalloprotein models, this can occur if metal centers are coordinated to ligands containing flexible alkyl chains, or, otherwise, if such fragments are present in the second or third coordination sphere of the QM model. Also, weakly encumbered torsions

feature such floppy modes of vibration. Being at the denominator of the generic NR step expression, such low value numbers can cause the NR step to go far beyond the so-called Trust Radius, i.e., the hypersphere region where the second order Taylor polynomial is still valid.

18. One of the issues which is most frequently encountered when searching transition state structures is the convergence of calculations to undesired saddle points of order higher than one. This corresponds to an Hessian structure featuring two (or more) normal modes vibrating with imaginary frequency (or associated with negative eigenvalues): usually only one of the two is associated with the desired reaction coordinate, whereas the other represents a different transformation. Large sized systems, i.e., composed by many degrees of freedom, such as metalloenzyme active sites containing rotatable bonds/groups (such as phenyl and/or alkyl moieties) are quite often affected by such problem. This means that very often the undesired negative eigenvalue is associated to a normal mode featuring phenyl/methyl rotations. The way of getting rid of the "secondary transition modes" implies to distort the structure corresponding to the high order saddle point, in order to locate one of the two minima that are separated by such "secondary transition structure." In this step, one has to preserve the quantitative parameters associated with the reaction coordinate in the high order saddle point, i.e., maintain all those degrees of freedom that were originally kept fixed during the energy optimization in the same frozen disposition. Concomitantly only the degrees of freedom associated with the secondary (undesired) transition modes must be altered in order to relax them toward one of the two minima they put in mutual connection. If, for example, one is dealing with an unwanted methyl rotation, the torsions centered on the C–X (rotatable) bond of the moiety ($H_{1,2,3}$–C)–(X–Y) (where X and Y are topologically connected atoms which the methyl group is linked to) should be varied by approximately 40–45°. This is expected to cause the transition structure associated with the rotation to collapse to one of the connected minima.

19. Even though the rigorous (mathematic) definition of transition state structure implies the presence of only one negative eigenvalue in the Hessian matrix (at the TS point), it must be underscored that relaxing secondary transition modes of vibration may not lead to significant variations in terms of energy and conformation. This is just the case of low-lying energy processes, associated to low-frequency vibrations, such as sloppy vibrational modes, often encompassing large parts of the entire molecular structure. Even in the case of small group rotation (such as the aforementioned methyl issue), the gain in energy

respect to the true transition structure may be very limited (fraction of kcal/mol) and the resulting conformational change negligible.

20. Since the SHE is actually a conceptual electrode (i.e., only hypothetical, in which, among other assumptions, no ionic interaction is present), one may want to refer the computed absolute reduction potential to some other real electrode, such as the saturated calomel, which has a potential of +0.244 V versus SHE. Other reference electrodes, commonly employed in electrochemistry are the silver chloride electrode (+0.222 V vs. SHE, see IUPAC. Compendium of Chemical Terminology) and the Fc+/Fc based electrode (+0.64 V vs. SHE, see IUPAC. Compendium of Chemical Terminology).
21. The comparison between experimental and theoretical spectra should be carried out using some caution. First of all, the oscillators strengths depend on the coupling between the ground state and the excited states, which is very sensitive to inaccuracies due to the functional, the basis set, the molecular geometry, and the model used to describe the solvent. Regarding the excitation energies, many benchmark studies showed that TDDFT values calculated using standard GGA functionals are affected by errors in the range 0.1–0.5 eV [176]. However, it is important to recall that TDDFT poorly predicts excitation energies for charge-transfer (CT) states, which are very common in transition metal complexes. In this case the quality of TDDFT predictions can be sensibly improved by adopting hybrid functional, such as PBE0 [177]. Moreover, the solvatochromic effect (i.e., the dependence of the excitation energy by the solvent polarity) can induce a significant shift of the experimental CT absorptions, which can be simulated using an implicit solvent model such as the Polarized-continuous Model [178].

References

1. Rokob TA, Srnec M, Rulisek L (2012) Theoretical calculations of physico-chemical and spectroscopic properties of bioinorganic systems: current limits and perspectives. Dalton Trans 41:5754–5768
2. Acevedo O, Jorgensen WL (2010) Advances in quantum and molecular mechanical (QM/MM) simulations for organic and enzymatic reactions. Accounts Chem Res 43:142–151
3. Cramer CJ, Truhlar DG (2009) Density functional theory for transition metals and transition metal chemistry. Phys Chem Chem Phys 11:10757
4. Deeth RJ (2004) Recent developments in computational bioinorganic chemistry. Principles and applications of density functional theory in inorganic chemistry II. Springer, Berlin, pp 37–70
5. Neese F (2006) A critical evaluation of DFT, including time-dependent DFT, applied to bioinorganic chemistry. J Biol Inorg Chem 11:702–711
6. Siegbahn PEM, Borowski T (2006) Modeling enzymatic reactions involving transition metals. Acc Chem Res 39:729–738
7. Noodleman L, Han W-G (2006) Structure, redox, pK_a, spin. A golden tetrad for

understanding metalloenzyme energetics and reaction pathways. J Biol Inorg Chem 11:674–694

8. Bertini L, Bruschi M, De Gioa L, Fantucci P, Greco C, Zampella G (2007) Quantum chemical investigations of reaction paths of metalloenzymes and biomimetic models – the hydrogenase example. Topics in Current Chemistry 268:1
9. Ranaghan KE, Mulholland AJ (2010) Investigations of enzyme-catalysed reactions with combined quantum mechanics/molecular mechanics (QM/MM) methods. Int Rev Phys Chem 29:65–133
10. Kohn W, Becke AD, Parr RG (1996) Density functional theory of electronic structure. J Phys Chem 100:12974–12980
11. Hohenberg P, Kohn W (1964) Inhomogeneous electron gas. Phys Rev 136: B864–B871
12. Dirac PAM (1930) Note on exchange phenomena in the thomas atom. Math Proc Camb Phil Soc 26:376–385
13. Slater JC (1951) Magnetic effects and the Hartree-Fock equation. Phys Rev 82:538–541
14. Becke AD (1988) Density-functional exchange-energy approximation with correct asymptotic behavior. Phys Rev A 38:3098–3100
15. Perdew JP (1986) Density-functional approximation for the correlation energy of the inhomogeneous electron gas. Phys Rev B 33:8822–8824
16. Lee C, Yang W, Parr RG (1988) Development of the Colle-Salvetti correlation-energy formula into a functional of the electron density. Phys Rev B 37:785–789
17. Perdew JP (1991) In: Ziesche P, Eschrig H (eds) Electronic structure of solids '91. Akademie, Berlin
18. Perdew JP, Burke K, Ernzerhof M (1996) Generalized gradient approximation made simple. Phys Rev Lett 77:3865–3868
19. Adamo C, Barone V (1998) Exchange functionals with improved long-range behavior and adiabatic connection methods without adjustable parameters: the mPW and mPW1PW models. J Chem Phys 108:664–675
20. Perdew JP, Ruzsinszky A, Tao J, Staroverov VN, Scuseria GE, Csonka GI (2005) Prescription for the design and selection of density functional approximations: more constraint satisfaction with fewer fits. J Chem Phys 123:062201–062209
21. Becke AD (1998) A new inhomogeneity parameter in density-functional theory. J Chem Phys 109:2092–2098
22. Stephens PJ, Devlin FJ, Chabalowski CF, Frisch MJ (1994) Ab Initio calculation of vibrational absorption and circular dichroism spectra using density functional force fields. J Phys Chem 98:11623–11627
23. Adamo C, Barone V (1997) Toward reliable adiabatic connection models free from adjustable parameters. Chem Phys Lett 274:242–250
24. Ernzerhof M, Scuseria GE (1999) Assessment of the Perdew–Burke–Ernzerhof exchange-correlation functional. J Chem Phys 110:5029–5036
25. Lynch BJ, Fast PL, Harris M, Truhlar DG (2000) Adiabatic connection for kinetics. J Phys Chem A 104:4811–4815
26. Jensen F (2006) Introduction to computational chemistry, 2nd edn. Wiley, New York
27. Cramer CJ (2006) Essentials of computational chemistry: theories and models, 2nd edn. Wiley, West Sussex
28. Barone V (1996) Structure, magnetic properties and reactivities of open-shell species from density functional and self-consistent hybrid methods. In: Chong DP (ed) Recent advances in density functional methods part I. World Scientific Publishing Company, Singapore, pp 287–334
29. Frisch MJ, Trucks GW, Schlegel HB, Scuseria GE, Robb MA, Cheeseman JR, Scalmani G, Barone V, Mennucci B, Petersson GA, Nakatsuji H, Caricato M, Li X, Hratchian HP, Izmaylov AF, Bloino J, Zheng G, Sonnenberg JL, Hada M, Ehara M, Toyota K, Fukuda R, Hasegawa J, Ishida M, Nakajima T, Honda Y, Kitao O, Nakai H, Vreven T, Montgomery Jr, JA, Peralta JE, Ogliaro F, Bearpark M, Heyd JJ, Brothers E, Kudin KN, Staroverov VN, Kobayashi R, Normand J, Raghavachari K, Rendell A, Burant JC, Iyengar SS, Tomasi J, Cossi M, Rega N, Millam JM, Klene M, Knox JE, Cross JB, Bakken V, Adamo C, Jaramillo J, Gomperts R, Stratmann RE, Yazyev O, Austin AJ, Cammi R, Pomelli C, Ochterski JW, Martin RL, Morokuma K, Zakrzewski VG, Voth GA, Salvador P, Dannenberg JJ, Dapprich S, Daniels AD, Farkas Ö, Foresman JB, Ortiz JV, Cioslowski J, Fox DJ (2009) Gaussian 09 (Revision A.1); Gaussian, Inc., Wallingford CT.
30. Guerra CF, Snijders JG, te Velde G, Baerends EJ (1998) Towards an order-N DFT method. Theor Chem Acc 99:391–403

31. Te Velde G, Bickelhaupt FM, Baerends EJ, Fonseca Guerra C, van Gisbergen SJA, Snijders JG, Ziegler T (2001) Chemistry with ADF. J Comput Chem 22:931–967
32. Baerends EJ, Ziegler T, Autschbach J, Bashford D, Bérces A, Bickelhaupt FM, Bo C, Boerrigter PM, Cavallo L, Chong DP, Deng L, Dickson RM, Ellis DE, van Faassen M, Fan L, Fischer TH, Fonseca Guerra C, Ghysels A, Giammona A, van Gisbergen SJA, Götz AW, Groeneveld JA, Gritsenko OV, Grüning M, Gusarov S, Harris FE, van den Hoek P, Jacob CR, Jacobsen H, Jensen L, Kaminski JW, van Kessel G, Kootstra F, Kovalenko A, Krykunov MV, van Lenthe E, McCormack DA, Michalak A, Mitoraj M, Neugebauer J, Nicu VP, Noodleman L, Osinga VP, Patchkovskii S, Philipsen PHT, Post D, Pye CC, Ravenek W, Rodríguez JI, Ros P, Schipper PRT, Schreckenbach G, Seldenthuis JS, Seth M, Snijders JG, Solà M, Swart M, Swerhone D, te Velde G, Vernooijs P, Versluis L, Visscher L, Visser O, Wang F, Wesolowski TA, van Wezenbeek EM, Wiesenekker G, Wolff SK, Woo TK, Yakovlev AL. DF2012, SCM, theoretical chemistry, Vrije Universiteit, Amsterdam, The Netherlands, http://www.scm.com.
33. Schmidt MW, Baldridge KK, Boatz JA et al (1993) General atomic and molecular electronic structure system. J Comput Chem 14:1347–1363
34. Gordon MS, Schmidt MW (2005) In: Dykstra CE, Frenking G, Kim KS, Scuseria GE (eds) Theory and applications of computational chemistry: the first forty years, Elsevier, Amsterdam, p 1167
35. Karlström G, Lindh R, Malmqvist P-Å et al (2003) MOLCAS: a program package for computational chemistry. Comput Mater Sci 28:222–239
36. Ahlrichs R, Bär M, Häser M, Horn H, Kölmel C (1989) Electronic structure calculations on workstation computers: the program system turbomole. Chem Phys Lett 162:165–169
37. Helgaker T, Jensen HJA, Jørgensen P, Olsen J, Ruud K, Agren H, Auer AA, Bak KL, Bakken V, Christiansen O, Coriani S, Dahle P, Dalskov EK, Enevoldsen T, Fernandez B, Hattig C, Hald K, Halkier A, Heiberg H, Hettema H, Jonsson D, Kirpekar S, Kobayashi R, Koch H, Mikkelsen KV, Norman P, Packer MJ, Pedersen TB, Ruden TA, Salek P, Sanchez A, Saue T, Sauer SPA, Schimmelpfennig B, Sylvester-Hvid KO, Taylor PR, Vahtras O. Dalton, a molecular electronic structure program, Release 2.0, 2005.
38. Petrenko T, Kossmann S, Neese F (2011) Efficient time-dependent density functional theory approximations for hybrid density functionals: analytical gradients and parallelization. J Chem Phys. doi:10.1063/1.3533441
39. Dziedzic J, Fox SJ, Fox T, Tautermann CS, Skylaris C-K (2013) Large-scale DFT calculations in implicit solvent – a case study on the T4 lysozyme L99A/M102Q protein. Int J Quant Chem 113:771–785
40. Siegbahn PEM, Himo F (2009) Recent developments of the quantum chemical cluster approach for modeling enzyme reactions. J Biol Inorg Chem 14:643–651
41. Sousa SF, Fernandes PA, Ramos MJ (2012) Computational enzymatic catalysis – clarifying enzymatic mechanisms with the help of computers. Phys Chem Chem Phys 14:12431–12441
42. Nemukhin AV, Grigorenko BL, Lushchekina SV, Varfolomeev SD (2012) Quantum chemical modelling in the research of molecular mechanisms of enzymatic catalysis. Russ Chem Rev 81:1011–1025
43. Siegbahn PEM, Himo F (2011) The quantum chemical cluster approach for modeling enzyme reactions. Wiley Interdiscip Rev Comput Mol Sci 1:323–336
44. Marino T, Russo N, Toscano M (2005) A comparative study of the catalytic mechanisms of the zinc and cadmium containing carbonic anhydrase. J Am Chem Soc 127:4242–4253
45. Leopoldini M, Marino T, del Michelini MC, Rivalta I, Russo N, Sicilia E, Toscano M (2007) The role of quantum chemistry in the elucidation of the elementary mechanisms of catalytic processes: from atoms, to surfaces, to enzymes. Theor Chem Acc 117:765–779
46. Amata O, Marino T, Russo N, Toscano M (2011) A proposal for mitochondrial processing peptidase catalytic mechanism. J Am Chem Soc 133:17824–17831
47. Alberto ME, Marino T, Russo N, Sicilia E, Toscano M (2012) The performance of density functional based methods in the description of selected biological systems and processes. Phys Chem Chem Phys 14:14943–14953
48. Klamt A, Schuurmann G (1993) Cosmo – a new approach to dielectric screening in solvents with explicit expressions for the screening energy and its gradient. J Chem Soc-Perkin Trans 2:799–805
49. Klamt A (1995) Conductor-like screening model for real solvents – a new approach to

the quantitative calculation of solvation phenomena. J Phys Chem 99:2224–2235
50. Andzelm J, Kolmel C, Klamt A (1995) Incorporation of solvent effects into density-functional calculations of molecular-energies and geometries. J Chem Phys 103:9312–9320
51. Klamt A, Jonas V (1996) Treatment of the outlying charge in continuum solvation models. J Chem Phys 105:9972–9981
52. Barone V, Cossi M (1998) Quantum calculation of molecular energies and energy gradients in solution by a conductor solvent model. J Phys Chem A 102:1995–2001
53. Cossi M, Barone V, Mennucci B, Tomasi J (1998) Ab initio study of ionic solutions by a polarizable continuum dielectric model. Chem Phys Lett 286:253–260
54. Rega N, Cossi M, Barone V, Pomelli CS, Tomasi J (1999) Toward an effective and reliable representation of solvent effects in the study of biochemical systems. Int J Quantum Chem 73:219–227
55. Cossi M, Rega N, Scalmani G, Barone V (2003) Energies, structures, and electronic properties of molecules in solution with the C-PCM solvation model. J Comput Chem 24:669–681
56. Tomasi J, Persico M (1994) Molecular-interactions in solution – an overview of methods based on continuous distributions of the solvent. Chem Rev 94:2027–2094
57. Tomasi J, Mennucci B, Cammi R (2005) Quantum mechanical continuum solvation models. Chem Rev 105:2999–3093
58. Chen S-L, Fang W-H, Himo F (2009) Reaction mechanism of the binuclear zinc enzyme glyoxalase II – a theoretical study. J Inorg Biochem 103:274–281
59. Liao R-Z, Himo F, Yu J-G, Liu R-Z (2010) Dipeptide hydrolysis by the dinuclear zinc enzyme human renal dipeptidase: mechanistic insights from DFT calculations. J Inorg Biochem 104:37–46
60. Sousa SF, Fernandes PA, Ramos MJ (2009) The search for the mechanism of the reaction catalyzed by farnesyltransferase. Chem Eur J 15:4243–4247
61. Amata O, Marino T, Russo N, Toscano M (2011) Catalytic activity of a zeta-class zinc and cadmium containing carbonic anhydrase. Compared work mechanisms. Phys Chem Chem Phys 13:3468–3477
62. Sevastik R, Himo F (2007) Quantum chemical modeling of enzymatic reactions: the case of 4-oxalocrotonate tautomerase. Bioorganic Chem 35:444–457
63. Hopmann KH, Himo F (2008) Quantum chemical modeling of the dehalogenation reaction of haloalcohol dehalogenase. J Chem Theor Comput 4:1129–1137
64. Warshel A, Levitt M (1976) Theoretical studies of enzymic reactions – dielectric, electrostatic and steric stabilization of carbonium-ion in reaction of lysozyme. J Mol Biol 103:227–249
65. Ranaghan KE, Ridder L, Szefczyk B, Sokalski WA, Hermann JC, Mulholland AJ (2003) Insights into enzyme catalysis from QM/MM modelling: transition state stabilization in chorismate mutase. Mol Phys 101:2695–2714
66. Senn HM, Thiel W (2007) QM/MM studies of enzymes. Curr Opin Chem Biol 11:182–187
67. Olsson MHM, Parson WW, Warshel A (2006) Dynamical contributions to enzyme catalysis: critical tests of a popular hypothesis. Chem Rev 106:1737–1756
68. Warshel A (2003) Computer simulations of enzyme catalysis: methods, progress, and insights. Annu Rev Biophys Biomolec Struct 32:425–443
69. Claeyssens F, Harvey JN, Manby FR, Mata RA, Mulholland AJ, Ranaghan KE, Schuetz M, Thiel S, Thiel W, Werner H-J (2006) High-accuracy computation of reaction barriers in enzymes. Angew Chem Int Edit 45:6856–6859
70. Cornell WD, Cieplak P, Bayly CI, Gould IR, Merz KM, Ferguson DM, Spellmeyer DC, Fox T, Caldwell JW, Kollman PA (1995) A second generation force field for the simulation of proteins, nucleic acids, and organic molecules. J Am Chem Soc 117:5179–5197
71. Duan Y, Wu C, Chowdhury S et al (2003) A point-charge force field for molecular mechanics simulations of proteins based on condensed-phase quantum mechanical calculations. J Comput Chem 24:1999–2012
72. Brooks BR, Bruccoleri RE, Olafson BD, States DJ, Swaminathan S, Karplus M (1983) CHARMM: a program for macromolecular energy, minimization, and dynamics calculations. J Comput Chem 4:187–217
73. Schmid N, Eichenberger AP, Choutko A, Riniker S, Winger M, Mark AE, van Gunsteren WF (2011) Definition and testing of the GROMOS force-field versions 54A7 and 54B7. Eur Biophys J 40:843–856
74. Jorgensen WL, Tirado-Rives J (1988) The OPLS (optimized potentials for liquid simulations) potential functions for proteins, energy minimizations for crystals of cyclic peptides

and crambin. J Am Chem Soc 110:1657–1666
75. Fox SJ, Pittock C, Fox T, Tautermann CS, Malcolm N, Skylaris C-K (2011) Electrostatic embedding in large-scale first principles quantum mechanical calculations on biomolecules. J Chem Phys 135:224107–224107–13.
76. Wallrapp FH, Guallar V (2011) Mixed quantum mechanics and molecular mechanics methods: looking inside proteins. Wiley Interdiscip Rev Comput Mol Sci 1:315–322
77. Hagiwara Y, Tateno M (2010) Recent advances in jointed quantum mechanics and molecular mechanics calculations of biological macromolecules: schemes and applications coupled to ab initio calculations. J Phys Condes Matter. doi:10.1088/0953-8984/22/41/413101
78. Zhang R, Lev B, Cuervo JE, Noskov SY, Salahub DR (2010) A guide to QM/MM methodology and applications. In: Sabin JR, Brandas E, Canuto S (eds) Advances in quantum chemistry, vol 59, Combining quantum mechanics and molecular mechanics – some recent progresses in Qm/Mm methods. Elsevier Academic Press Inc, San Diego, pp 353–400
79. Hirao H, Morokuma K (2011) Recent progress in the theoretical studies of structure, function, and reaction of biological molecules. Yakugaku Zasshi J Pharm Soc Jpn 131:1151–1161
80. Ryde U (1996) The coordination of the catalytic zinc in alcohol dehydrogenase studied by combined quantum-chemical and molecular mechanics calculations. J Comput Aided Mol Des 10:153–164
81. Seminario JM (1996) Calculation of intramolecular force fields from second-derivative tensors. Int J Quant Chem 60:1271–1277
82. Reuter N, Dejaegere A, Maigret B, Karplus M (2000) Frontier bonds in QM/MM methods: a comparison of different approaches. J Phys Chem A 104:1720–1735
83. Olsson MHM, Hong G, Warshel A (2003) Frozen density functional free energy simulations of redox proteins: computational studies of the reduction potential of plastocyanin and rusticyanin. J Am Chem Soc 125:5025–5039
84. Rod TH, Ryde U (2005) Quantum mechanical free energy barrier for an enzymatic reaction. Phys Rev Lett 94:138302
85. Reiher M, Salomon O, Hess BA (2001) Reparameterization of hybrid functionals based on energy differences of states of different multiplicity. Theor Chem Acc 107:48–55
86. Becke A (1993) Density-functional thermochemistry. 3. The role of exact exchange. J Chem Phys 98:5648–5652
87. Geerlings P, De Proft F, Langenaeker W (2003) Conceptual density functional theory. Chem Rev 103:1793–1873
88. Siegbahn PEM, Blomberg MRA (2000) Transition-metal systems in biochemistry studied by high-accuracy quantum chemical methods. Chem Rev 100:421–437
89. Reiher M (2002) Theoretical study of the Fe (phen)(2)(NCS)(2) spin-crossover complex with reparametrized density functionals. Inorg Chem 41:6928–6935
90. Wolny JA, Paulsen H, Trautwein AX, Schuenemann V (2009) Density functional theory calculations and vibrational spectroscopy on iron spin-crossover compounds. Coord Chem Rev 253:2423–2431
91. Heisenberg W (1928) Z Phys 49:619
92. Dirac PAM (1929) Proc R Soc London, Ser A 123:714
93. Van Vleck JH (1945) Rev Mod Phys 17:27
94. Noodleman L (1981) Valence bond description of anti-ferromagnetic coupling in transition-metal dimers. J Chem Phys 74:5737–5743
95. Noodleman L, Norman J (1979) X-alpha valence bond theory of weak electronic coupling – application to the low-lying states of Mo2cl84. J Chem Phys 70:4903–4906
96. Noodleman L, Peng C, Case D, Mouesca J (1995) Orbital interactions, electron delocalization and spin coupling in iron-sulfur clusters. Coord Chem Rev 144:199–244
97. Mouesca J, Chen J, Noodleman L, Bashford D, Case D (1994) Density-functional Poisson-Boltzmann calculations of redox potentials for iron-sulfur clusters. J Am Chem Soc 116:11898–11914
98. Torres RA, Lovell T, Noodleman L, Case DA (2003) Density functional-and reduction potential calculations of Fe(4)S(4) clusters. J Am Chem Soc 125:1923–1936
99. Noodleman L, Norman J, Osborne J, Aizman A, Case D (1985) Models for ferredoxins – electronic-structures of iron sulfur clusters with one, 2, and 4 iron atoms. J Am Chem Soc 107:3418–3426
100. Aizman A, Case D (1982) Electronic-structure calculations on active-site models for 4-Fe, 4-S iron sulfur proteins. J Am Chem Soc 104:3269–3279
101. Papaefthymiou V, Millar M, Munck E (1986) Mossbauer and electron-paramagnetic-Res studies of a synthetic analog for the Fe4s4

core of oxidized and reduced high-potential iron proteins. Inorg Chem 25:3010–3014
102. Carney M, Papaefthymiou G, Spartalian K, Frankel R, Holm R (1988) Ground spin state variability in (fe4s4(sr)4)3- – synthetic analogs of the reduced clusters in ferredoxins and other iron sulfur proteins – cases of extreme sensitivity of electronic state and structure to extrinsic factors. J Am Chem Soc 110:6084–6095
103. Auric P, Gaillard J, Meyer J, Moulis J (1987) Analysis of the high-spin states of the 2(4fe-4se) + ferredoxin from Clostridium pasteurianum by Mossbauer spectroscopy. Biochem J 242:525–530
104. Champion P, Munck E, Debrunner P, Moss T, Lipscomb J, Gunsalus I (1975) Magnetic-susceptibility of reduced cytochrome P-450cam. Biochimica Et Biophysica Acta 376:579–582
105. Tang S, Spiro T, Antanaitis C, Moss T, Holm R, Herskovits T, Mortensen L (1975) Resonance Raman spectroscopic evidence for structural variation among bacterial ferredoxin, Hipip, and Fe4s4(sch2ph)42-. Biochem Biophys Res Commun 62:1–6
106. Middleton P, Dickson D, Johnson C, Rush J (1980) Interpretation of the Mossbauer spectra of the high-potential iron protein from Chromatium. Eur J Biochem 104:289–296
107. Middleton P, Dickson D, Johnson C, Rush J (1978) Interpretation of Mossbauer spectra of 4-iron ferredoxin from Bacillus stearothermophilus. Eur J Biochem 88:135–141
108. Blondin G, Girerd J (1990) Interplay of electron exchange and electron-transfer in metal polynuclear complexes in proteins or chemical-models. Chem Rev 90:1359–1376
109. Noodleman L, Baerends E (1984) Electronic-structure, magnetic-properties, electron-spin-resonance, and optical-spectra for 2-Fe ferredoxin models by Lcao-X-alpha valence bond theory. J Am Chem Soc 106:2316–2327
110. Papaefthymiou V, Girerd J, Moura I, Moura J, Munck E (1987) Mossbauer study of D-gigas ferredoxin-ii and spin-coupling model for the Fe3s4 cluster with valence delocalization. J Am Chem Soc 109:4703–4710
111. Noodleman L, Case D (1992) Density-functional theory of spin polarization and spin coupling in iron-sulfur clusters. Adv Inorg Chem 38:423
112. Borshch S, Bominaar E, Blondin G, Girerd J (1993) Double exchange and vibronic coupling in mixed-valence systems – origin. J Am Chem Soc 115:5155–5168
113. Pereira AS, Tavares P, Moura I, Moura JJG, Huynh BH (2001) Mossbauer characterization of the iron-sulfur clusters in Desulfovibrio vulgaris hydrogenase. J Am Chem Soc 123:2771–2782
114. Popescu CV, Munck E (1999) Electronic structure of the H cluster in (Fe)-hydrogenases. J Am Chem Soc 121:7877–7884
115. Bruschi M, Greco C, Fantucci P, De Gioia L (2008) Structural and electronic properties of the (FeFe) hydrogenase H-cluster in different redox and protonation states. A DFT investigation. Inorg Chem 47:6056–6071
116. Fiedler AT, Brunold TC (2005) Computational studies of the H-cluster of Fe-only hydrogenases: geometric, electronic, and magnetic properties and their dependence on the (Fe(4)S(4)) cubane. Inorg Chem 44:9322–9334
117. Peters JW, Lanzilotta WN, Lemon BJ, Seefeldt LC (1998) X-ray crystal structure of the Fe-only hydrogenase (Cpl) from Clostridium pasteurianum to 1.8 angstrom resolution. Science 282:1853–1858
118. Zampella G, Neupane KP, De Gioia L, Pecoraro VL (2012) The importance of stereochemically active lone pairs for influencing PbII and AsIII protein binding. Chem Eur J 18:2040–2050
119. Johnson MK, Rees DC, Adams MWW (1996) Tungstoenzymes. Chem Rev 96:2817–2840
120. Huheey JE (2006) Inorganic chemistry: principles of structure and reactivity. Pearson Education
121. Zampella G, Fantucci P, De Gioia L (2009) DFT characterization of the reaction pathways for terminal- to μ-hydride isomerisation in synthetic models of the (FeFe)-hydrogenase active site. Chem Commun 46:8824–8826
122. Bruschi M, Zampella G, Fantucci P, De Gioia L (2005) DFT investigations of models related to the active site of (NiFe) and (Fe) hydrogenases. Coord Chem Rev 249:1620–1640
123. Chen JL, Noodleman L, Case DA, Bashford D (1994) Incorporating solvation effects into density functional electronic structure calculations. J Phys Chem 98:11059–11068
124. Connors KA (1990) Chemical kinetics: the study of reaction rates in solution. Wiley, New York
125. Bigeleisen J, Mayer MG (1947) Calculation of equilibrium constants for isotopic exchange reactions. J Chem Phys 15:261–267

126. Zhao Y, Truhlar DG (2008) Density functionals with broad applicability in chemistry. Acc Chem Res 41:157–167
127. Zhao Y, Truhlar DG (2008) The M06 suite of density functionals for main group thermochemistry, thermochemical kinetics, noncovalent interactions, excited states, and transition elements: two new functionals and systematic testing of four M06-class functionals and 12 other functionals. Theor Chem Account 120:215–241
128. Ziegler T (1991) Approximate density functional theory as a practical tool in molecular energetics and dynamics. Chem Rev 91:651–667
129. Cramer CJ (2005) Essentials of computational chemistry: theories and models. Wiley, New York
130. Koch PDW, Holthausen DMC (2001) A chemist's guide to density functional theory, 2nd edn.
131. Becke AD (2000) Simulation of delocalized exchange by local density functionals. J Chem Phys 112:4020–4026
132. Siegbahn PEM, Borowski T (2006) Modeling enzymatic reactions involving transition metals. Accounts Chem Res 39:729–738
133. Evans MG, Polanyi M (1935) Some applications of the transition state method to the calculation of reaction velocities, especially in solution. Trans Faraday Soc 31:875–894
134. Eyring H (1935) The activated complex in chemical reactions. J Chem Phys 3:107–115
135. Leach AR (2001) Molecular modelling: principles and applications. Pearson Education
136. Hughes TF, Friesner RA (2012) Development of accurate DFT methods for computing redox potentials of transition metal complexes: results for model complexes and application to cytochrome P450. J Chem Theory Comput 8:442–459
137. Moens J, De Proft F, Geerlings P (2010) A density functional theory study on ligand additive effects on redox potentials. Phys Chem Chem Phys 12:13174–13181
138. Galstyan A, Knapp E-W (2009) Accurate redox potentials of mononuclear iron, manganese, and nickel model complexes. J Comput Chem 30:203–211
139. Izutsu K (2009) Electrochemistry in nonaqueous solutions. Wiley, New York. doi: 10.1002/3527600655
140. Liptak MD, Gross KC, Seybold PG, Feldgus S, Shields GC (2002) Absolute pKa determinations for substituted phenols. J Am Chem Soc 124:6421–6427
141. Bartmess JE (1994) Thermodynamics of the electron and the proton. J Phys Chem 98:6420–6424
142. Namazian M, Coote ML (2007) Accurate calculation of absolute one-electron redox potentials of some para-quinone derivatives in acetonitrile. J Phys Chem A 111:7227–7232
143. Szaciłowski K, Macyk W, Drzewiecka-Matuszek A, Brindell M, Stochel G (2005) Bioinorganic photochemistry: frontiers and mechanisms. Chem Rev 105:2647–2694
144. Rose MJ, Betterley NM, Mascharak PK (2009) Thiolate S-oxygenation controls nitric oxide (NO) photolability of a synthetic iron nitrile hydratase (Fe-NHase) model derived from mixed carboxamide/thiolate ligand. J Am Chem Soc 131:8340–8341
145. Dunietz BD, Dreuw A, Head-Gordon M (2003) Initial steps of the photodissociation of the CO ligated heme group. J Phys Chem B 107:5623–5629
146. Chen Z, Lemon BJ, Huang S, Swartz DJ, Peters JW, Bagley KA (2002) Infrared studies of the CO-inhibited form of the Fe-only hydrogenase from clostridium pasteurianum i: examination of its light sensitivity at cryogenic temperatures. Biochemistry 41:2036–2043
147. Lodowski P, Jaworska M, Andruniów T, Kumar M, Kozlowski PM (2009) Photodissociation of Co–C bond in methyl- and ethylcobalamin: an insight from TD-DFT calculations. J Phys Chem B 113:6898–6909
148. Dreuw A, Head-Gordon M (2005) Single-reference ab initio methods for the calculation of excited states of large molecules. Chem Rev 105:4009–4037
149. Bringmann G, Bruhn T, Maksimenka K, Hemberger Y (2009) The assignment of absolute stereostructures through quantum chemical circular dichroism calculations. Eur J Org Chem 2009:2717–2727
150. Santoro F, Lami A, Improta R, Barone V (2007) Effective method to compute vibrationally resolved optical spectra of large molecules at finite temperature in the gas phase and in solution. J Chem Phys 126:184102–184111

151. Bertini L, Greco C, De Gioia L, Fantucci P (2009) DFT/TDDFT exploration of the potential energy surfaces of the ground state and excited states of Fe2(S2C3H6)(CO)6: a simple functional model of the (FeFe) hydrogenase active site. J Phys Chem A 113:5657–5670
152. Dyall KG, Faegri K (2007) Introduction to relativistic quantum chemistry. Oxfrod University Press, New York, USA
153. Almlöf J, Gropen O (1996) In: Lipkowitz KB, Boyd DB (eds) Relativistic effects in chemistry. VCH, Weinheim, Germany, pp 203–244
154. Dolg M, Stoll H, Preuss H (1989) Energy-adjusted ab initio pseudopotentials for the rare earth elements. J Chem Phys 90:1730–1734
155. Cundari TR, Stevens WJ (1993) Effective core potential methods for the lanthanides. J Chem Phys 98:5555–5565
156. Ross RB, Gayen S, Ermler WC (1994) Ab initio relativistic effective potentials with spin–orbit operators V. Ce through Lu. J Chem Phys 100:8145–8155
157. Dolg M, Stoll H, Preuss H (1989) Energy-adjusted pseudopotentials for the rare earth elements. Theor Chim Acta 75:173–194
158. Merbach AE, Toth E (2001) The chemistry of contrast agents in medical magnetic resonance imaging. Wiley, Chichester, UK
159. Davidson G, Choudhury SB, Gu ZJ, Bose K, Roseboom W, Albracht SPJ, Maroney MJ (2000) Structural examination of the nickel site in Chromatium vinosum hydrogenase: redox state oscillations and structural changes accompanying reductive activation and CO binding. Biochemistry 39:7468–7479
160. Fan HJ, Hall MB (2001) Recent theoretical predictions of the active site for the observed forms in the catalytic cycle of Ni-Fe hydrogenase. J Biol Inorg Chem 6:467–473
161. Stein M, Lubitz W (2002) Quantum chemical calculations of (NiFe) hydrogenase. Curr Opin Chem Biol 6:243–249
162. Dole F, Fournel A, Magro V, Hatchikian EC, Bertrand P, Guigliarelli B (1997) Nature and electronic structure of the Ni-X dinuclear center of Desulfovibrio gigas hydrogenase. Implications for the enzymatic mechanism. Biochemistry 36:7847–7854
163. Wang C, Franco R, Moura J, Moura I, Day E (1992) The Nickel site in active desulfovibrio-baculatus (nifese) hydrogenase. J Biol Chem 267:7378–7380
164. Stein M, Lubitz W (2001) DFT calculations of the electronic structure of the paramagnetic states Ni-A, Ni-B and Ni-C of (NiFe) hydrogenase. Phys Chem Chem Phys 3:2668–2675
165. De Gioia L, Fantucci P, Guigliarelli B, Bertrand P (1999) Ab initio investigation of the structural and electronic differences between active-site models of (NiFe) and (NiFeSe) hydrogenases. Int J Quant Chem 73:187–195
166. De Gioia L, Fantucci P, Guigliarelli B, Bertrand P (1999) Ni-Fe hydrogenases: a density functional theory study of active site models. Inorg Chem 38:2658–2662
167. Volbeda A, Charon M, Piras C, Hatchikian E, Frey M, Fontecillacamps J (1995) Crystal-structure of the nickel-iron hydrogenase from Desulfovibrio-gigas. Nature 373:580–587
168. Wang HX, Ralston CY, Patil DS et al (2000) Nickel L-edge soft X-ray spectroscopy of nickel-iron hydrogenases and model compounds – Evidence for high-spin nickel(II) in the active enzyme. J Am Chem Soc 122:10544–10552
169. Khangulov SV, Gladyshev VN, Dismukes GC, Stadtman TC (1998) Selenium-containing formate dehydrogenase H from Escherichia coli: a molybdopterin enzyme that catalyzes formate oxidation without oxygen transfer. Biochemistry 37:3518–3528
170. Tiberti M, Papaleo E, Russo N, De Gioia L, Zampella G (2012) Evidence for the formation of a Mo-H intermediate in the catalytic cycle of formate dehydrogenase. Inorg Chem 51:8331–8339
171. Lewin JL, Heppner DE, Cramer CJ (2007) Validation of density functional modeling protocols on experimental bis(μ-oxo)/μ-η 2:η2-peroxo dicopper equilibria. J Biol Inorg Chem 12:1221–1234
172. Schultz NE, Zhao Y, Truhlar DG (2005) Density functionals for inorganometallic and organometallic chemistry. J Phys Chem A 109:11127–11143
173. Baerends EJ, Ellis DE, Ros P (1973) Self-consistent molecular Hartree—Fock—Slater calculations I. The computational procedure. Chem Phys 2:41–51
174. Dunlap BI, Connolly JWD, Sabin JR (1979) On some approximations in applications of Xα theory. J Chem Phys 71:3396–3402
175. Vahtras O, Almlöf J, Feyereisen MW (1993) Integral approximations for LCAO-SCF calculations. Chem Phys Lett 213:514–518
176. Jacquemin D, Mennucci B, Adamo C (2011) Excited-state calculations with TD-DFT: from benchmarks to simulations in complex

environments. Phys Chem Chem Phys 13:16987–16998

177. Adamo C, Barone V (2000) Inexpensive and accurate predictions of optical excitations in transition-metal complexes: the TDDFT/PBE0 route. Theor Chem Acc 105:169–172

178. Scalmani G, Frisch MJ, Mennucci B, Tomasi J, Cammi R, Barone V (2006) Geometries and properties of excited states in the gas phase and in solution: theory and application of a time-dependent density functional theory polarizable continuum model. J Chem Phys 124:094107–094115

Chapter 15

Density Functional Theory–Broken Symmetry (DFT–BS) Methodology Applied to Electronic and Magnetic Properties of Bioinorganic Prosthetic Groups

Jean-Marie Mouesca

Abstract

The goal of this "how to" chapter is to present in a way as simple and practical as possible some of the concepts, key issues, and practices behind the so-called broken symmetry (BS) state which is widely used within the density functional theory (DFT) (for a very nice but thoughtful introduction to DFT (without equations!), read Perdew et al. (J Chem Theory Comput 5:902–908, 2009)) community to compute energetic as well as spectroscopic properties pertaining to (poly-)radicals, bioinorganic clusters (especially those containing transition metal ions), etc. Such properties encompass exchange coupling constants J (molecular magnetism) but also (among other things) **g**-tensors and hyperfine coupling tensors **A** (from electron paramagnetic resonance), isomer shifts $\boldsymbol{\delta}$ and quadrupolar tensors $\boldsymbol{\Delta E_Q}$ (from Mössbauer), etc.

Hopefully, this chapter will appeal to those DFT practitioners who would like to understand the basics behind the BS state and help them "demystify" some of the issues involved with them. More technical issues will only be alluded to, and appropriate references will be given for those interested to go beyond this mere introduction. This chapter is however not a review of the field. Consequently, it will be primarily based on my own experience.

The goal here (in the spirit of a "how to" chapter) is to accompany the readers' thoughts in a progressive way along increasingly complex issues rather than encumbering the same thoughts with too complicate mathematical details (the few derivations which are given will therefore be explicit). Moreover, I will emphasize in this chapter the interplay between the *computation* of BS states on the one hand, and the derivation of *phenomenological models* on the other hand, whose parameters can be supplied from appropriate BS states.

Finally, this chapter is dedicated to Louis Noodleman (Scripps Research Institute, CA, USA), pioneer (Noodleman, J Chem Phys 74:5737–5743, 1981; Noodleman, Chem Phys 109:131–143, 1986) and leading practitioner (for those interested, I invite them to look for his papers online (via Web of Knowledge for example) as they are simply too numerous to be cited here) of the BS methodology described below in the rich field of (bio)inorganic (transition metal) complexes and prosthetic groups and also a mentor (since my 2-year postdoctoral fellowship with him, back in 1992–1993) and a friend.

Key words Density functional theory (DFT), Broken symmetry (BS) state, Electronic and magnetic properties, Monomer and dimer, Valence bond (VB) theory, Spin coupling, Exchange coupling constants, Clebsch–Jordan coefficients, Hyperfine coupling constants

Juan C. Fontecilla-Camps and Yvain Nicolet (eds.), *Metalloproteins: Methods and Protocols*, Methods in Molecular Biology, vol. 1122, DOI 10.1007/978-1-62703-794-5_15,

1 Introduction

1.1 Broken Symmetry States and Energetics

1.1.1 What Is "Broken" in a Broken Symmetry State?

The first point to consider is what is a BS state [1–4]. And what is "broken" in such a BS state? In order to answer that question, let us consider the simplest case of two spin 1/2 bearers within a molecular system, for example a hydrogen molecule (H_2) or a Cu(II) (d^9) dimer. Let us write each individual bearer's spin state as $|s,ms>$ (lower case) where *s* is the bearer's unpaired electronic spin and *ms* its magnetic quantum number (+1/2 for the so-called *alpha* spin, and −1/2 for a *beta* spin, depending on how they are aligned in the presence of a magnetic field ... and also as they appear classified in any standard nonrelativistic density functional theory (DFT) output file). To each unpaired electron and spin state will be ascribed a spatial (i.e., orbital) extension called a(r) for the first one and b(r) for the second one (*see* **Note 1**). We thus have defined two *local* monomer states $|s_a,ms_a>$ and $|s_b,ms_b>$.

Let us then consider the determinant Ψ_{ab} constructed from these two *local* monomer states $|s_a,ms_a>$ and $|s_b,ms_b>$. We consider the spin arrangement for which one monomer is with spin α (upward arrow), $|s_a,ms_a> = |1/2,+1/2>$, and the other monomer is with spin β (downward arrow), $|s_b,ms_b> = |1/2,-1/2>$. I will explain below how to construct such an initial state Ψ_{ab} of total $Ms = ms_a + ms_b = 0$.

If, after SCF convergence, the α spin electron of orbital extension a(r) turns out to be *spatially* localized, say on the left site of the molecule, and the β spin electron of orbital extension b(r) is localized on its right site, we have what we commonly call a broken symmetry state (*see* Fig. 1). Of course, it will also be noticed that, for symmetrical molecules, Ψ_{ab} is energetically degenerate with $\Psi_{ba} = |1/2,-1/2>\otimes|1/2,+1/2>$.

Can we then construct such BS determinants for all molecules? The answer is negative. Such BS $\{\Psi_{ab}, \Psi_{ba}\}$ determinants are in fact constructed for *chemically* valence-bond (VB) systems, i.e., for systems in which, to a first approximation, the dimer (trimer, etc.) can *really* be considered as the sum of its two (or more) weakly interacting monomers. In other words, the BS state corresponds to one of the VB resonant structures characterized each by a specific distribution of *localized* electrons (here, one on the left side, and the other on the right side of the molecule). In a VB dimer, this

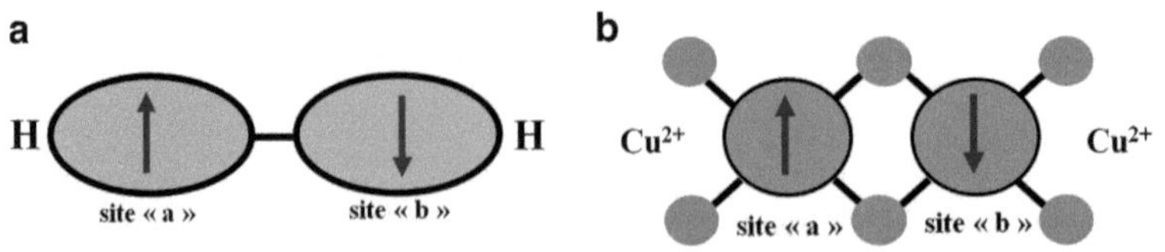

Fig. 1 BS states $\Psi_{ab} = |1/2,+1/2>\otimes|1/2,-1/2>$ for H_2 (**a**) and Cu(II) dimer (**b**, where bridging and terminal ligands are there depicted schematically)

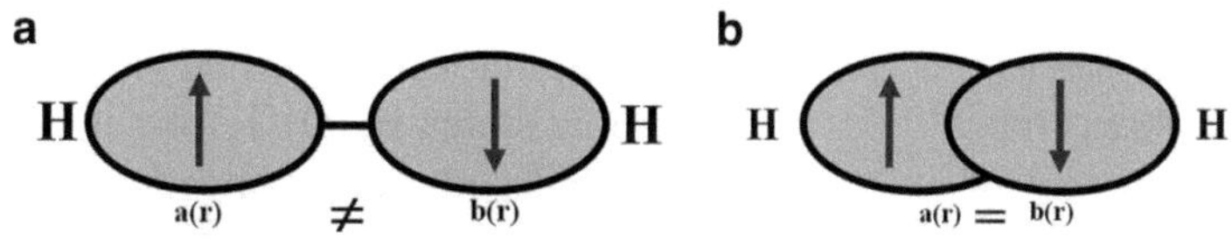

Fig. 2 Valence-bond limit (**a**) versus molecular orbital limit (**b**) for H_2

allows for the lateral localization via DFT of both spins onto both ends of the dimer (*see* **Note 2**).

For example, in the simple case of the H_2 molecule, this occurs when the two hydrogen atoms are sufficiently far away from one another (i.e., during dissociation): Fig. 2a. In the case of the Cu(II) (d^9) dimer, this usually occurs when the two metal sites are indirectly linked via intermediate bridges (*see* Fig. 1b).

As a starting point to study molecules, this valence-bond limit is conceptually opposite to the molecular orbital (MO) limit where both α and β spins couple together to form a true (Lewis-type) bond as in H_2 in its *equilibrium* state. In such a limit, it is no longer possible to spatially localize both spins: they belong to the same *delocalized* molecular orbital, that is, $a(r) = b(r)$ (*see* Fig. 2b). From a theoretical point of view, full VB and MO derivations converge ultimately, showing that both approaches only differ by their viewpoints and starting points [5]. Simply, in Nature, some chemical systems are best described to first approximation within VB-type theories and others within MO-type theories.

To make it transparent that, from now on, we will deal with *localized* (VB) monomer functions combined within dimers' BS $\{\Psi_{ab}, \Psi_{ba}\}$ determinants, I will represent such determinants as *outer products* (I chose the symbol ⊗ to represent it) of monomers' states: $|s_a,ms_a\rangle \otimes |s_b,ms_b\rangle$ (*see* **Note 3**). In the present case $\Psi_{ab} = |1/2,+1/2\rangle \otimes |1/2,-1/2\rangle$. We have what is called a *spatially* "broken symmetry" state Ψ_{ab}. Notice at once that Ψ_{ab} is not a pure spin state: Ψ_{ab} is not an eigenvalue of the operator $\mathbf{S}^2$. Only the value of Ms ($=ms_a + ms_b$) is controlled (here $Ms = 0$ when $s_a = s_b$). But Ψ_{ab} is related to pure spin states (singlet and triplet in the present case) in a manner that will be shown below.

The conceptual point here is that we have constructed a *global* VB state (Ψ_{ab}) for the dimer in terms of *local* monomer states ($|s_a,ms_a\rangle$ and $|s_b,ms_b\rangle$). This is a game (local versus global) we will often play in the following presentation.

1.1.2 The Broken Symmetry State as a Hybrid State of Two Worlds . . .

Broken symmetry (mono-determinant) states are very interesting from a computational point of view. In effect, in the construction of electronic states by DFT, one is often limited by the fact that standard DFT codes are mono-determinantal. In other words, they are designed to construct the electronic density of the system, which is the real quantity with physical meaning at the heart of DFT, as the square of a (*N*-electrons) Slater determinant. This determinant is supposedly flexible enough to yield the "true" density of the system or, in practice, the "best" density, i.e., the one minimizing the total energy of the system. Within the DFT world, such a determinant is therefore just an ***intermediate*** computational quantity with, a priori, no physical or chemical meaning per se attached to it.

This situation contrasts with that of the various wave function theories (WFT: Hartree–Fock and beyond), approaches for which determinants are the very elementary elements allowing for a rationale construction of *N*-electronic wave functions, usually a mixture of many determinants whose respective weights are optimized so as to minimize the energy of the system. The true variable in the WFT world is therefore the (multi-determinant) wave function, not the density as in DFT.

This last fact means that, for systems which can be represented, to first order, by a single (Hartree–Fock) determinant within WFT, the corresponding DFT determinant (a mathematical intermediate allowing for the construction of the "real" variable, the density) can be expected to be close enough to that WFT determinant.

> This raises an interesting paradox to which I will come back at the end: when dealing with BS states, we deal with ***determinants*** (a WFT concept) within DFT (density based)!

1.1.3 How Are Broken Symmetry States Related to Pure Spin States?

Let us now introduce pure spin states in order to see how they are related to BS states. From a WFT point of view, a pure spin state can be constructed by mixing various determinants in such a way as to obtain an eigenfunction of the operator $\mathbf{S}^2$. For example, from the two $s = 1/2$ monomer spin bearers considered above, one can construct two simple spin-coupled magnetic states of the general form $\Psi_{S,Ms} = |S,Ms\rangle$ (upper case). First, there is the ***singlet*** state (called Σ) written as $\Psi_{0,0}$ or $|0,0\rangle$, that is, $\mathbf{S}^2|0,0\rangle = S(S+1) = 0$ and $S_z|0,0\rangle = Ms = 0$. This singlet state is a mixture of the two previously defined BS states $\Psi_{ab} = |1/2,+1/2\rangle \otimes |1/2,-1/2\rangle$ and $\Psi_{ba} = |1/2,-1/2\rangle \otimes |1/2,+1/2\rangle$: *see* Fig. 3 and Eq. 1(I) below.

We can also construct the three magnetic components $Ms = +1, 0, -1$ of the ***triplet*** state (called T) of eigenfunction $\Psi_{1,Ms}$ or $|1,Ms\rangle$, that is, $\mathbf{S}^2|1,Ms\rangle = S(S+1) = 2$ and $S_z|1,Ms\rangle = Ms$. As it turns out, $\Psi_{1,0}$ can also be expanded in terms of the same localized BS states Ψ_{ab} and Ψ_{ba} (*see* Fig. 3 and Eq. 2(III)).

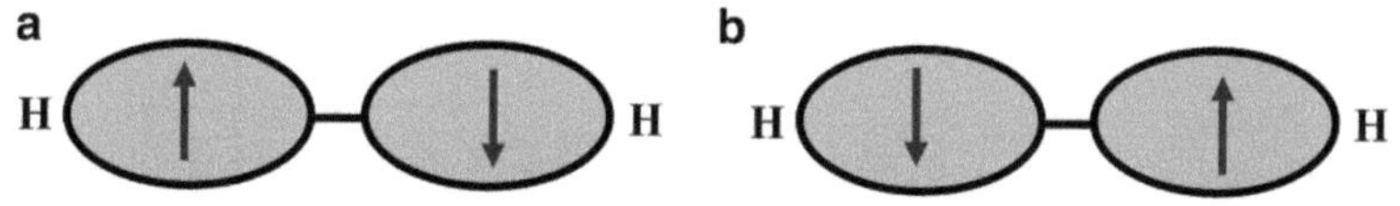

Fig. 3 The two BS states Ψ_{ab} and Ψ_{ba} combined to yield both the singlet state (**a**: "−" combination; cf. Eq. 1) and the triplet (*Ms* = 0) state (**b**: "+" combination; cf. Eq. 2, III)

For the singlet state:

$$\mathrm{I}: \Psi_{0,0} = |0,0> = \left(2^{-1/2}\right)(\Psi_{ab} - \Psi_{ba}) \quad (1)$$

For the triplet state:

$$\begin{aligned} \mathrm{II}: \Psi_{1,+1} &= |1,+1> = |1/2,+1/2> \otimes |1/2,+1/2> \\ \mathrm{III}: \Psi_{1,0} &= |1,0> = \left(2^{-1/2}\right)(\Psi_{ab} + \Psi_{ba}) \\ \mathrm{IV}: \Psi_{1,-1} &= |1,-1> = |1/2,-1/2> \otimes |1/2,-1/2> \end{aligned} \quad (2)$$

It will be noticed that combinations II ($\Psi_{1,+1}$) and IV ($\Psi_{1,-1}$) are mono-determinants where both spins are parallel (both "up" or both "down"). This means that the energies of both triplet state's components II and IV can be directly computed by DFT: $E(\mathrm{II}) = E(\mathrm{IV})$. This energy is none but the energy of the triplet state $E(\Psi_{1,Ms})$ called E_T. Moreover, the three magnetic components of the triplet state (II–IV) are energetically degenerate in the absence of magnetic field. Incidentally, we thus have *indirectly* computed the energy of the two-determinant state $\Psi_{1,0} = |1,0>$: $E(\mathrm{III}) = E_T$ (*see* **Note 4**).

The next question is the following: How do we compute the energy of the (two-determinant) singlet state $E(\Psi_{0,0}) = E_\Sigma$ within a mono-determinant DFT code? The answer comes from mixing both I (singlet $Ms = 0$) and III (triplet $Ms = 0$) states yielding the BS state Ψ_{ab} whose energy we can compute:

$$\Psi_{ab} = \left(2^{-1/2}\right)\left(\Psi_{0,0} + \Psi_{1,0}\right) \quad (3)$$

Consequently, with $E(\Psi_{ab}) = E_{BS}$:

$$E_{BS} = (1/2)(E_\Sigma + E_T) \rightarrow E_\Sigma = 2 \cdot E_{BS} - E_T \quad (4)$$

If one desires to compute the value of the energy difference called J between triplet (T) and singlet (Σ) states, it can be seen at once from Eq. 4 that

$$J = E_\Sigma - E_T = 2 \cdot [E_{BS} - E_T] \quad (5)$$

This singlet–triplet energy difference is the exchange coupling constant J (*see* Fig. 4) corresponding to a (valence-bond)

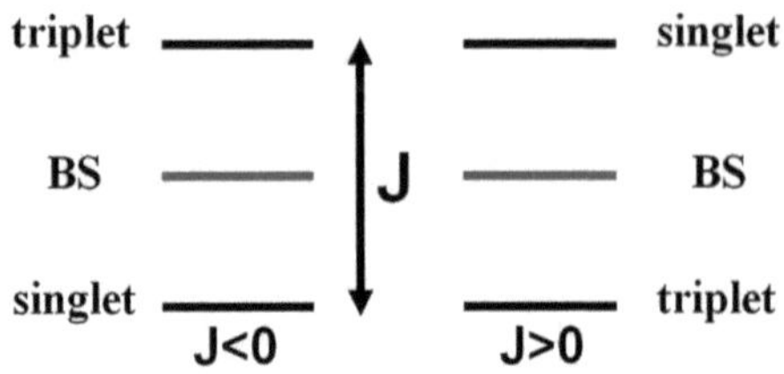

Fig. 4 Relative energies of the singlet (Σ), triplet (T), and BS states for both $J < 0$ and $J > 0$

Heisenberg spin Hamiltonian of the form $-J\mathbf{s_a}\cdot\mathbf{s_b}$. It turns out to be twice the BS-triplet energy difference (*see* **Note 5**). What is important here is that, indeed, the singlet–triplet energy gap J can be indirectly estimated from the computation of two DFT (mono-determinant) states (triplet and broken symmetry).

From this section, we learned that, although most pure spin states, being many-determinantal, cannot be ***directly*** computed within DFT, the energies of some of them (here the singlet state Σ) can be ***indirectly*** computed through appropriate combinations of pure spin states (here the triplet state) to yield broken symmetry states accessible by DFT.

1.1.4 Generalization for Local Spins Larger Than 1/2

This last statement can now be generalized to dimers whose local monomer spins are larger than 1/2. Without being too complicate, let us consider first a *symmetrical* dimer made of two monomers of local spin states $|s_a,ms_a>$ and $|s_b,ms_b>$, for example a di-ferric dimer (of the type $[2Fe2S](SCys)_4$ (as encountered in iron–sulfur proteins) made of two ferric monomers ($s_a = s_b = 5/2$) (*see* Fig. 5). We would like to compute the energy difference between the high-spin (HS) state $S = 5$: $|5/2,+5/2>\otimes|5/2,+5/2>$ (typical input file *2fe2s_hsox.inp* given in Subheading 3) and the singlet state (Σ) of spin $S = 0$ (explicit expression given below). The problem within DFT will be again how do we compute the energy of Σ. The energy of the (mono-determinant) HS state (equivalent to the triplet T above) can be approximated at once (i.e., there is only one way, and therefore one determinant, to represent a configuration with all spins up).

From an analytical/phenomenological point of view, and first without recourse to BS states, let us again assume that a (valence-bond) Heisenberg spin Hamiltonian of the form $H_{\text{Heis}} = -J\mathbf{s_a}\cdot\mathbf{s_b}$ describes the magnetic interaction between the two ferric ions of spins $\mathbf{s_a}$ and $\mathbf{s_b}$, respectively. A pure spin state of spin S $(0 \leq S \leq 5)$ has the following energy (*see* **Note 6**):

$$E_S = (-J/2)[S(S+1) - s_a(s_a+1) - s_b(s_b+1)] \qquad (6)$$

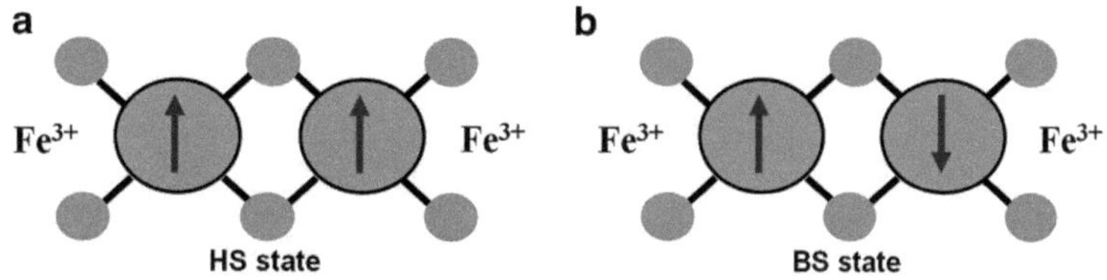

Fig. 5 HS (**a**) and BS (**b**) states for a [2Fe2S](SCys)$_4$ dimer where bridging (S) and terminal ligands (SCys) are depicted schematically

with $s_a = s_b = 5/2$ for (locally high-spin) ferric ions. Notice that the sum of the last two terms is constant ($35J/4$). The energy of the HS state of pure spin $S = 5$ is $E_{HS} = -25J/4$ whereas that of the singlet state $\Sigma(S = 0)$ is given by $E_\Sigma = 35J/4$. Therefore

$$E_\Sigma - E_{HS} = 15J \rightarrow J = [E_\Sigma - E_{HS}]/15 \qquad (7)$$

The "only" problem is, again, how to compute E_Σ within DFT. For that, we use as above (cf. Eqs. 1–2) an indirect method via an appropriate BS state of the form $\Psi_{ab} = |s_a, ms_a\rangle \otimes |s_b, ms_b\rangle = |5/2, +5/2\rangle \otimes |5/2, -5/2\rangle$ with $Ms = 0$ (typical input file *2fe2s_bsox.inp* given in Subheading 3).

> In practice, in the present case of this [2Fe2S](Scys)$_4$ dimer, not all $|s_a, ms_a\rangle \otimes |s_b, ms_b\rangle$ outer products (i.e., not all pairs of ms_a, ms_b values) are suitable for DFT calculations. Here, the *chemical* system (i.e., ferric ions within a weak sulfur ligand field) exhibits locally high-spin ions (i.e., $|ms_a| = |ms_b| = 5/2$). Therefore, the only two BS states which can be constructed/monitored by DFT are $\Psi_{ab} = |5/2, +5/2\rangle \otimes |5/2, -5/2\rangle$ and $\Psi_{ba} = |5/2, -5/2\rangle \otimes |5/2, +5/2\rangle$ (*see* Fig. 5) of energy E_{BS}. Had we two ferric ions imbedded within strong local ligand fields, inducing $s_a = s_b = 1/2$ for example (we thus have low-spin ions), the only two BS states we could compute would be $\Psi_{ab} = |1/2, +1/2\rangle \otimes |1/2, -1/2\rangle$ and $\Psi_{ba} = |1/2, -1/2\rangle \otimes |1/2, +1/2\rangle$, and we go back to the previous section of two $s = 1/2$ spins.

The key relation to compute either $E_\Sigma - E_{HS}$ or J (cf. Eq. 7) is the following general relation which conceptually links *global* (pure spin) states $\Psi_{S,Ms} = |S, Ms\rangle$ to *localized* (broken symmetry) states $|s_a, ms_a\rangle \otimes |s_b, ms_b\rangle$:

$$|S, Ms\rangle = \Sigma_{msa,msb} C_{S \rightarrow BS}(ms_a, ms_b) \cdot |s_a, ms_a\rangle \otimes |s_b, ms_b\rangle \qquad (8)$$

where $ms_a + ms_b$ is constrained to be equal to Ms and $\Sigma_{msa,msb}[C_{S \rightarrow BS}(ms_a, ms_b)]^2 = 1$. The $\{C_{S \rightarrow BS}\}$ are Clebsch–Jordan coefficients whose values are tabulated [6] (*see* **Note 7**).

$$
\begin{array}{l}
|S,Ms> \\
|5,0> = \\
|4,0> = \\
|3,0> = \\
|2,0> = \\
|1,0> = \\
|0,0> =
\end{array}
\begin{bmatrix}
\frac{1}{\sqrt{252}} & \frac{5}{\sqrt{252}} & \frac{5}{\sqrt{63}} & \frac{5}{\sqrt{63}} & \frac{5}{\sqrt{252}} & \frac{1}{\sqrt{252}} \\
\frac{1}{\sqrt{28}} & \frac{3}{\sqrt{28}} & \frac{1}{\sqrt{7}} & -\frac{1}{\sqrt{7}} & -\frac{3}{\sqrt{28}} & -\frac{1}{\sqrt{28}} \\
\sqrt{\frac{5}{36}} & \frac{7}{\sqrt{180}} & -\frac{2}{\sqrt{45}} & -\frac{2}{\sqrt{45}} & -\frac{7}{\sqrt{180}} & \sqrt{\frac{5}{36}} \\
\sqrt{\frac{25}{84}} & \sqrt{\frac{1}{84}} & -\sqrt{\frac{4}{21}} & \sqrt{\frac{4}{21}} & -\sqrt{\frac{1}{84}} & -\sqrt{\frac{25}{84}} \\
\sqrt{\frac{5}{14}} & -\sqrt{\frac{9}{70}} & \sqrt{\frac{1}{70}} & \sqrt{\frac{1}{70}} & -\sqrt{\frac{9}{70}} & \sqrt{\frac{5}{14}} \\
\frac{1}{\sqrt{6}} & -\frac{1}{\sqrt{6}} & \frac{1}{\sqrt{6}} & -\frac{1}{\sqrt{6}} & \frac{1}{\sqrt{6}} & -\frac{1}{\sqrt{6}}
\end{bmatrix}
\begin{array}{l}
|s_a,ms_a>\otimes|s_b,ms_b> \\
|5/2,+5/2>\otimes|5/2,-5/2> \\
|5/2,+3/2>\otimes|5/2,-3/2> \\
|5/2,+1/2>\otimes|5/2,-1/2> \\
|5/2,-1/2>\otimes|5/2,+1/2> \\
|5/2,-3/2>\otimes|5/2,+3/2> \\
|5/2,-5/2>\otimes|5/2,+5/2>
\end{array}
$$

Fig. 6 Clebsch–Jordan coefficients corresponding to Eqs. 8 and 9 relating pure spin states |S,*Ms*> (with $Ms = 0$) to outer product states $|s_a,ms_a>\otimes|s_b,ms_b>$ (with $ms_a + ms_b = 0$) for a di-ferric dimer of the type [2Fe2S]$(SCys)_4$

For the di-ferric dimer of the type [2Fe2S]$(SCys)_4$, one can express the Clebsch–Jordan coefficients in a matrix form (*see* Fig. 6).

For a given pure spin state $|S,Ms>$ (say, |5,0>), the coefficients of Eq. 8 are read in the corresponding row of the matrix of Fig. 6 (here, first row). Among the six outer products with $Ms = 0$ appearing on the right side of Fig. 6, only two (first and last) can be computed by DFT. Notice finally that for the HS state (with $S = 5$, $Ms = 5$) (not appearing in Fig. 6), that is $\Psi_{5,5}$, there is only one determinant, $|5/2,+5/2>\otimes|5/2,+5/2>$, and therefore $C_{S\to BS}(+5/2,+5/2) = 1$. All other pure spin states are multi-determinantal.

The relation of Eq. 8 can be reversed:

$$|s_a, ms_a > \otimes |s_b, ms_b >= \Sigma_S C_{BS\to S}(S, Ms) \cdot |S, Ms > \qquad (9)$$

where Ms is equal to $ms_a + ms_b$ and $\Sigma_S[C_{BS\to S}(S,0)]^2 = 1$. These $\{C_{BS\to S}\}$ are the same Clebsch–Jordan coefficients as above (*see* Fig. 6): for a given outer product state $|s_a,ms_a>\otimes|s_b,ms_b>$: only, the coefficients of Eq. 9 are read in the corresponding column of the matrix (*see* Fig. 6) (example below) (*see* **Note 8**). The idea is therefore to apply Eq. 9 to the specific BS state which can be constructed by DFT, that is, $\Psi_{ab} = |5/2,+5/2>\otimes|5/2,-5/2>$ (first top right on the list appearing in Fig. 6). Practically (with the sum over S running from $S = 0$ up to $S = 5$)

$$
\begin{aligned}
\Psi_{ab} &= |5/2, +5/2 > \otimes |5/2, -5/2 > \\
&= C_0 \cdot \Psi_{0,0} + C_1 \cdot \Psi_{1,0} + C_2 \cdot \Psi_{2,0} + C_3 \cdot \Psi_{3,0} \\
&\quad + C_4 \cdot \Psi_{4,0} + C_5 \cdot \Psi_{5,0}
\end{aligned} \qquad (10)
$$

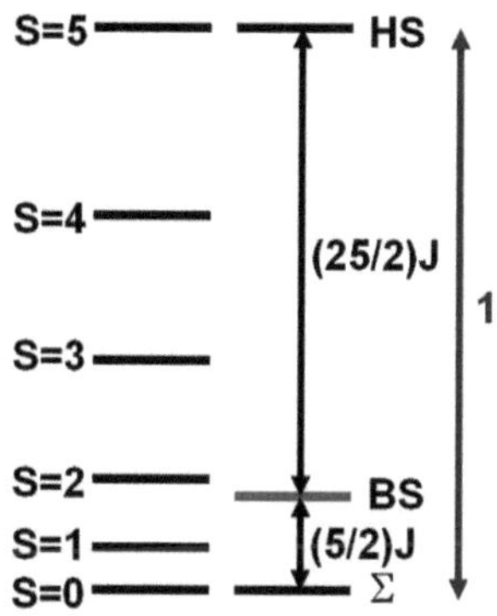

Fig. 7 Magnetic spin ladder constructed for an antiferromagnetic ferric dimer ($J < 0$). Energy differences between HS, BS, and Σ states are represented

where (cf. first column of the matrix in Fig. 6) $C_0 = (1/6)^{1/2}$, $C_1 = (5/14)^{1/2}$, $C_2 = (25/84)^{1/2}$, $C_3 = (5/36)^{1/2}$, $C_4 = (1/28)^{1/2}$, and $C_5 = (1/252)^{1/2}$. The corresponding BS energy is computed as

$$E_{BS} = \Sigma_S [C_S]^2 \cdot E_S \quad (11)$$

that is, $E_{BS} = [42E_0 + 90E_1 + 75E_2 + 35E_3 + 9E_4 + E_5]/252$. The energy E_{BS} appears to be a weighed average of pure spin states' energies! We know moreover E_S from Eq. 6. Numerically, for the ferric dimer $E_{BS} = +25J/4$. As the energy of the HS state of pure spin 5 is $E_{HS} = -25J/4$, we obtain (alternatively to Eq. 7) (*see* **Note 9**)

$$E_{BS} - E_{HS} = (25J/2) \rightarrow J = 2 \cdot [E_{BS} - E_{HS}]/25 \quad (12)$$

> The point here is that both energies E_{BS} and E_{HS} can be computed by DFT and the whole magnetic spin ladder can be reconstructed from there (*see* Fig. 7) with Eq. 6.

Finally, by combining Eq. 7 and Eq. 12, we derive

$$E_\Sigma = E_{BS} + [E_{BS} - E_{HS}]/5 \quad (13)$$

For an antiferromagnetic spin-coupled dimer $E_\Sigma < E_{BS} < E_{HS}$, whereas for a ferromagnetic dimer $E_{HS} < E_{BS} < E_\Sigma$, similar to what we had in Fig. 4. Although both the singlet state Σ and the BS state have in common the fact that $Ms = 0$, Eq. 13 tells us quantitatively by how much E_Σ and E_{BS} differ in energy, as BS is a weighed average of pure spin states' $Ms = 0$ components (cf. Eqs. 10 and 11). Visually (Fig. 7), E_{BS} is close to $E_S = 2$, in agreement with the fact that the largest coefficients of the expansion in Eq. 10 are those of the lower spins $S = 0$ to $S = 3$ (i.e., C_0 to C_3).

Now that we have seen how to compute BS states and how to relate them to HS states and derive from their comparison exchange coupling constants, let us now apply these BS states to the computations of electronic and magnetic properties.

1.2 Broken Symmetry States and Spectroscopies

In what follows, I will distinguish between spectroscopic quantities derived from *charge* properties (for example, isomer shifts and quadrupolar tensors measured by Mössbauer, etc.) and those magnetic properties derived from *spin* quantities (**g**-tensors derived from electron paramagnetic resonance (EPR), hyperfine coupling tensors measurable by both EPR and Mössbauer techniques, etc.). In fact, the main characteristic of a BS state is to be spin uncoupled. In other words, a dimer (or a polymer) is conceptually seen as a sum of magnetically non-interacting monomers. As a consequence, one would naively expect that *charge*-derived quantities are not affected directly by this BS property (I will show below that this is not always the case) whereas spin-derived quantities for pure spin states will be strongly affected by the fact that the corresponding quantities are computed for spin-uncoupled BS states.

1.2.1 Quantities Derived from Charge Properties

In order to compute a *charge*-derived quantity, such as a Mössbauer isomer shift or a quadrupolar tensor, for example, the main issue will be to decide if the *electronic* structure converged for a given DFT state (HS or BS) represents to a first approximation the correct *electronic* structure of the real system. This is not a trivial question, as electronic structures are in part determined by monomer spin alignments as will be shown below. As a consequence, a *charge* property can be indirectly impacted by a spin property.

To illustrate this point, let us again consider the same structurally symmetrical $[2Fe2S](SCys)_4$ dimer (*see* Fig. 5). In the cluster redox state with two ferric ions, we constructed three DFT states: HS ($Ms = 5$) ($|5/2,+5/2\rangle \otimes |5/2,+5/2\rangle$) and two BS ($Ms = 0$) ($|5/2,+5/2\rangle \otimes |5/2,-5/2\rangle$ and its $\pm ms_i$ symmetrical), as already mentioned above. *Electronically*, the ferric sites remain identical (to first order) in these three states upon local spin flip. The same is true, were the two iron ions ferrous (HS: $Ms = 4$, and BS: $Ms = 0$).

In the intermediate case however, where one ferric ion and one ferrous ion forming the so-called mixed-valence pair, the local *electronic* iron states will depend on that of the spin state! In the BS state ($Ms = 1/2$) ($|5/2,+5/2\rangle \otimes |2,-2\rangle$), the sixth ferrous electron will be fully localized on the second site (*see* Fig. 8a). We thus start from formal Fe^{3+} and Fe^{2+} ions which are antiferromagnetically coupled ($J < 0$).

In the HS state however ($Ms = 9/2$), for structurally *symmetrical* dimers, that same electron will be equally delocalized between both iron sites (*see* Fig. 8b) resulting into a $Fe^{2.5+}$–$Fe^{2.5+}$ pair, an equal mixture of $|5/2,+5/2\rangle \otimes |2,+2\rangle$ and $|2,+2\rangle \otimes |5/2,+5/2\rangle$.

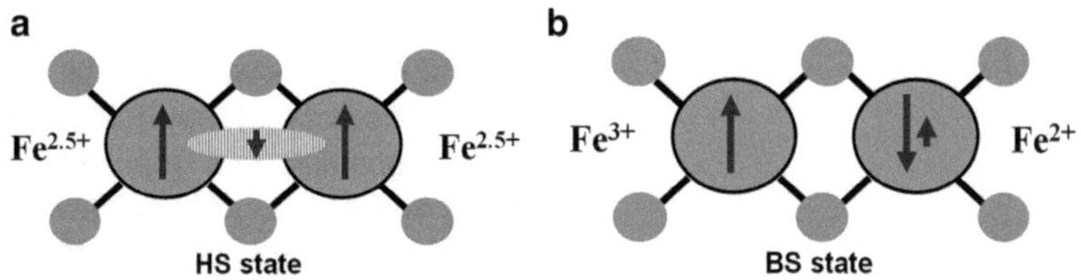

Fig. 8 HS (**a**) (with $Ms = 9/2$) and BS (**b**) states (with $Ms = 1/2$) for a symmetrical mixed-valence [2Fe2S](SCys)$_4$ dimer where bridging (*S*) and terminal ligands (SCys) are depicted schematically

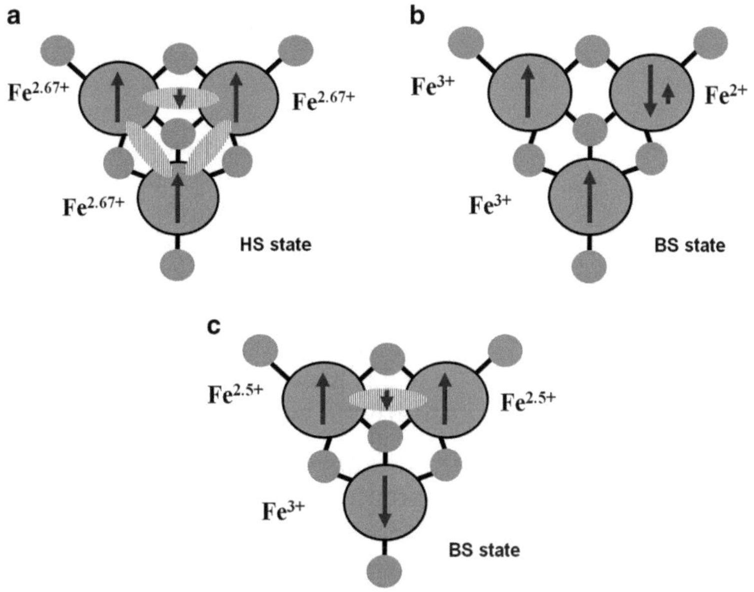

Fig. 9 DFT states constructed for a [3Fe4S](SCys)$_3$ trimer formally composed of one ferrous and two ferric ions. (**a**) Fully delocalized HS state (with $Ms = 14/2$); (**b**) fully localized BS state (with $Ms = 6/2$); (**c**) partially delocalized BS state (with $Ms = 4/2$)

This delocalization occurs via the so-called double-exchange mechanism [7] which favors ferromagnetism. We can now see that, depending on the spin alignments (HS versus BS), resulting from the competition between antiferromagnetic super-exchange and ferromagnetic double-exchange, the final *electronic* structures differ. The intermediate case between fully trapped and fully delocalized has been treated in full details elsewhere [8].

The previous interplay between charge and spin properties can be generalized to more complicated clusters. Let us consider for example the next simplest case of a [3Fe4S](SCys)$_3$ trimer (cf. Fig. 9a–c representing such a trimer for specific charge and spin distributions discussed below). In general, for such a trimer, one can in principle construct one HS state and three BS$_i$ states of

Table 1
HS and BS states which can be constructed for a trimer such as a [3Fe4S](SCys)$_3$ cluster. Local spin alignments are represented by arrows (up: ↑ and down: ↓, red downward arrows). BS states are labelled according to which spin site (*i* = a, b, c) is flipped (BS$_i$)

BS states	Fe #a	Fe #b	Fe #c	Total Ms
HS	↑	↑	↑	maximal
BSa	↓	↑	↑	intermediate
BSb	↑	↓	↑	intermediate
BSc	↑	↑	↓	intermediate

intermediate *Ms* value, that is, four states total (see Table 1), irrespective of the local redox states of the iron ions and not counting the states obtained by "up" ↔ "down" spin permutation (generalization to four sites and beyond follows along the same lines). These four states all assume the form $|s_a,ms_a\rangle \otimes |s_b,ms_b\rangle \otimes |s_c,ms_c\rangle$ (this is by the way where outer products are conceptually simple and nice representations of determinants).

At this general level, three practical and conceptually related points have to be noted:

(i) *Monomer transferability*: To a first approximation, a charge property for one iron site can be computed from either HS or BS state as long as the formal charge of that ion is the same. For example, in Fig. 5a, b both iron ions of the dimer are formally ferric, as is the left iron in Fig. 8b. For the trimer, among the three states of Fig. 9, one finds ferric ions in b (upper left and lower) and c (lower). The same reasoning applies for whatever ions as long as they carry the *same* formal valence charge (see the two following points below).

In practice though, even if such a condition is satisfied, an electronic site property computed from different states may still vary a little because monomers within dimers or trimers are mutually, though weakly, interacting and, therefore, not exactly identical.

(ii) *Cluster/electronic symmetry*: The higher the structural geometry of the cluster, the more BS states of the same *Ms* may become equivalent (i.e., energetically degenerate). For example, for a C_3 cluster geometry as depicted in Fig. 9, but for a cluster redox state for which all iron ions are formally identical (all ferrous or all ferric), it is sufficient to construct HS and one of the three BS states. The number of distinct BS states will again increase upon relaxing the structural symmetry and/or the electronic symmetry (*see* next point iii).

(iii) *Electronic localization/delocalization*: Usually, especially in the case of mixed-valence systems, differential electron flow may occur between iron sites. Such flows may be contained, for example within a single mixed-valence pair (*see* Fig. 8a). But when the cluster/electronic symmetry is lowered, this is not always the case, and this has to be checked carefully.

To illustrate this point, let us consider again the previous structurally symmetrical [3Fe4S](SCys)$_3$ cluster formally containing two ferric and one ferrous ions. In HS ($Ms = 14/2$), one could construct either one of the three *initial* DFT states localizing the ferrous ion on any of the three monomer sites a, b, or c: **|2,+2>**⊗|5/2,+5/2>⊗|5/2,+5/2>, |5/2,+5/2>⊗**|2,+2>**⊗|5/2,+5/2>, or |5/2,+5/2>⊗|5/2,+5/2>⊗**|2,+2>**, respectively. After SCF convergence however, and whatever the initial HS state, the (ferrous) β (spin down) electron will be fully delocalized among the *three* equivalent iron sites, each formally $Fe^{2.67+}$ (↑): Fig. 9a.

Moreover, two types of BS states of intermediate MS value can be constructed: either two Fe^{3+} (↑) and one (fully localized) Fe^{2+} (↓) of $Ms = 6/2$ (there are three such BS states: |5/2,+5/2>⊗|5/2,+5/2>⊗**|2,−2>** and the other two resulting from site permutations: Fig. 9b) or one Fe^{3+} (↓) and two $Fe^{2.5+}$ (↑) ($Ms = 4/2$) where electron delocalization occurs between *two* sites: Fig. 9c. Both initial |5/2,−5/2>⊗|5/2,+5/2>⊗**|2,+2>** and |5/2,−5/2>⊗**|2,+2>**⊗|5/2,+5/2> will yield such ($Ms = 4/2$) BS states (+ permutations).

> This illustrates that, for the *same initial* set of formal ions, the iron valences can be trapped on their respective sites, or *electronic* delocalization may occur (fully or partially) between two or three sites depending on how one constructs the *magnetic* states!

To conclude this first subsection, let us state that the interplay between spin and electronic properties thus appears to be quite subtle [7]. But, on the bright side, this means that different BS states will exhibit different electronic properties, and a proper comparison of computed versus experimentally measured quantities (typically Mössbauer properties) [9] may help decide which BS state best represents the real system *from an electronic point of view*, especially with respect to electronic (de)localization and charge flow.

1.2.2 Quantities Derived from Spin Properties

The computation of spin-dependent properties such as *local* (i.e., site) hyperfine coupling constants {**A**} (to be compared to experimentally measured coupling of given *local* nonzero nuclear spins) or *global* (i.e., cluster) **g**-tensors characterizing the *global* cluster spin is (I think) more involved than those computed for *charge* properties.

> The basic point to understand here is that the HS state is a pure (spin-coupled) spin state whereas BS states are by nature spin uncoupled (only *Ms* is controlled: cf. Eq. 9). As a consequence, a spin-coupling procedure has to be defined in each case to transform a raw (spin-uncoupled) property computed for monomer sites within a given (artificial) BS state into the corresponding property as it is measured in the real spin-coupled state.

(a) *Dimers.* To illustrate this essential point, let us consider hyperfine coupling constants computed for the paramagnetic ($S = 1/2$) mixed-valence [2Fe2S]$(SCys)_4$ dimer (HS = 9/2 and BS = 1/2; cf. Fig. 8) comprising initially a ferrous ion and a ferric ion. The hyperfine couplings can be computed for the (^{57}Fe) iron ions or for some peripheral nuclei with non-zero nuclear spin (^{1}H, ^{13}C, etc.). Let us for arguments' sake treat the case of ^{57}Fe nuclei (of nuclear spin I = 1/2). Experimentally, two (average) **A** values are measured for such [2Fe2S]$(SCys)_4$ dimers in the $S = 1/2$ pure spin state, +22.1 and −48.2 MHz, which begs the following question: Which value corresponds to which iron ion?

In the BS state $\Psi_{ab} = |5/2,+5/2\rangle \otimes |2,-2\rangle$ (*see* Fig. 8b), the raw hyperfine coupling constant computed for the ferric site, $\mathbf{a_a(3+)}$, will be (as a first approximation: cf. point (i) of the previous subsection) the same as that computed or measured [10] for the corresponding isolated ferric monomer of the (rubredoxin-type) [Fe]$(SCys)_4$ (experiment: −21.8 MHz). The same is expected for the other ferrous site, yielding $\mathbf{a_b(2+)}$ (experiment: −22.2 MHz). The task at hand is now to relate these (spin-uncoupled) monomer quantities $\{\mathbf{a_i}\}_{i=a,b}$ to spin-coupled dimer ones $\{\mathbf{A_i}\}_{i=a,b}$. The $\{\mathbf{a_i}\}_{i=a,b}$ are either measured or computed for the BS state which, as is the case for the "real" $S = 1/2$ spin state, is both valence trapped (two couplings of opposite signs) and shares with the "real" $S = 1/2$ state the fact that $Ms = 1/2$. The BS state is thus a good *electronic* model of the pure $S = 1/2$ state.

> The simplest way to understand it is to realize that we have to link two descriptions (i.e., two spin Hamiltonians) of the *same* reality (here, the hyperfine couplings) by establishing a mathematical correspondence between two equivalent formalisms, a *local* one suited to describe the monomers' spins and a *global* one suited to describe the whole cluster spin.

The first formalism is expressed in terms of the *local* spins $\mathbf{s_a}$ and $\mathbf{s_b}$: $H_{\mathrm{local}} = \mathbf{s_a}\cdot\mathbf{a_a}\cdot\mathbf{I_a} + \mathbf{s_b}\cdot\mathbf{a_b}\cdot\mathbf{I_b}$ where $\mathbf{I_a}$ (resp. $\mathbf{I_b}$) is the nuclear spin belonging to the magnetic site "a" (resp. "b") for which

$\lvert S,Ms\rangle$						$\lvert s_a,ms_a\rangle\otimes\lvert s_b,ms_b\rangle$
$\lvert 9/2,1/2\rangle =$	$\frac{1}{\sqrt{126}}$	$\sqrt{\frac{10}{63}}$	$\sqrt{\frac{10}{21}}$	$\sqrt{\frac{20}{63}}$	$\sqrt{\frac{10}{252}}$	$\lvert 5/2,+5/2\rangle\otimes\lvert 2,-2\rangle$
$\lvert 7/2,1/2\rangle =$	$\sqrt{\frac{4}{63}}$	$\frac{11}{\sqrt{315}}$	$\sqrt{\frac{4}{105}}$	$-\sqrt{\frac{14}{45}}$	$-\sqrt{\frac{64}{315}}$	$\lvert 5/2,+3/2\rangle\otimes\lvert 2,-1\rangle$
$\lvert 5/2,1/2\rangle =$	$\sqrt{\frac{3}{14}}$	$\sqrt{\frac{6}{35}}$	$\sqrt{\frac{8}{35}}$	0	$\sqrt{\frac{27}{70}}$	$\lvert 5/2,+1/2\rangle\otimes\lvert 2,0\rangle$
$\lvert 3/2,1/2\rangle =$	$\sqrt{\frac{8}{21}}$	$-\sqrt{\frac{2}{105}}$	$-\sqrt{\frac{2}{35}}$	$\sqrt{\frac{5}{21}}$	$-\sqrt{\frac{32}{105}}$	$\lvert 5/2,-1/2\rangle\otimes\lvert 2,+1\rangle$
$\lvert 1/2,1/2\rangle =$	$\frac{1}{\sqrt{3}}$	$\frac{-2}{\sqrt{15}}$	$\frac{1}{\sqrt{5}}$	$-\sqrt{\frac{2}{15}}$	$\sqrt{\frac{1}{15}}$	$\lvert 5/2,-3/2\rangle\otimes\lvert 2,+2\rangle$

Fig. 10 Clebsch–Jordan coefficients corresponding to Eqs. 8 and 9 relating pure spin states $|S,Ms\rangle$ (here with $Ms = 1/2$) to outer product states $|s_a,ms_a\rangle\otimes|s_b,ms_b\rangle$ (with $ms_a + ms_b = 1/2$) for a mixed-valence dimer of the type $[2Fe2S](SCys)_4$

the coupling $\mathbf{a}_a$ (resp. $\mathbf{a}_b$) is computed in the spin-uncoupled BS state.

The second formalism is expressed in terms of the *total* cluster spin S: $H_{global} = \mathbf{S}\cdot\mathbf{A}_a\cdot\mathbf{I}_a + \mathbf{S}\cdot\mathbf{A}_b\cdot\mathbf{I}_b$ where the couplings $\mathbf{A}_a$ and $\mathbf{A}_b$ are now measured for the spin-coupled state. The nuclear spins are in common to both formalisms.

What follows is not a complete derivation, but only some steps showing how one arrives at the correct solution. Moreover, as the exchange coupling constant J (to be added to H_{global}) in such systems is supposed to energetically dominate the (local/global) hyperfine terms (*see* **Note 10**), we can define the z-axis along $\mathbf{S}$ and project the local spins onto that axis.

> To relate $\mathbf{a}_a$ (local/BS hyperfine coupling) to $\mathbf{A}_a$ (global/experimental coupling), there are two ways to proceed. Both are analytical. But the first one is interesting because it can be easily generalized to systems with more than two local spins (as illustrated below in the case of a three spins' system). The second one is much simpler to derive but only works for two spins (*see* **Note 11**).

(b) *Solution for dimers.* We can first equate the local hyperfine term $\mathbf{s}_a\cdot\mathbf{a}_a$ with the global hyperfine term $\mathbf{S}\cdot\mathbf{A}_a$ (same for **b** terms) by computing their average value over $|S,Ms\rangle$:

$$\mathbf{a}_a < S, Ms|\mathbf{s}_{az}|S, Ms > \equiv \mathbf{A}_a < S, Ms|\mathbf{S}_z|S, Ms > \qquad (14)$$

The second bracket term of Eq. 14 is straightforward to compute (i.e., $<S,Ms|\mathbf{S}_z|S,Ms> = Ms$, here 1/2). To compute the first bracket term, we use Eq. 8 with the appropriate coefficients of the matrix of Fig. 10 derived for the mixed-valence $[2Fe2S](SCys)_4$ dimer (the first column expresses $|1/2,1/2\rangle$ as a function of the $|s_a,ms_a\rangle\otimes|s_b,ms_b\rangle$ states) with $<s_a,ms_a|\otimes<s_b,ms_b|\mathbf{s}_{az}|$

Table 2
Spin-coupling coefficients $\{K_i\}$ computed for all spin values ($1/2 \leq S \leq 9/2$) in the case of a (valence-localized) mixed-valence [2Fe2S](SCys)$_4$ dimer (*see* Fig. 8b)

S	K_a (Fe^{3+})	K_b (Fe^{2+})
1/2	+7/3	−4/3
3/2	+13/15	+2/15
5/2	+23/35	+12/35
7/2	+37/63	+26/63
9/2	+5/9	+4/9

$s_a,ms_a\!>\otimes|s_b,ms_b\!> = ms_a$. We find numerically $<S,Ms|\mathbf{s_{az}}|S,Ms> = +7/6$ and $<S,Ms|\mathbf{s_{bz}}|S,Ms> = -2/3$.

We then define the spin-coupling coefficient K_a from $\mathbf{A_a} = K_a \cdot \mathbf{a_a}$ which is the expression needed to relate spin-uncoupled quantities (here $\mathbf{a_a}$) to spin-coupled ones (here $\mathbf{A_a}$). K_b is defined in the same way (exchanging "a" and "b" labels in Eq. 14), with moreover $K_a + K_b = 1$ since $\mathbf{s_a} + \mathbf{s_b} = \mathbf{S}$ (*see* moreover Eq. 14). In the case of the mixed-valence [2Fe2S](SCys)$_4$ (*see* Fig. 8b), $\mathbf{s_a} = 5/2$, $\mathbf{s_b} = 2$, and the different sets of spin-coupling coefficients for all cluster spins *S* from 1/2 to 9/2 are given in Table 2.

For $S = 1/2$, from $\mathbf{a}(\mathbf{Fe^{2+}}) = -22.2$ MHz and K(Fe^{2+}) $= -4/3$ (*see* Table 2), one would predict $\mathbf{A}(\mathbf{Fe^{2+}}) = +29.6$ MHz (against +22.1 experimentally), whereas from $\mathbf{a}(\mathbf{Fe^{3+}}) = -21.8$ MHz and $K(Fe^{2+}) = +7/3$ (*see* Table 2), one predicts $\mathbf{A}(\mathbf{Fe^{2+}}) = -50.9$ MHz (against −48.2 experimentally). Thus, starting from (here measured or else computed) monomer hyperfine coupling constants for both ferrous and ferric monomers (experimentally, both around −22 MHz: similar magnitudes and same sign), the spin-coupling procedure explains (to first order) both signs and magnitudes of the experimentally measured values (*see* **Note 12**).

Notice that, in Table 2, the spin-coupling coefficients have been computed in the case of full valence *localization* (i.e., even when both iron spins are parallel). In case of full valence *delocalization* (as depicted in Fig. 8a), both K_a and K_b coefficients are averaged to +1/2 each. The case of intermediate (de)localization has been treated in great details elsewhere [8].

(c) *Trimers and more.* More complicate cases can be treated and require attention. In effect, so far, we only considered explicitly "isolated" (i.e., non-degenerate) pure spin states.

Table 3
List of all pure spin states S ($1/2 \leq S \leq 15/2$) which can be constructed from three mutually interacting 5/2 monomer spins $\{s_a, s_b, s_c\}$. The coupling order is first $S' = s_a + s_b$ (first column: $0 \leq S' \leq 5$) and then $S = S' + s_c$ (second column)

$S' = \lvert s_a + s_b \rvert$	$S = \lvert S' + s_c \rvert$
0	5/2
1	3/2, 5/2, 7/2
2	1/2, 3/2, 5/2, 7/2, 9/2
3	1/2, 3/2, 5/2, 7/2, 9/2, 11/2
4	3/2, 5/2, 7/2, 9/2, 11/2, 13/2
5	5/2, 7/2, 9/2, 11/2, 13/2, 15/2

In the previous dimer, there was only one pure spin state with $S = 1/2$ for which we directly computed spin-projection coefficients $\{K_i\}$ linking magnetic properties computed for the spin-uncoupled ($Ms = 1/2$) BS state to those predicted for the spin-coupled state $S = 1/2$, $Ms = 1/2$. Let us now consider the same trimer as above, a [3Fe4S]$(SCys)_3$ cluster (*see* Fig. 9) but with all ferric ions [11]. The goal is just to "feel" what is the problem at hand and how to solve it.

Let us start with a list of all pure spin states which can be constructed from three $S_i = 5/2$ monomer spins. Calling first $\mathbf{S}'$ the intermediate sum $\mathbf{s_a} + \mathbf{s_b}$ (with $|s_a - s_b| \leq S' \leq s_a + s_b$), we then couple $\mathbf{S}'$ to $\mathbf{s_c}$ yielding the total cluster spin $\mathbf{S}$ (with $1/2 \leq S \leq 15/2$). With such a coupling scheme (**Note 13**), the cluster spin state can be written as $|(S')S>$ (Table 3). Remember that, within mono-determinant DFT codes, one can only construct the states listed in Table 1.

It can be seen at once that, assuming (as is experimentally the case) a $S = 1/2$ ground spin state, there are a priori two theoretical candidates: $|(S' = 2)S = 1/2>$ and $|(S' = 3) S = 1/2>$. In fact, the experimental $S = 1/2$ ground spin state will be a mixture of these two "isolated" $S = 1/2$ spin states: $|S_x, Ms> = x|(2)1/2> + y|(3)1/2>$ with $x^2 + y^2 = 1$. The question is then how do we get the value of x.

The logics behind the whole procedure is the following: The ground state energy and wave function of the system are determined by its magnetic spin ladder (analogous to those depicted in Fig. 4 or 7). The character and ordering of these magnetic states result from the spin system (*see* Table 3) and from the appropriate Heisenberg spin Hamiltonian of the form $H_{Heis} = -J_{ab}(\mathbf{s_a}\cdot\mathbf{s_b}) - J_{ac}(\mathbf{s_a}\cdot\mathbf{s_c}) - J_{bc}(\mathbf{s_b}\cdot\mathbf{s_c})$.

The ground state $|S_x, Ms>$, and therefore the value of x, depends on the relative values of the three exchange coupling constants $\{J_{ab}, J_{bc}, J_{ac}\}$. Once we know the value of x, or its expression as a function of computed or measured $\{J_{ab}, J_{bc}, J_{ac}\}$ values, one knows $|S_x, Ms>$ and can compute from it spectroscopic quantities, for example hyperfine coupling constants whose signs and magnitudes are determined by x.

As a consequence, let us apply this procedure in two different ways. A first way, empirically based, consists in deriving from $|S_x, Ms>$ an analytical model for some experimental spectroscopic quantity (here, hyperfine couplings) as a function of the variable x (or y) and fitting the experimental data to obtain a "best" value for x.

A second application, here entirely theoretical, consists in deriving the expression of x as a function of $\{J_{ab}, J_{bc}, J_{ac}\}$ which can be computed by DFT–BS methodology (via HS and BS states of Table 1).

(d) *Hyperfine coupling constants.* The first way consists in estimating the value of x from, for example, experimental hyperfine coupling constants measured for each site, for example ^{57}Fe hyperfine couplings. We again equate $\mathbf{s_a}\cdot\mathbf{a_a}$ (local) with $\mathbf{S}\cdot\mathbf{A_a}$ (global) (same for b and c terms). For the site **a**, we may thus write (cf. Eq. 14)

$$\mathbf{a_a} < S_x, Ms|\mathbf{s_{az}}|S_x, Ms > \equiv \mathbf{A_a} < S_x, Ms|\mathbf{S_z}|S_x, Ms > \qquad (15)$$

The trick here is to use iteratively Eq. 8 linking coupled states to uncoupled ones. To show how it works in principle, but without using too complicated numerical expressions [12], let us start with $|(S' = 2)S = 1/2>$. Starting from the uncoupled spins s_a and s_b, we first couple them into $S' = 2$ (cf. Eq. 8):

$$|S', M's >= \Sigma_{\text{msa,msb}} C'_{S'}(ms_a, ms_b)\cdot|5/2, ms_a > \otimes|5/2, ms_b > \qquad (16)$$

(with $ms_a + ms_b = M's$). Iteratively, for $S' = 2$ and $s_c = 5/2$ coupled into $S = 1/2$

$$|(S')S, Ms >= \Sigma_{M's,\text{msc}} C''_S(M's, ms_c)\cdot|S' = 2, M's > \otimes|5/2, ms_c \qquad (17)$$

(with $M's + ms_c = Ms$). Finally, Eq. 16 can be inserted into Eq. 17. One obtains an equation of the form

$$|(S')S, Ms > = \Sigma Q'(ms_a, ms_b, ms_c) \cdot |5/2, ms_a > \otimes|5/2, ms_b > \otimes|5/2, ms_c > \qquad (18)$$

where the sum runs over $M's$ (with $M's = ms_a + ms_b$) and ms_c (with $Ms = ms_a + ms_b + ms_c$) and the coefficients $\{Q'\}$ are functions of tabulated $C'_{S'}(ms_a,ms_b)$ (Eq. 16) and $C''_S(M's,ms_c)$ (Eq. 17) coefficients. A similar expression can be written for $|(S'=3)S,Ms>$ as a function of $\{Q''\}$ coefficients. The mathematics is rather involved and tedious, but it is conceptually simple. As a consequence, and from Eq. 15,

$$< S_x, Ms|\mathbf{s_{az}}|S_x, Ms > \\ = x^2\left[\Sigma(Q')^2 ms_a\right] + y^2\left[\Sigma(Q'')^2 ms_a\right] + 2xy\left[\Sigma(Q'Q'')ms_a\right] \quad (19)$$

The computation of the second bracket term of Eq. 15 is again straightforward as $<S_x,Ms|\mathbf{S_z}|S_x,Ms> = 1/2$. Consequently (with the sites' correspondence a ≡ 2, b ≡ 3, c ≡ 1) [11]

$$\begin{aligned} < \mathbf{s_{az}} > &= y^2 - (1/3) - \left(3^{1/2}\right)xy \\ < \mathbf{s_{bz}} > &= y^2 - (1/3) + \left(3^{1/2}\right)xy \\ < \mathbf{s_{cz}} > &= (7/6) - 2y^2 \end{aligned} \quad (20)$$

It remains to fit the theoretical expressions of $\mathbf{A_i} = 2\mathbf{a_i}<\mathbf{s_{iz}}>$ against the experimental values to estimate a best value for the parameter y (or x).

(e) *Exchange coupling constants.* Alternatively, we may try to determine from theoretical grounds alone the value of the mixing coefficient x (or y) in the expression of the ground spin state $|S_x, Ms> = x \cdot |(2)1/2> + y \cdot |(3)1/2>$. Recalling that the ground spin state and energy of this tri-ferric [3Fe4S] $(SCys)_4$ system are determined by the spin Hamiltonian, $H_{Heis} = -J_{ab}(\mathbf{s_a}\cdot\mathbf{s_b}) - J_{ac}(\mathbf{s_a}\cdot\mathbf{s_c}) - J_{bc}(\mathbf{s_b}\cdot\mathbf{s_c})$, x will result from the relative values of the three exchange coupling constants $\{J_{ab}, J_{bc}, J_{ac}\}$.

The mathematics linking x to $\{J_{ab}, J_{bc}, J_{ac}\}$ is rather involved. I will only give here its outlines as, for the general case in which all three exchange coupling constants are different, there is no analytical solution. In the present case, there are only two $S = 1/2$ spin states to consider: $|(2)1/2>$ and $|(3)1/2>$. Both can be expanded according to Eq. 18. As a consequence, it is relatively easy (though tediously painful) to compute the elements of the 2 × 2 matrix of the Heisenberg spin Hamiltonian $H_{Heis} = -J_{ab}(\mathbf{s_a}\cdot\mathbf{s_b}) - J_{ac}(\mathbf{s_a}\cdot\mathbf{s_c}) - J_{bc}(\mathbf{s_b}\cdot\mathbf{s_c})$.

After diagonalization of this matrix (see Table 4), a relatively simple expression relates the mixing coefficient x (or y) to the exchange coupling constants $\{J_{ab}, J_{bc}, J_{ac}\}$:

$$\begin{aligned} x^2 &= (1/2)\left(1 + 1/\left(1+m^2\right)^{1/2}\right) \\ y^2 &= (1/2)\left(1 - 1/\left(1+m^2\right)^{1/2}\right) \end{aligned} \quad (21)$$

Table 4
Matrix elements of the general three-spin Hamiltonian over the two possible $S = 1/2$ spin states $|(2)1/2>$ and $|(3)1/2>$ (see main text)

$<(2)1/2\|H_{\mathrm{Heis}}\|(2)1/2>$	$<(2)1/2\|H_{\mathrm{Heis}}\|(3)1/2>$
$<(2)1/2\|H_{\mathrm{Heis}}\|(3)1/2>$	$<(3)1/2\|H_{\mathrm{Heis}}\|(3)1/2>$

with (again with the sites' correspondence $a \equiv 2, b \equiv 3, c \equiv 1$) [11]: $m = (3)^{1/2}(J_{ac} - J_{bc})/(2J_{ab} - J_{ac} - J_{bc})$.

The final step of the theoretical procedure consists in computing these three exchange coupling constants $\{J_{ab}, J_{bc}, J_{ac}\}$. We therefore need the energies of four states. Then we found in Table 1 actualized for the problem at hand. We have one DFT-HS state with $S = Ms = 15/2$ and three BS states (BSa, BSb, and BSc) of each $Ms = 5/2$ (intermediate) value. If the three iron sites are strictly equivalent, then $J_{ab} = J_{bc} = J_{ac}$ ($=J$). Conveniently, the three BS states then become energetically degenerate, and J can be derived from E_{HS} and E_{BS} alone. In the general case, we want to compute energies for (outer product) states of the form $|5/2, ms_a> \otimes |5/2, ms_b> \otimes |5/2, ms_c>$ with local $|ms_i| = 5/2$ ($i =$ a, b, c). As a consequence, for each scalar product of the type $\mathbf{s_i} \cdot \mathbf{s_j}$ in H_{Heis}, only the $s_{iz}s_{jz}$ term will yield nonzero values. From the four states

$$\begin{aligned} \mathrm{HS} &\equiv |5/2, +5/2 > \otimes |5/2, +5/2 > \otimes |5/2, +5/2 > \\ \mathrm{BSa} &\equiv |5/2, \mathbf{-5/2} > \otimes |5/2, +5/2 > \otimes |5/2, +5/2 > \\ \mathrm{BSb} &\equiv |5/2, +5/2 > \otimes |5/2, \mathbf{-5/2} > \otimes |5/2, +5/2 > \\ \mathrm{BSc} &\equiv |5/2, +5/2 > \otimes |5/2, +5/2 > \otimes |5/2, \mathbf{-5/2} > \end{aligned} \tag{22}$$

we have

$$\begin{aligned} E_{\mathrm{HS}} &= < \mathrm{HS}| -J_{ab}(s_{az}s_{bz}) - J_{ac}(s_{az}s_{cz}) - J_{bc}(s_{bz}s_{cz})|\mathrm{HS} > \\ &= -(25/4)(+J_{ab} + J_{ac} + J_{bc}) \\ E_{\mathrm{BSa}} &= < \mathrm{BSa}| -J_{ab}(s_{az}s_{bz}) - J_{ac}(s_{az}s_{cz}) - J_{bc}(s_{bz}s_{cz})|\mathrm{BSa} > \\ &= -(25/4)(-J_{ab} - J_{ac} + J_{bc}) \\ E_{\mathrm{BSb}} &= < \mathrm{BSb}| -J_{ab}(s_{az}s_{bz}) - J_{ac}(s_{az}s_{cz}) - J_{bc}(s_{bz}s_{cz})|\mathrm{BSb} > \\ &= -(25/4)(-J_{ab} + J_{ac} - J_{bc}) \\ E_{\mathrm{BSc}} &= < \mathrm{BSc}| -J_{ab}(s_{az}s_{bz}) - J_{ac}(s_{az}s_{cz}) - J_{bc}(s_{bz}s_{cz})|\mathrm{BSc} > \\ &= -(25/4)(+J_{ab} - J_{ac} - J_{bc}) \end{aligned} \tag{23}$$

It remains to extract values for the three exchange coupling constants from these four energies computed by DFT by combining the four expressions of Eq. 23, for example:

$$\begin{aligned} E_{\mathrm{HS}} - E_{\mathrm{Bsa}} &= -(25/2)(+J_{\mathrm{ab}} + J_{\mathrm{ac}}) \\ E_{\mathrm{BSc}} - E_{\mathrm{BSb}} &= -(25/2)(+J_{\mathrm{ab}} - J_{\mathrm{ac}}) \end{aligned} \quad (24)$$

yielding J_{ab} and J_{ac}, etc. These J's values could then be inserted into Eq. 21 which, in turn, determines the values of the predicted hyperfine coupling constants $\{\mathbf{A}_i\}$ via Eq. 20.

As a final remark, in some magnetic systems, there are more DFT states which can be constructed than exchange coupling constants. In these cases, one can extract the "best" set of exchange coupling constants through a least-square procedure (one example is provided in ref. 13).

It remains now to see what are some of the caveats and drawbacks of the broken symmetry methodology.

1.3 Caveats of Broken Symmetry States

In the following, I will expand on two caveats of the BS methodology. The first one is conceptual and is related to the very existence of BS states within DFT. After all, within a "perfect" DFT world and, say, for an antiferromagnetic spin-coupled Cu(II) dimer, the ground state with DFT-constrained $Ms = 0$ should be directly the (spatially symmetrical) singlet state ($S = 0$) and not the (spatially localized) BS state [14].

The second one is more practical and is related to the quantitative failure of the DFT–BS in predicting exchange coupling constants. The main reason has been made explicit only recently by confronting actual computed DFT–BS states to their analytical/phenomenological models (as they should be, and as they really are).

1.3.1 Neither Bird Nor Mouse . . .

As already stated above, the broken symmetry concept is a sort of hybrid (a bat) between the WFT (a bird) and the DFT (a mouse). It is constructed for valence-bond systems allowing for a left-versus-right localization of α and β electrons, respectively. The whole reasoning behind such BS states is based on mono-determinants and their astute use to circumvent the problem of computing energies of multi-determinant pure spin states with mono-determinant DFT codes.

The paradox is therefore clear: were we in possession of the "exact" exchange-correlation potential at the heart of all DFT codes, we could compute directly the energy of, say, a singlet state from the density minimizing the energy with $Ms = 0$. Of course, the corresponding DFT mono-determinant would have nothing to do with the exact singlet wave function determined by sophisticated wave function approaches (WFT). But its optimized DFT *density* would be the same as that computed by WFT. Of course, it would still be possible to compute a ($Ms = 0$) BS state with an "exact" DFT code, but its energy will most probably be higher than that obtained for the corresponding singlet state ($S = 0$, $Ms = 0$).

In other words, the BS state is doubly artificial. First, so to speak, it does not exist in Nature. Although its energy will be

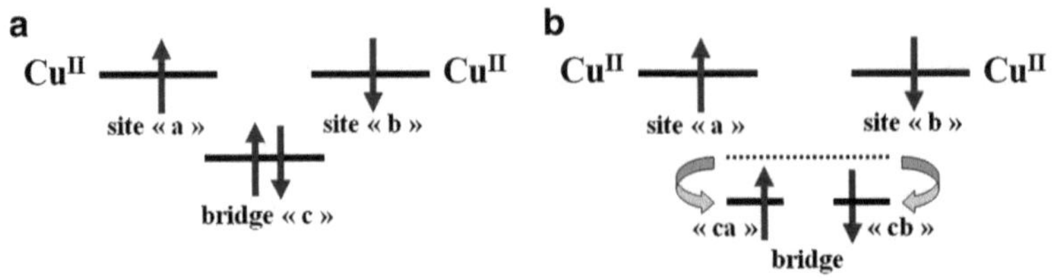

Fig. 11 Three sites-four electrons schematic diagrams. (**a**) Magnetic (a,b) and bridge (c) orbitals as they should be (i.e. all mutually orthogonal); (**b**) As they appear in a DFT-BS state (i.e. unrealistic broken bridge orbitals ca and cb)

representative of that of monomers upon dissociation, it is not the "real" wave function of such a dissociating system. Second, it is useful as long as our lack of knowledge of the "true" DFT lasts. Happily for DFT-NS practitioners (or not from a purist's point of view), this situation may still endure for a very long time . . . unless getting closer to "reality" comes from hybridizing both DFT and WFT worlds [15].

1.3.2 Artificial (i.e., Nonrealistic) Breaking of the Bridges' Orbitals

In order to explain in simple terms what is the point here, let us go back to the case of two weakly interacting $S = 1/2$ monomers such as a Cu(II) d^2 dimer depicted in Fig. 1b, right. So far, in all this presentation about BS states, we considered only the two magnetic orbitals bearing each an unpaired (i.e., magnetic) electron. This situation corresponds to the so-called 2 sites–2 electrons case, the two sites being "a" and "b" introduced above. If we now take into account the diamagnetic bridges (sites "c"), one would expect their electrons to remain paired in a "real" system. In other words, the spatial symmetry breaking leading to the emergence of BS states should affect only the *magnetic* orbitals a and b (*see* Fig. 11a) and not the diamagnetic bridge orbital c. This would be the correct "3 sites–4 electrons" model, the additional site "c" being identical for each pair of α/β electrons.

However, upon constructing the DFT–BS state, i.e., upon spatially localizing the two magnetic electrons on each end of the dimer, the DFT code will also break the symmetry of the bridging orbitals! This unwanted artefact will have two important consequences:

1. Qualitatively, the description of the bridge orbitals is no longer correct, not being chemically sound. We now have two bridge orbitals "ca" and "cb" which are different (*see* Fig. 11b) instead of one doubly occupied bridge orbital "c."
2. Quantitatively, by thus removing a symmetry constraint at the level of the bridge orbitals, i.e., by breaking their symmetry and identity, their energy is artificially lowered, as is that of the whole BS state! Therefore, the magnitude of the DFT-computed exchange coupling constant J for antiferromagnetic systems (*see* Figs. 4—left and 7) becomes too large! This last problem had been a long-standing issue for decades and found here its

source. The mathematical demonstration is rather involved, but the remedial solution is clear: it consists in constraining the bridge orbitals in order to avoid their symmetry breaking [16].

To conclude this "how to" chapter devoted to the use of broken symmetry states within DFT, we realize that, beyond computing mere energies, these BS states can become integral parts of increasingly more sophisticated *phenomenological* models, both analytical and/or numerical, which have to be developed for each particular problem at hand to compute various (electronic and magnetic) spectroscopic quantities. In the present work, I only gave a few examples in order to show as simply as possible but, at the same time, as precisely as possible (in the spirit of the Methods in Molecular Biology volume) some of the (simple or complex) calculations which can be done.

2 Materials

Adequate computer power will be required to run the programs described in this chapter.

3 Methods

As "material" for the present contribution, I will provide the reader with two typical minimalist DFT-ADF input files for a simple problem (within the scope of this chapter), that is, that of the study of a di-ferric [2Fe2S](SCys)$_4$ dimer. These input files should be self-explanatory, especially for those who have already run DFT calculations. I will be referring to these two files according to their respective titles: *2fe2s_hsox.inp* and *2fe2s_bsox.inp*.

3.1 2fe2s_hsox.inp

The goal is to converge the di-ferric [2Fe2S] dimer in the HS state of $Ms = 10/2$. In ADF, this is indicated by $n_\alpha - n_\beta = 10$, the difference number between alpha (spin up) and beta (spin down) electrons, here 10 as each (locally high-spin) ferric ion bears 5 unpaired electrons. The total charge count (here −2) results from +3 for each iron ion (26 − 3 electrons each), −2 for each bridging inorganic sulfur (16 + 2 electrons each), and −1 for each cysteinyl ligand (here modelled as SCH_3: 16 + 1 + 12 + 3 = 32 electrons). We thus have for the cluster $(+3) \times 2 + (-2) \times 2 + (-1) \times 4 = -2$ and a total of $(23) \times 2 + (18) \times 2 + (32) \times 4 = 210$ electrons, separated into 110 alpha electrons and 100 beta electrons. This is an all-electrons' calculation because we use the b3lyp exchange-correlation potential. The SCF convergence of this first file generates an output file TAPE13 renamed as tape13_2fe2s_hsox.out (procedure not shown here). It does contain a spatial representation of the alpha and beta spin densities on a basis of atomic fit functions. This will be our restart file for the convergence of the BS state:

```
$ADFBIN/adf<<eof
title 2fe2s_hsox.inp file

# blocs (most closed by "end") describing the molecular system

units
  length angstrom
end
symmetry NOSYM
atoms cartesian
Fe   0.000   0.000   0.000
Fe   0.000   0.000   2.747
S    0.104   1.781   1.428
S    0.110  -1.735   1.402
S   -1.995  -0.402  -1.199
S    1.582   0.188  -1.659
S    1.650  -0.084   4.367
S   -1.973  -0.236   3.899
C   -2.347   0.806  -2.490
C    1.256  -1.144  -2.826
C    2.374   1.585   4.463
C   -2.038   1.268   4.933
H    1.955  -1.103  -3.608
H    0.378  -1.103  -3.400
H    1.426  -2.133  -2.517
H   -3.246   0.553  -2.970
H   -1.723   0.833  -3.334
H   -2.594   1.786  -2.205
H    3.132   1.596   5.189
H    1.787   2.367   4.845
H    2.938   1.936   3.650
H   -2.916   1.259   5.509
H   -2.176   2.197   4.463
H   -1.346   1.371   5.716
end
charge -4 10
occupations
A 110 // 100
end

# blocs (most closed by «end») describing the DFT options

basis
type TZ2P
core none
end
unrestricted
scf
```

```
   iterations  200
   converge  0.00003
   mixing     0.06
end
integration 6.0
xc
   hybrid b3lyp
end
savefile TAPE13
end input
eof
```

3.2 2fe2s_bsox.inp

The only changes in this BS input file concern $n_\alpha - n_\beta$, now equal to 0. Moreover, the total number of electrons, 210, will be now separated into 105 alpha electrons and 105 beta electrons. The important bloc is written in bold: it does indicate that the atom #2, that is, Fe #2, will have its alpha and beta coefficients exchanged, which will create a suitable initial BS electronic structure. The SCF convergence of this first file generates an output file TAPE13 which will be renamed as tape13_2fe2s_bsox.out (procedure not shown here):

```
$ADFBIN/adf<<eof
title 2fe2s_bsox.inp file

# blocs (most closed by "end") describing the molecular system

units
   length angstrom
end
symmetry NOSYM
atoms cartesian
Fe     0.000    0.000    0.000
Fe     0.000    0.000    2.747
S      0.104    1.781    1.428
S      0.110   -1.735    1.402
S     -1.995   -0.402   -1.199
S      1.582    0.188   -1.659
S      1.650   -0.084    4.367
S     -1.973   -0.236    3.899
C     -2.347    0.806   -2.490
C      1.256   -1.144   -2.826
C      2.374    1.585    4.463
C     -2.038    1.268    4.933
H      1.955   -1.103   -3.608
H      0.378   -1.103   -3.400
H      1.426   -2.133   -2.517
H     -3.246    0.553   -2.970
H     -1.723    0.833   -3.334
H     -2.594    1.786   -2.205
H      3.132    1.596    5.189
```

```
H    1.787   2.367   4.845
H    2.938   1.936   3.650
H   -2.916   1.259   5.509
H   -2.176   2.197   4.463
H   -1.346   1.371   5.716
end
charge -4 0
occupations
A 105 // 105
end

# blocs (most closed by «end») describing the DFT options
basis
type TZ2P
core none
end
unrestricted
restart tape13_2fe2s_hsox.out &
spinflip 2
end
scf
   iterations  200
   converge  0.00003
   mixing     0.06
end
integration 6.0
xc
   hybrid b3lyp
end
savefile TAPE13
end input
eof
```

4 Notes

1. The combination of one unpaired spin with its spatial extension (a(r), etc.) is what is called a magnetic orbital.
2. In practice, with the DFT code I am using, I first converge the high-spin state (here the triplet state of spin $S = 1$) and then restart the calculation from the triplet output file after having flipped one of the two monomer spins and set $Ms = 0$. If I get after SCF convergence nonzero spin populations of opposite signs on the main spin bearers (for example, positive on site a and negative on site b: *see* Fig. 1a, b), then I have successfully reached a broken symmetry state. If both atomic spin populations turn out to be zero (as it will be the case in Fig. 2b for example), it means that the system is (chemically) not of the valence-bond type suited for this kind of DFT–BS calculation.

3. This notation of determinants as outer products of monomer spin states is what will make the extension to three spins (i.e., $|s_a,ms_a\rangle \otimes |s_b,ms_b\rangle \otimes |s_c,ms_c\rangle$) and more quite easy and transparent in terms of notations.
4. This illustrates the kind of game we will play, i.e., get indirectly the energies of some pluri-determinant states from those of specific mono-determinants (pure spin or BS).
5. We thus neglect the magnetic orbital overlap $\langle a|b\rangle^2 \ll 1$. If not, BS gets closer to Σ as $\langle a|b\rangle^2$ tends to unity as Eq. 5 becomes $J = E_\Sigma - E_T = 2 \cdot [E_{BS} - E_T]/(1 + \langle a|b\rangle^2)$.
6. Computed from $\mathbf{S}^2 = (\mathbf{s_s} + \mathbf{s_b})^2 \rightarrow \mathbf{s_a}\cdot\mathbf{s_b} = [\mathbf{S}^2 - (\mathbf{s_a})^2 - (\mathbf{s_b})^2]/2$.
7. Notice that Eq. 8 generalizes the previous Eq. 1 derived for two weakly interacting (spin 1/2) monomers for which $[C_{S\rightarrow BS}(+1/2,-1/2)]^2 = [C_{S\rightarrow BS}(-1/2,+1/2)]^2 = 1/2$.
8. Again, Eq. 9 generalizes the previous Eq. 3 written for two weakly interacting (spin 1/2) monomers, for which $[C_{BS\rightarrow S}(0,0)]^2 = [C_{BS\rightarrow S}(1,0)]^2 = 1/2$.
9. Notice that, in Eq. 5, we had $J = 2 \cdot [E_{BS} - E_T]$, which is formally the same result as in Eq. 12 besides the factor 25. This last factor corresponds to $n_a n_b$ where n_a (resp. n_b) is the number of unpaired spins on site a (resp. b). Hence, $n_a = n_b = 1$ for a Cu(II) dimer, whereas $n_a = n_b = 5$ for a (locally high-spin) ferric dimer. In general, for symmetrical dimers $J = 2 \cdot [E_{BS} - E_{HS}]/(n_a n_b)$.
10. Typical J values for iron–sulfur clusters are of the order of few tens or hundreds per centimeter, whereas hyperfine coupling constants are in the range of 20–40 MHz (1 MHz ↔ $3.3 \times 10-^5$ $cm-^1$).
11. Alternatively to the method presented in the main text, we can equate as $\mathbf{s_a}\cdot\mathbf{a_a}$ (local) with $\mathbf{S}\cdot\mathbf{A_a}$ (global) (same for **b** terms) and multiply each of them by **S**, yielding $\mathbf{a_a}(\mathbf{s_a}\cdot\mathbf{S})$ and $\mathbf{A_a}(\mathbf{S}\cdot\mathbf{S})$, respectively. We then derive eigenvalues for both equivalent expressions, that is,

$$\mathbf{a}_a < S, Ms|\mathbf{s_a} \cdot \mathbf{S}|S, Ms > \equiv \mathbf{A_a} < S, Ms|\mathbf{S} \cdot \mathbf{S}|S, Ms >$$

where $\langle S,Ms|\mathbf{S}\cdot\mathbf{S}|S,Ms\rangle = S(S + 1)$. It remains to compute the first bracket term $\langle S,Ms|\mathbf{s_a}\cdot\mathbf{S}|S,Ms\rangle$ using Eq. 8 expressing $|S,Ms\rangle$ as a weighed average of $|s_a,ms_a\rangle \otimes |s_b,ms_b\rangle$ states, by making use of the identity $\mathbf{s_a}\cdot\mathbf{S} = (\mathbf{S}^2 + \mathbf{s_a}^2 - \mathbf{s_b}^2)/2$ yielding

$$\mathbf{a_a}[S(S+1) + s_a(s_a+1) - s_b(s_b+1)]/2 = \mathbf{A_a} \cdot S(S+1).$$

In other words, we have derived in a simple way closed expressions for the spin-coupling coefficients K_a and K_b, linking (local) $\mathbf{a_a}$ to (global) $\mathbf{A_a} = K_a\cdot\mathbf{a_a}$: $K_a = [S(S + 1) + s_a(s_a + 1) - s_b(s_b + 1)]/2S(S + 1)$ and $K_b = [S(S + 1) +$

$s_b(s_b + 1) - s_a(s_a + 1)]/2S(S + 1)$, yielding the same values found in Table 2. For more than two sites, the generalization of such a procedure linking local to global properties can only proceed along numerical lines, as illustrated in the main text.

12. If one has to compute a global cluster **g**-tensor, one would proceed in the same way. First, compute the local monomer $\mathbf{g_i}$-tensors (this **g**-tensor option is nowadays standard in DFT codes and works in most cases) and then spin-couple them in this way to obtain the total **g**-tensor of the cluster: $\mathbf{g} = \Sigma_i K_i \mathbf{g_i}$.
13. One could have started with $\mathbf{S''} = \mathbf{s_a} + \mathbf{s_c}$ or $\mathbf{S'''} = \mathbf{s_b} + \mathbf{s_c}$. The three resulting coupling schemes are equivalent, though not identical, and lead to the same final solution.

References

1. Perdew JP, Ruzsinsky A, Constantin LA et al (2009) Some fundamental issues in ground-state Density Functional Theory: a guide for the perplexed. J Chem Theory Comput 5:902–908
2. Noodleman L (1981) Valence bond description of antiferromagnetic coupling in transition metal dimers. J Chem Phys 74:5737–5743
3. Noodleman L, Davidson ER (1986) Ligand spin polarization and antiferromagnetic coupling in transition metal dimer. Chem Phys 109:131–143
4. For those interested, I invite them to look for his papers online (via Web of Knowledge for example) as they are simply too numerous to be cited here
5. Hoffmann R, Shaik S, Hiberty PC (2003) A conversation on VB vs MO theory : a never-ending rivalry? Acc Chem Res 36:750–756
6. Brink DM, Satchler GR (1968) Angular momentum. Clarendon, Oxford
7. Noodleman L, Peng CY, Case DA, Mouesca J-M (1995) Orbital interactions, electron delocalization and spin-coupling in iron-sulfur clusters. Coord Chem 144:199–244
8. Orio M, Mouesca J-M (2008) Variation of average g values and effective exchange coupling constants among [2Fe-2S] clusters: a Density Functional Theory study of the impact of localization (trapping forces) versus delocalization (double-exchange) as competing factors. Inorg Chem 47:5394–5416
9. Volbeda A, Amara P, Darnault C et al (2012) X-ray crystallographic and computational studies of the O_2-tolerant [NiFe]-hydrogenase 1 from *Escherichia coli*. Proc Natl Acad Sci USA 109:5305–5310
10. Mouesca J-M, Noodleman L, Case DA et al (1995) Spin densities and spin coupling in Iron-Sulfur clusters: a new analysis of hyperfine coupling constants. Inorg Chem 34:4347–4359
11. Kent TA, Huynh BH, Münck E (1980) Iron-sulfur proteins: spin-coupling model for three-iron clusters. Proc Natl Acad Sci USA 77:6574–6576
12. Griffith JS (1972) On the general theory of magnetic susceptibilities of polynuclear transition-metal compounds. In: Hemmerich P (et al) (eds) Structure and bonding, volume 10. Springer, Springer Verlag, pp 87–126
13. Li L, Liao D, Jiang Z et al (2006) An unprecedented asymmetric end-on azido-bridged copper(II) imino nitroxide complex: structure, magnetic properties, and Density Functional Theory analysis. Inorg Chem 45:7665–7670
14. Rudra I, Wu Q, Voorhis V (2010) Accurate magnetic exchange couplings in transition-metal complexes from constrained density-functional theory. J Chem Phys 124:024103-1–024103-3
15. Grabowski I, Lotrich HS (2010) *Ab initio* DFT—the seamless connection between WFT and DFT. Mol Phys 108:3313–3322
16. Onofrio N, Mouesca J-M (2011) Analysis of the singlet-triplet splitting computed by the Density Functional Theory–Broken Symmetry method: is it an exchange coupling constant? Inorg Chem 50:5577–5586

INDEX

Juan C. Fontecilla-Camps and Yvain Nicolet (eds.), *Metalloproteins: Methods and Protocols*, Methods in Molecular Biology, vol. 1122, DOI 10.1007/978-1-62703-794-5,

MIX
Papier aus verantwortungsvollen Quellen
Paper from responsible sources
FSC® C105338

If you have any concerns about our products,
you can contact us on
ProductSafety@springernature.com

In case Publisher is established outside the EU,
the EU authorized representative is:
Springer Nature Customer Service Center GmbH
Europaplatz 3, 69115 Heidelberg, Germany

Printed by Libri Plureos GmbH
in Hamburg, Germany